The publisher gratefully acknowledges the generous support of the Authors Imprint Endowment Fund of the University of California Press Foundation, which was established to support exceptional scholarship by first-time authors.

SAVANNAS OF OUR BIRTH

Robin S. Reid   ·   SAVANNAS OF
OUR BIRTH

*People, Wildlife, and Change
in East Africa*

University of California Press

Berkeley   Los Angeles   London

University of California Press, one of the most distinguished university presses in the United States, enriches lives around the world by advancing scholarship in the humanities, social sciences, and natural sciences. Its activities are supported by the UC Press Foundation and by philanthropic contributions from individuals and institutions. For more information, visit www.ucpress.edu.

University of California Press
Berkeley and Los Angeles, California

University of California Press, Ltd.
London, England

Library of Congress Cataloging-in-Publication Data

Reid, Robin Spencer.
  Savannas of our birth : people, wildlife, and change in east Africa / Robin S. Reid.
    p. cm.
  Includes bibliographical references and index.
  ISBN 978-0-520-27355-9 (cloth)
  1. Savanna ecology—Africa, east. 2. Land use—Environmental aspects—Africa, east.
  3. Pastoral systems—Environmental aspects—Africa, east. 4. Savannas—Africa, east.
  I. Title.
  QH195.A23R45 2012
  577.4'80967—dc23

                                                        2012004372

21 20 19 18 17 16 15 14 13 12
10 9 8 7 6 5 4 3 2 1

CONTENTS

ACKNOWLEDGMENTS

Often while researching this book I felt as if I were on an intellectual treasure hunt laid down three decades ago by my long-term mentor and friend Jim Ellis. Jim died in an avalanche in 2002, leaving so many of us poorer in both friendship and learning. At the same time, I drew courage from his fine example, for he taught me that good ideas come only from very hard thinking and scrupulous scholarship. I hope to honor his tradition with this book.

Jim introduced me to my most important teachers about African pastoralists and wildlife: pastoralists themselves. My first teachers were from Turkana, who regularly endure insecurity and drought in their remote pastures in northern Kenya. My Turkana graduate committee included Angerot, Nakwawi, Amarie, Lopeyone, Eliud Loweto, Achukwa Ewoi, and Judy Lokudo. My Maasai teachers were John Rakwa, David Nkedianye, Dickson Kaelo, Samson Lenjirr, Meole Sananka, Joseph Temut, James Kaigil, Charles Matankory, Ogeli Makui, Nicholas Kamuaro, Leonard Onetu, Stephen Kiruswa, Moses Neselle, John Siololo, James Turere, Nickson Parmisa, Joseph Kimiti, and Moses Koriata. My Samburu teachers were Pakuo Lesorogol, Lerali Lesorogol, and Sauna Lemiruni.

I learned a great deal from other scholars who have tried to reach the middle ground by integrating people, wildlife, and their wider ecosystems. Here I am indebted to Kathleen Galvin, Helen Gichohi, Katherine Homewood, Terrence McCabe, David Nkedianye, Michael Rainy, and David Western.

This book is written for a wide audience from pastoralists to academics to lay readers. As a result, I purposely do not describe or adhere to a particular theoretical framework, partly to make the book widely accessible, but also to allow me to cross disciplinary lines easily. Also, I took the references out of the text and put them in endnotes so that the text is cleaner, even though that approach may frustrate some scholars.

For my pastoralist readers, I hope this book holds some truth for you and inspires you to tell your own story better than I ever can.

This book was co-created with the superb help of Russell Kruska, Susan Mac-Millan, and Cathleen Wilson. Russell created the maps with great humor and the highest skill; Cathleen provided her excellent photographs and research skills; and Susan edited many of the chapters, helping me untangle my words and thoughts. I am so grateful to you three and will never forget our time working together at Kapiti Ranch in Kenya.

Russell Kruska got mapping support from Shem Kifugo, John Owuor, and Meshack Nyabenge. John Curry digitized George Murdock's 1959 map of the ethnic groups of Africa. Leah Ng'ang'a, Cathleen Wilson, Jeffrey Worden, and Joana Roque de Pinho conducted early literature surveys. I thank them all.

Many of the scholars whom I most admire thoughtfully reviewed all or part of this book. These fine folk include Stanley Ambrose, Jayne Belnap, Shauna Burn-Silver, William DiMichele, Douglas Frank, Kathleen Galvin, Patricia Kristjanson, Russell Kruska, Susan MacMillan, Fiona Marshall, Terrence McCabe, James McCann, Patricia Moehlman, Cynthia Moss, David Nkedianye, Joseph Ogutu, Bill Reid, Gwen Reid, Phillip Thornton, Cathleen Wilson, Jackie Wolf, Jeffrey Worden, Truman Young, and one anonymous reviewer. Other scholars clarified key concepts and gave encouragement; these include Andrew Ash, Randall Boone, Ruth DeFries, Lisa Graumlich, Andrew Hanson, N. Thompson Hobbs, Richard Lamprey, Thomas Mölg, Fortunate Msoffe, Andrew Muchiru, Onesmo Olengurumwa, and Phillip Thornton. Their comments improved the book immensely. All remaining annoyances, errors, and misinterpretations are mine alone.

I thank John McDermott, Shirley Tarawali, and Carlos Sere for granting me the short sabbatical at the International Livestock Research Institute (ILRI) that kicked this book off. They and Simon Kiberu kindly allowed me the time and space to write at ILRI's Kapiti Ranch, while Mohammed Said stepped in and took over my responsibilities leading our team. At Colorado State University, I thank Ed Warner and Joyce Berry for creating the Center for Collaborative Conservation, and Stacy Lynn, Jill Lackett, Bev Johnson, Ch'aska Huayhuaca, Patrick Flynn, and Kim Skyelander for making the Center real, and especially for taking up the slack when I stepped out to finish this book.

Russell Kruska and I thank George Murdock for creating his 1959 ethnic groups map, Mahesh Sankaran for graciously allowing us to use and revise his map of nutrient-rich and -poor savannas, and James Newman and Yale University Press for permitting us to adapt Newman's map of human population movements in east Africa.

I thank Blake Edgar for encouraging me to write this book. He, Lynn Meinhardt, and Dore Brown of the University of California Press skillfully shepherded me through the entire book production process. Anne Canright did excellent copyediting, and Naomi Linzer provided a thorough index.

For great friendship and encouragement in east Africa, I thank John Edwards, Mario Herrero, Dickson Kaelo, Patti Kristjanson, Russ Kruska, Jan Low, Susan Macmillan, Margaret Morehouse, David Nkedianye, Joseph Ogutu, Mike and Judy Rainy, Mohammed Said, Phil Thornton, Cathy Wilson, Jeff Worden and many other fine people. In Colorado, Gillian Bowser, Kathleen Galvin, Joshua Goldstein, Corrine Grim and Robert Dearing, Tom Hobbs, Heather Knight, Richard Knight, Lee Scharf, and Karl and Lucy Stefan are all great friends with ready support. In the Pacific Northwest, I thank George and Hildegard Dengler, Jackie Wolf, Christine Langley, Claudia Elwell, Pamela Pauly, Suzanne Berry, Karen Gilbert, Terry Marshall, Marty Clark, Tim Clark, Angie Ponder, Scott Morris, Laura Morris, and Kitty Harmon. And a special thanks to Daphne Morris for lending me a cabin during that first blustery winter of writing (and Cathy for firewood and fires!). Many thanks also to all those not mentioned here who boosted me along in the writing of this book.

Most of all, of course, I want to thank my family. Every author tests the patience and good will of his or her immediate family, over and over again as the writing goes on. Holly, Rich, Indy, and Lionel Reid-Shaw kept me going with their enthusiasm, support, and winter ice cream runs. Scott, Kris, Eliza, and Marjorie Reid and Nancy and Dexter Anderson provided support and good humor. My parents, Gwen and Bill Reid, who were the first to read this book cover to cover, represent a lifelong source of love, inspiration, and strength. Lastly, I thank Cathy, Robi, Kazi, and Kibo for putting up with many missed family weekends and vacations, and for their boundless encouragement, warmth, and love.

Finally, I ask each of you kind readers to join me in an effort to stitch together our divided world. Our problems are too large and too formidable, from all perspectives, to allow us to fight and argue about one side against the other. If there is an underlying message to this book, it is that we have far more in common, on almost every issue, than we have in difference.

CHAPTER ONE

Searching for the Middle Ground

Wildlife are an invaluable renewable resource that developing nations must learn to utilise judiciously, so that their benefits accrue today but in such a manner that future generations shall inherit with pride the legacy of previous generations. This will not be achieved by banishing the indigenous wildlife guardians from the land of their birth, and relegating them to marginal areas where impoverishment and deprivation will become their lot.

MORINGE PARKIPUNY, *former Tanzanian Minister of Parliament, 1991*[1]

Nature preserves . . . are not places to be saved to be used at a later stage when an ever-growing human population claims more land because of lack of economic development.

HERBERT PRINS, *biologist, 1992*[2]

Why can our animals not go there [to the Park] while the Park animals can come here?

MAASAI ELDER, *Tanzania, interview with Kadzo Kangwana and Rafael Ole Mako, 2001*[3]

SAVANNAS OF OUR BIRTH

This is the story of where all our human ancestors probably came from: Africa, particularly the savannas of Africa. Over millions of years, these savannas gradually replaced forests and woodlands in some places, and hominins (including ancestors of humans) took advantage of these new environments. In these savannas are the oldest known footprints of hominins walking upright, a habit that freed our ancestors' hands for carrying, digging, and throwing. Here, our ancestors probably crafted their first tools and hunted and scavenged wildlife. And as far as we know, it is from here, millennium after millennium, that our forebears repeatedly left Africa and populated other parts of the world, eventually becoming the ancestors of all humans. To the best of our knowledge, every person who reads these words has australopithecine ancestors who lived in Africa.[4]

Remarkably, this ancient home of humankind is also the setting for a unique story of people and wildlife who have lived side by side throughout thousands, perhaps millions, of years. Africa alone more or less escaped the massive extinctions of large mammals that occurred in the Pleistocene epoch. Today, only African savannas continue to maintain a startling diversity of large mammals. These large mammals thus thrive best not where they have been separated from people over the millennia, whether by accident of geography or history, but rather where people have lived on this planet the longest. This book is partly about why this is so.

Since the late 1800s, this ancient coexistence of people and wildlife in African savannas has often been ignored, often replaced by a modern practice of conservation which assumes that wildlife are best conserved in landscapes with no people. Colonists from southern Africa and the West imported this idea into east Africa to stem wildlife losses caused by colonial and African hunting. Rooted in a Western philosophy of the Middle Ages that separated people from nature, this new belief contrasted sharply with the belief in many cultures of Africa, Asia, and the Americas, which viewed humans as part of nature, interconnected and interdependent. Most colonial governments, and some African governments that followed, carved out large pieces of the savannas to create wildlife parks and reserves, most of which banned people from their traditional lands within the parks. Even though these parks today cover only about 9 percent of savannas across east Africa, they were often created where most species congregate in the dry season and in drought, places critical to the survival of wildlife and people alike.[5]

Despite the creation of parks to protect wildlife, many large animals in Kenya and some other countries today still live in savannas outside parks, where they share their habitat with people. Why is this so? One obvious reason is that there is about ten times more savanna land outside than inside parks. But another, less appreciated reason is that some communities located outside parks have chosen to live with rather than exterminate wildlife. This is particularly true where low and variable rainfall make settlement and farming difficult; as a consequence, people often use the land in a way that keeps the land open for wildlife. Remarkably, the people who coexist with wildlife in these savannas are not just wealthy wildlife ranchers or tax-funded national governments with significant funds to support those wild populations. Many are herders, far poorer than nearby farmers, pursuing a pastoral lifestyle that has been common in these savannas for thousands of years (Figure 1).[6]

A central irony in this story is that the peoples who, amid tremendous change, still coexist with wildlife are those most often held responsible for overgrazing,

FIGURE I.
Wildebeest and cattle grazing together in Kitengela,
Kenya, a common sight in east African savannas. (Photo by
Cathleen J. Wilson.)

the spread of deserts, and generally poor environmental management: pastoral
herders (also called pastoralists). Herders themselves recognize that their lifestyle
is not always compatible with wildlife. More than two decades ago, for example,
I asked a group of herders in northern Kenya why there were so few gazelles in
their pastures. They pointed to a herder leaning against an acacia tree next to his
rifle, who served as a home guard against cattle raiders, and said: "100 bullets, 100
dead gazelles." Borana herders in southern Ethiopia see clearly that they are overstocking their land, causing woody plants to spread (though grazing might not be
the only cause), which makes the land less hospitable for grazing livestock and
wildlife. Even in places where traditional practices are compatible with wildlife,
as in Maasailand, the land is changing rapidly with the spread of wells, settlements,
and farming. These herders, who rarely hunt and consider themselves the guardians of wildlife, now see their age-old compatibility with wildlife under threat,
even disappearing in some places. (See Maps 1 and 2 for the place names used in
this book. Map 3 shows linguist Joseph Greenberg's distribution of language

phyla; Maps 4–6 show anthropologist George Murdock's historical distribution of some of the major ethnic groups in east Africa.)[7]

These contradictions have created two camps of outside observers, one supporting pastoral development and one supporting wildlife conservation. On the one hand, many social scientists and development practitioners see efforts to conserve wildlife through establishment of parks as a violation of both human and land rights and one cause of poverty. Across Africa many people still live in these parks, and an estimated 1 to 14 million people would have to be displaced to enforce the rules governing human populations within these protected areas. In African savannas, these parks are usually on land that hunter-gatherers and herders have used for thousands of years. Even in conservation areas where pastoral people are allowed to live, such as Ngorongoro in Tanzania, research shows that herders become impoverished when conservation policy requires they live mostly off their livestock, by restricting their ability to grow crops. And local residents, particularly those who are poor, often experience most of the costs and few of the benefits of living near parks and conservation areas. The creation of protected areas denies local people the option of using this land in other ways in the future, which could limit their economic growth. These realities create critical problems in African savannas.[8]

On the other hand, many conservationists see parks and reserves as critical to conserving wildlife in Africa and as ecological baselines for measuring how people affect wildlife and savannas. As the human footprint grows in savannas outside protected areas, they are pointing to a long-term loss of wildlife and a rise in human-wildlife conflicts. In Kenya, for example, wildlife populations fell by as much as 80 percent in some savannas between the 1970s and 1990s. Around Kenya's Amboseli National Park, some elephants die at the hands of spear-wielding herders, especially during drought when elephants and cattle compete for limited food and water. In the wildlife reserves of Karamojong, Uganda, armed herders are a major cause of wildlife loss. In Tanzania, aerial surveys suggest that wildlife declined in the decade from 1990 to 2000 both in and around many protected areas, although elephant populations grew. In all of these cases, people, and sometimes herders, are a likely cause of wildlife declines. These realities also create critical problems in African savannas.[9]

Underlying these views are conflicting values. From a human (or anthropo-) centric perspective, the health and welfare of people are most important and their needs should be served first. Nature has value because it provides people with material benefits; thus, conserving natural resources has value because it affects

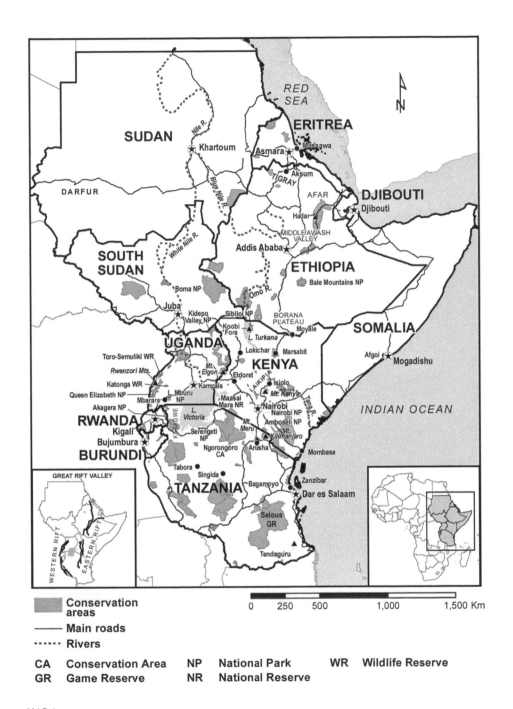

MAP 1.
Locations described in this book across the east African
region. (Map adapted from DMA 1992 and WPDA 2009 by
Russell L. Kruska.)

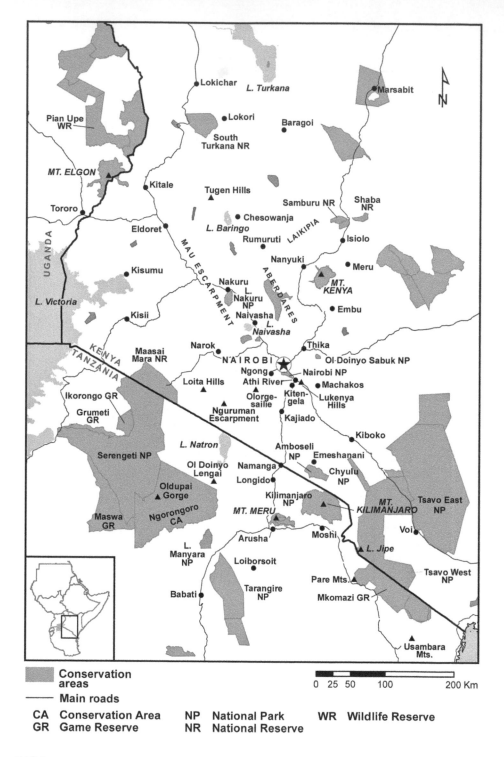

MAP 2.

Locations described in this book from northern Kenya to northern Tanzania. (Map adapted from DMA 1992 and WPDA 2009 by Russell L. Kruska.)

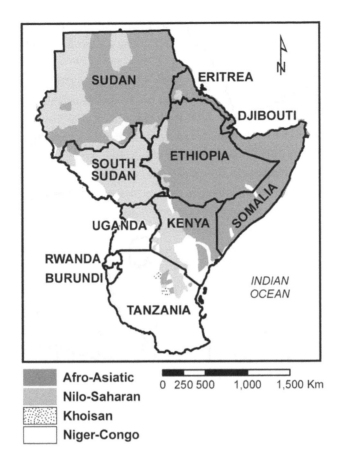

MAP 3.
Broad language groups in east Africa based on the historical
tribal group map of Murdock (1959) and using the language
classification of Greenberg (1963). (Map by Russell L. Kruska.)

people's well-being. In Western societies, this view typically emphasizes the individual, whereas in many African societies it focuses on the clan, community, village, tribe, or nation. A contrasting, ecocentric point of view maintains that nature has value for its own sake, apart from how nature affects humans, and deserves moral consideration as such. People holding this view can appear indifferent to human welfare, seeming to consider conserving nature more important than feeding people. When people argue either for the development of pastoralism or for nature conservation, the argument often is about who should come first, people or nature. The Western philosophy of people's separation from nature typically underlies these dichotomies, which may explain why the divide is both large and seemingly unbridgeable.[10]

What do pastoralists in east Africa think about the separation of people and nature—about people first or environment first? Scholars, whether from Africa or elsewhere, have written very little about this subject. The Oromo herders of Ethiopia (included with the Borana and Arusi on Map 4) depend on the land for food, shelter, and fuel, but they also view the environment spiritually and morally. Philosopher Workineh Kelbessa describes it this way: "For them, land is not only a resource for humans' utilitarian ends, but also it has its own inherent value given to it by *Waaqa* (God). For the Oromo, *Waaqa* is the guardian of all things, and nobody is free to destroy natural things to satisfy his or her needs." Anthropologist David Turton describes how Mursi herders in Ethiopia (included with the Suri in Map 4) regard wild animals: "They kill them to obtain economically useful products and, when necessary, to protect their cattle, but otherwise their disposition is, as Evans Pritchard . . . writes of the Nuer, 'to live and let live.'" I have had many conversations with Maasai about this, and they take pride in their long-term stewardship of savannas; many clearly think that people and wildlife are meant to live together. Animal characters feature prominently in Maasai oral history and stories, as both bad and good examples of behavior. Maasai schoolchildren see parks as places that attract tourists, provide income, and protect people from wildlife and wildlife from people. Thus, while pastoralists are certainly human-centric, they also respect, or at least tolerate, wildlife and consider themselves a part of the life of the savannas around them.[11]

In a sense, the pastoral view falls somewhere in the middle ground between support of human needs on the one hand and wildlife conservation on the other. Recently, the pastoral middle ground has been growing in influence among non-pastoralists, as those occupying this terrain attempt to move beyond an either-or view of people and wildlife and try to understand, rather, how people and wildlife coexist in the savannas and how they do not. The solution does not lie in a compromise of the extremes, however, but in the creation of a productive third way. Instead, it is critical that this third view recognize and address the real difficulties that pastoral families and wildlife face living with each other, and the inherent contradictions of this course: the fact that people have lived in savannas for millennia with wildlife *and* that wildlife are now fast disappearing; the fact that wildlife conservation policy can impoverish local communities *and* that agricultural development policy can harm wildlife populations; the fact that pastoral herders

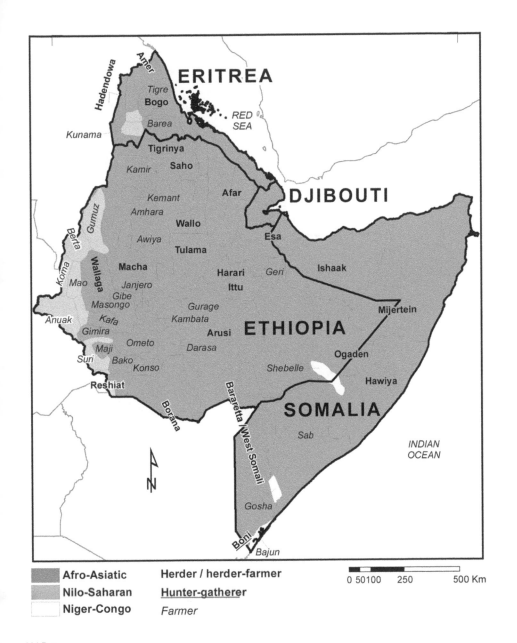

Afro-Asiatic   Herder / herder-farmer

Nilo-Saharan   Hunter-gatherer

Niger-Congo   Farmer

0 50 100   250   500 Km

MAP 4.
Historical tribal groups in Eritrea, Ethiopia, Djibouti, and
Somalia. Groups are distinguished by lifestyle (hunter-gatherer,
pastoral, farmer) and linguistic group. (Map adapted from
Murdock 1959 by Russell L. Kruska.)

Afro-Asiatic | Herder / herder-farmer
Nilo-Saharan | <u>Hunter-gatherer</u>
Khoisan | *Farmer*
Niger-Congo

0 50 100 250 500 Km

MAP 5.

Historical tribal groups in Uganda, Kenya, Tanzania, Rwanda, and Burundi. (Map adapted from Murdock 1959 by Russell L. Kruska.)

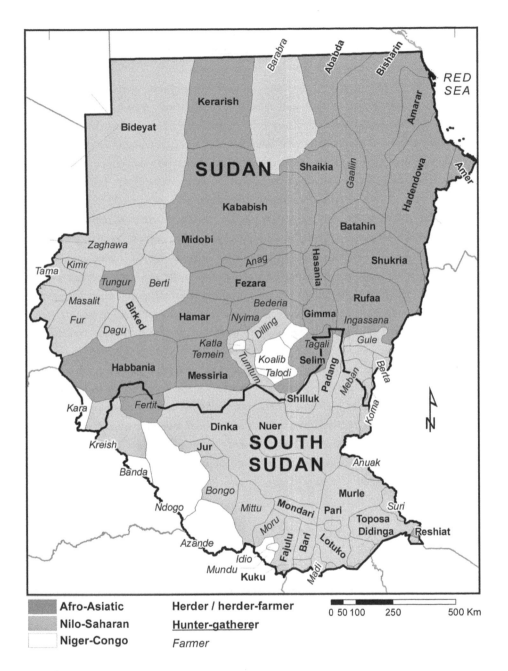

MAP 6.
Historical tribal groups in Sudan and South Sudan.
(Map adapted from Murdock 1959 by Russell L. Kruska.)

can coexist with wildlife *and* that they can extinguish wildlife. This book is about making sense of these contradictions—understanding how they came about in the first place and how they may shape the future of savannas.[12]

But why look to a middle, more pastoral view for solutions? Aren't herders the problem, overgrazing pastures and causing desertification? As a matter of fact, new evidence suggests that the story is far more complicated, emblematic of the middle ground. Take the following examples of this shifting, and often conflicting, narrative. Once, scientists thought the Sahara Desert was marching south into the Sahel; now, that transitional zone appears to be greening again. In the past, scholars thought African pastoralists were irrational and even greedy to increase the size of their herds; now, many view periodic herd growth as a sound strategy for coping with recurrent drought. Once, overgrazing by herders was thought to cause bush to invade grasslands; now, evidence suggests that CO_2 generated by burning fossil fuels (mostly outside Africa) may also encourage bush to spread. In the past, scientists alone defined and measured savanna decline and degradation; recently, however, scientists have begun to include the observations of pastoralists in this undertaking. Whereas scholars once thought that communal grazing damaged rangelands, they have come to realize that herders create rules-of-use that in fact sustain rangelands; now, though, even some pastoral leaders are saying that their communities are overstocking their lands. Some people think that parks protect biodiversity, while others consider those same parks a violation of human land rights. To some, pastoralists destroy biodiversity; to others, they are the hope for conserving it. This book assesses the scientific merits of these points of view and attempts to take this discussion to a third place, somewhere in the elusive middle ground.[13]

THE LARGER STORY OUTSIDE SAVANNAS

The story of people and wildlife in African savannas is set within a much larger story of how people develop and live within their environment around the world. By some measures, the human condition is improving. The average child on Earth grows up in a family that is wealthier, better nourished, and better educated than that of their parents. There are now 100 million more adults who can read these words than could twenty-five years ago. Today, 2.1 million more children live past the age of five than they did just fifteen years ago. Since 1800, the percentage of people on Earth who participated in electing their governmental leaders rose from

2.5 to 45 percent. We eradicated two major diseases, smallpox and rinderpest, that used to cause sweeping epidemics in people and animals. We invented a worldwide system of parks to protect wildlife, plants, and ecosystems for present and future generations of people. A measure of our success, and a cause of our current undoing, is the sheer number of people alive today on Earth: about 7 billion—almost two people for every year the planet has existed.[14]

But not all people share in this wealth, health, and education. The world is now more inequitable than it was just one generation ago. Nearly half the people on Earth, some 2.8 billion people, live on less than $2 per day. Almost half the population of Africa, 300 million people, live on less than $1 per day, and incomes are stagnant or falling. Directors of major corporations are paid as much as 30,000 times more than some of their rural African clients who walk several kilometers to buy corporate products in small village shops across the continent. The world's five hundred richest people together make more money than the 416 million poorest people. In 2004, for every child that died before the age of five in the world's wealthiest countries, thirty children died in sub-Saharan Africa, more than double the difference that existed in 1980. Belgium has about 450 doctors for every 100,000 people in the country, Rwanda only 4—less than 1 percent of Belgium's number. These disparities tend to be even wider where women are concerned. For example, although there are fewer illiterate people in the world today, two-thirds of those are women. One might wonder, given these facts, how well people in the richest countries are able to understand what it is like to grow up in the poorest countries, and vice versa.[15]

We have accomplished many of these gains in human well-being by turning the Earth's riches—minerals, soil nutrients, water, animals, plants, fossil fuels—into food, fiber, and consumer goods. Over the last century, the industrialized nations focused on growing their economies and have done so with spectacular success, assuming they could figure out how to solve any social, cultural, or environmental problems later. When in the 1980s it became clear that development could be sustained only if it purposefully addressed environmental and social as well as economic problems, "sustainable development"—the idea that our future survival depends in part on a healthy environment—was born.[16]

Few contest this idea today, and the understanding of our dependence on the environment, and of how strongly we can affect it, has deepened considerably. For example, we have understood for millennia that most of our food, clothing, and other essential material goods depend on the environment: soil nutrients, water, fuels, a diverse and abundant plant and animal life, and other natural resources.

What has become clear only recently is the importance of a much wider range of "ecosystem services" that maintain the climate, provide clean water, pollinate crops, and even hold genetic information in store for future generations (in the form of biodiversity).[17]

This new understanding of ecosystem services reveals the tremendous economic cost of historical development. In fact, the value of the services provided to humans by all the Earth's ecosystems each year is larger, probably much larger, than the sum of all the gross national products (GNPs) of all the Earth's nations. For example, the services—such as timber production, flood control, and climate regulation—provided by an intact rain forest in Cameroon are worth more, in purely economic terms, never mind ecological, than forest cleared and planted for farming. Simply put, the foundation of human livelihoods is built on a diverse and resilient supply of the manifold services of nature. It turns out that biodiversity, like the wildlife discussed in this book, is the cornerstone of not only economic services but cultural, social, and aesthetic ones as well.[18]

This cornerstone that is biodiversity is eroding away. Humans now dominate most of the Earth's ecosystems, altering the Earth's life support system at scales from genes to species to ecosystems. Our best estimate, as indicated by the fossil record, is that species are disappearing a thousand times faster today than they did during the millions of years before our species migrated out of Africa. Most of this loss of species diversity is caused by converting land to agricultural production and settlements, introducing nonnative plant and animal species that often outcompete native species, and overharvesting plants and animals. As people convert land to more intensive farming and livestock raising, they inadvertently create ideal conditions for new infectious diseases to arise in wildlife, livestock, and people; these diseases, in turn, can kill yet other species. And global warming, caused by greenhouse gases emitted into the atmosphere by human activities, is shifting species into new habitats, some already heavily used by people for other purposes.[19]

Drylands, home to most of the Earth's savannas, seem to have more than their share of human and environmental problems. Of the major ecosystems around the globe, drylands, along with the cold lands near the poles, have the least rainfall and thus are the least productive. Mostly because of this, the gross domestic product (GDP) of global drylands is about $4,500 per person per year, lower than anywhere else on Earth (and African dryland GDP is even lower). Despite higher infant mortality, human populations still grew faster in drylands than in any other region between 1990 and 2000. Drylands and the people who inhabit them are expected to be most affected by future water scarcity and climate change. And the dryland countries ranking lowest in human development are in Africa.[20]

WHAT THIS BOOK IS ABOUT

This book is about relationships between people and wildlife in east African savannas, which can vary from synergy to measured tolerance to conflicted coexistence to full-blown conflict. I investigate the extremes of human-centric and eco-centric points of view and the important "eco-human-centric" middle ground that is opening up between them. I focus on the two-way interactions between pastoral people and wildlife but also make some comparisons with hunter-gatherer peoples, farmers, townspeople, and other groups. I also venture beyond Africa's well-known large wildlife to explore how herders and their livestock affect savanna insects, smaller mammals, plants, water, and soils.

Many people associate the term *East Africa* with the three countries of Uganda, Kenya, and Tanzania; I use the term *east Africa* in this book more broadly, however, to refer to eleven countries (see Map 1): Burundi, Djibouti, Eritrea, Ethiopia, Kenya, Rwanda, Somalia, Sudan, South Sudan, Tanzania, and Uganda. While all general references to east Africa in this book refer to this eleven-country region, four case-study chapters (8–11) focus on the wildlife-rich Maasailand of southern Kenya and northern Tanzania, where I have lived and worked for more than two and a half decades.

A fundamental assumption of this book is that people are part of ecosystems. Evidence from prehistory and history, as we will explore later on, suggests that people have had some role in creating and maintaining savannas over millennia. And people, by hunting, farming, and spreading diseases, affect savanna plants and animals deep inside "protected areas" today. I also assume that the rights of people living with wildlife are as important to consider as the rights of those who live outside pastoral lands within the same nation or in other parts of the world. I assume that conservation of wildlife, while good for the globe, should also be good for the local neighborhood. I assume that people are not the only species with rights, that nonhuman species have a right to thrive and survive also. And I assume that there exist elusive but practical win-win situations that can help local people to meet their changing aspirations and needs while conserving the wildlife around them—and further, that it is in the self-interest of all nations that this happens.

The rest of this book investigates the following fundamental questions:

1. How do savannas and pastoral societies work? Is there anything special about the region of east Africa in a story about wildlife, people, and savannas? These topics are explored in Chapters 2 and 3.

2. Chapter 3 starts by asking: How do people who hunt, herd, and farm affect African savannas today? The focus then narrows to herders and wildlife and explores the costs and benefits of their coexistence.

3. What does the past tell us about current relations, whether in balance or in conflict, between wildlife and pastoralists? Chapters 4 and 5 review the evolution of savannas, wildlife, and (proto-)humans throughout Earth's history, focusing on the last 4 million years, from the Pliocene epoch to the mid–twentieth century, in east Africa.

4. Do today's pastoral peoples do more than coexist with wildlife? Do they in fact enrich savannas for wildlife? Chapter 6 focuses on recent research which suggests that in some savannas, traditional herding not only is compatible with wildlife but may indeed attract wildlife to pastoral lands.

5. When and why does coexistence break down into conflict? This is the topic explored in Chapter 7. More generally, are Africa's savannas being irreversibly "degraded," as is widely believed? If so, who (or what) is most responsible?

6. Chapters 8–11 present case studies that explore how pastoralist-wildlife issues are being played out in savannas of east Africa today. The focus here is on four savanna ecosystems across Maasailand in Kenya and Tanzania where pastoral culture is changing at unprecedented rates, herding can no longer meet all family needs, and large wildlife, though in some areas on the decline, is still often abundant in both numbers and diversity.

7. The book concludes by proposing answers to two questions: What are the most likely future scenarios for pastoral people and wildlife in the savannas of east Africa? And is there a path to the middle ground that promises a more resilient future for both?

CHAPTER TWO

· Savannas of Our Birth

The shade under the "tree of the men" is cool, the light soft compared to the brightness of noon out in the sun in northern Kenya. "Cool" is relative here; the shade is almost body temperature (37°C, or 98.6°F), the sun beyond the leafy canopy a painful 110°F. It is the dry season. No grass grows on the plains beyond the line of riverine trees under which we sit. Upstream, a group of calves stands sleepily under a large acacia tree. Someone in Ewoi's family will come soon, unhook a long stick lodged in the tree's branches, and shake down the seedpods dangling from branches high over their heads. This is the last food available for the calves. Lying under blankets to ward off flies, the head herders of several Turkana families discuss where to take the animals next for food and water, and what places will be safe from cattle rustlers. Despite their relaxed body poses, this con-versation will determine how and if their families make it through this crushing dry season. They decide today that they must send the cattle up the rift escarpment into Uganda, where they hear there is still some grass. This decision is not made lightly, for they fear for the lives of the two brothers they will send to herd the cattle. They do not speak of it, but only last year Ugandan herders, their age-old rivals, killed their cousin in a midnight raid up on the escarpment.

Just over five hundred kilometers to the south, the short-grass plains of the Serengeti are black with wildebeest, the darkness broken only by an occasional tree, gray rock outcrop, or cluster of lighter colored zebras and gazelles. At the edge of the herd, lions stalk prey, feast on the latest kill, or lie full-belly up in the shade of a tree along a small stream. In the bush along the rivers to the west are snares set the previous night by villagers to harvest—illegally—bushmeat from the Serengeti. Today, the herd is on the move, like a river flowing north to Kenya, away from the drying water holes and browning plains in the southern Serengeti. How the herds know when to move is unknown. Perhaps the salty taste of the water or the "sour" taste of the grass tells them, or perhaps they feel and smell the rain in the north, and the calmness of the wind. Elephants in the Serengeti move instinctively for the same reasons, finding the last bits of dry-season food and water left in swamps, in shady spots along rivers, or in hillier, cooler places. Many elephants climb high up the nearby rift escarpment to cooler well-watered forests, away from the parched lowlands.

What the herders and wildlife don't know is that these diverse landscapes and the choices they offer are unusually abundant in the region where they live. In the same season but over four thousand kilometers to the west of Turkanaland, Fulani herders sitting under a similar acacia tree have a similar discussion, and one of the last remaining herds of elephants in west Africa moves from forest patch to forest patch to find more water and food. But there is no nearby escarpment and few large hills to climb to find life-sustaining resources. Instead, as the dry season deepens, the herders, and the elephants, must head farther south because that is where the food and water will be, not in any other direction. Along the way they may find a small forest enclave on a hill, or a wetland, but mostly they will find a flat, almost featureless landscape. They will not follow exactly the same route south every year, but they will head that way; they have no other choices.

Their east African contemporaries will scatter in several directions, taking advantage of the small hills and large escarpments that provide opportunities for sustenance. But even here they find their choices cut off, more and more, by expanding farms and settlements, filled with people often new to savanna living.[1]

WHERE, WHAT, WHO: A GLOBAL VIEW OF PASTORAL PEOPLES AND SAVANNAS

In our mind's eye, we might picture a savanna as a hot, bright, open plain of knee-high grass, with a flat-topped acacia in the distance and zebras, wildebeest, and giraffes grazing peacefully nearby. This is a very African-centric vision of a savanna. Elsewhere, in Asia, Latin America, and northern Australia, tropical savannas usually support fewer large animals, often have no acacias but many other kinds of trees, and are covered by grass that varies from wispy ankle-high annuals to robust chest-high perennials. Even in Africa, our mind's-eye view can be misleading. Although the most-visited game parks often do look like what we imagined, many African savannas look more like vast dry forests interrupted by occasional open stretches of grass; others have no trees at all, just grass and scattered shrubs. Many African savannas support few large animals or are so bushy that wildlife are hard to see. Most important, this vision excludes a species that likely evolved in African wooded savannas and whose ancestors eventually populated the entire world: us, *Homo sapiens.*

Scientists working in savannas often argue about what savannas are, partly because of their variability and partly because there is no stasis: a dry, patchy forest can become an open grassy savanna and vice versa, depending on how much it is burned, browsed, grazed, or tilled. The word *savanna,* which entered the English language from Spanish nearly 400 years ago, originally referred to grassland free of trees. According to the modern definition, savannas are grass–dominated ecosystems that include some woody plants (trees, shrubs), are restricted to the tropics or subtropics, and have alternating wet and dry seasons. Across Africa, areas where rainfall is less than about 100 millimeters (mm) per year (4 in) and there are no trees (except where there is groundwater) are considered deserts, not savannas (Map 7).[2] Even though people and wildlife do interact in deserts, our focus will be on savannas.

African savannas are of two types. The first, representing 43 percent of African savannas, exists in areas with about 100–650 mm (4–26 in) annual rainfall, where low rainfall suppresses tree growth, creating landscapes dominated by grass with scattered trees. These are called *rainfall-driven savannas.* A few forests grow in this rainfall range where there is abundant groundwater (along rivers, for example), but these are relatively rare.[3]

Above about 650 mm annual rainfall, where moisture is abundant enough to support vigorous tree growth in a forest, African savannas exist only where people, browsing animals, fires (at least once every ten years), or heavy, shallow, or waterlogged soils create them by stunting or killing trees. This second type, known as *disturbance-driven savanna* (ignoring soils), represents 57 percent of African savannas.

By these definitions, African savannas occupy the vast lands that stretch between the broad belt of rain forests spanning the Earth's equator and the dry deserts, such as the Sahara and Kalahari, that form two more belts at middle latitudes north and south of the equator (Map 7). The danger of defining the savanna this broadly is that it includes places as different as the miombo woodland of southern Tanzania, where sunlight dapples the forest floor under a high, dense tree canopy,

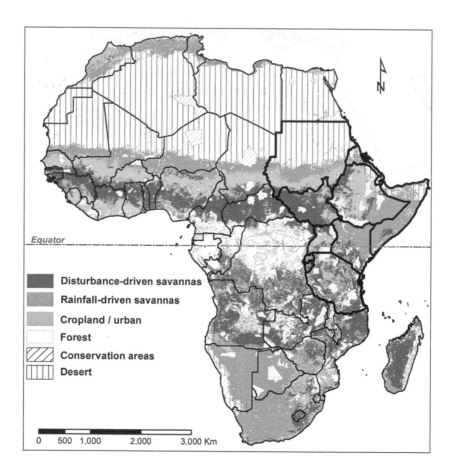

MAP 7.
Savannas (rainfall-driven and disturbance-driven), forests,
deserts, conservation areas, cropland, and urban areas in Africa.
(Map adapted from GLC 2003 and WPDA 2009 by Russell L.
Kruska.)

and the dry, sandy, dwarf bushlands of southeastern Ethiopia, where a newborn
goat is easy to spot from afar among short, leguminous shrubs and slender annual
grasses. Yet there is a benefit to defining African savannas so broadly, and that is
that this definition covers most of the areas where people and their livestock interact with wildlife on the African continent today.[4]

There are more savannas in Africa than on any other continent on Earth; in fact,
Africa could be called the savanna continent. About 44 percent of the world's tropi-

cal and subtropical savannas are in Africa (11.8 million km² or 4.6 million mi²), compared to 22 percent in Australia (5.9 million km² or 2.3 million mi²), 15 percent in Asia (3.9 million km² or 1.5 million mi²), 13 percent in South America (3.5 million km² or 1.4 million mi²), and 6 percent in North America (1.6 million km² or 0.6 million mi²). Europe's only savannas are on the southern tip of Crete. Africa has such extensive savannas by chance: it is the only continent with most of its land along the equator, or just north and south of it, between the heavy rainfall areas on the equator and the global belt of midlatitude deserts. In Africa, more than 75 percent of savannas are dry (arid and semiarid), as they are in North America and Australia. Strikingly different are the savannas in South America, which are two-thirds wet (subhumid or humid) and one-third dry. If you were to blindly parachute into Africa at a random point, 4 times out of 10 you would touch down in a savanna, 3 times out of 10 in a desert, and only 1.2 times out of 10 in a forest or cropland (Map 7). And once on foot in an African savanna, you would be walking within a vegetation type that covers more area than all of Europe, all of Australia, or all of the United States.

We think of African savannas as teeming with diverse mammals, but there are actually more mammalian species in the savannas of South America, which, though at just over half the size of Africa, swarms with small "gnawers": rodents and rabbits and other lagomorphs. Asia has relatively few large mammals but lots of bats. Africa is remarkable not because there are many types of mammals but because they are so big. Africa has the big plant eaters such as elephants, rhinos, giraffes, hippos, and relatives of cattle (bovids, including antelopes); some primates such as catarrhine monkeys and big apes (chimps, gorillas); and many smaller species such as hyraxes, galagos, and aardvarks. While only 40 percent of Africa's mammals live in savannas, these include the majority of the meat eaters (carnivores) and hooved plant eaters (grazers and browsers, both herbivores). No place else in the world supports the diversity of large mammals that live in Africa today, a remnant of the megafauna that thrived across the Earth only 20,000 years ago, during the Pleistocene.[5]

People today use Africa's savannas for everything imaginable: mining, charcoal production, wildlife conservation, tourism, settlement, and of course food production—hunting, crop farming, and horticulture. But the most common way for people to turn sunlight into food in the African savannas (and deserts) is to keep domesticated animals, in mobile herds on commonly owned land, that transform hard-to-digest plant matter into easy-to-digest animal tissue and fluids: meat, milk, and blood. This way of using the land is known as pastoralism, the herding of livestock.

Because the world's savannas and deserts are so vast, pastoralism is the most widespread way people use the land on Earth. But in these vast lands the absolute number of pastoralists in drylands worldwide is small: less than 250 million (or less than 4 percent of the total world population). Pastoral people are particularly abundant in Africa, which is home to more than half of the world's pastoral peoples, who in turn produce 75 percent of the continent's milk and more than half of its meat. And yet to most Africans, who grew up in cities, in forest villages, or on farms, as well as to most development officials, pastoralism is a foreign way of life. This minority status marginalizes pastoral peoples in many African countries (notable exceptions being pastoralist-dominated Somalia and Mauritania).[6]

Pastoralism is a fluid lifestyle, with people taking up different ways of living depending on circumstance, which makes it sometimes difficult to pinpoint just who is a "herder" or a "pastoralist." For example, herders often take refuge and settle with farmers when drought strikes and kills most of their livestock; in these trying circumstances, it can take years for herders to build up their herds sufficiently to allow their families to take up a moving lifestyle again. Many African families combine herding with other activities such as fishing and crop-raising. Are these people then herders, or are they farmers or fishers? The surest approach is to view people as people see themselves, as suggested by anthropologist Andrew Smith. Those who think of themselves as herders, even if they do not always herd livestock, are considered herders; those who value their cultivated gardens or river fishing more than their herding are considered farmers or fishers, not pastoralists.[7]

One fundamental way herders differ around the world is in their reasons for herding livestock. A Colorado cattle rancher, an Argentinean llama herder, a Somali camel herder, an outback Australian shepherd, and a horse-riding shepherd from Mongolia all have herding in common. Yet their strategies differ, in line with their goals. *Commercial* ranchers, such as the Coloradan, Argentinean, and Australian, have strong and available markets for their livestock, so they produce the most valuable and marketable product: meat, with some meat on the side for subsistence. In these operations, livestock graze on huge tracts of land, require considerable capital investment, and are tended by few people. Ranchers usually herd only one species of animal and raise male-dominated herds.

The largely *subsistence* herders, in contrast, such as the Somali or Mongolian, need to produce edible animal products—milk and meat—as many days of the year as possible as well as extra animals to sell for cash. To do this, they keep a herd with two or more species (cattle, sheep, and goats; sheep and yaks) to take advantage of their varying rangelands most efficiently and avoid the risk of disease wiping out a single species. Their herds are dominated by females for milking, creating a daily source of family food. In the 1980s, anthropologist Kathleen Galvin found that the diet of pastoralists in east and west Africa was 18–64 percent milk and only 0–18 percent meat, blood, and fat, with the remainder made up of grains. According to Roger Blench, a linguist and anthropologist, "Because pastoral production depends on human and animal energy rather than elaborate tools, pastoral economies require large inputs of labor but little capital other than the animals themselves." With large families providing lots of labor, subsistence herders can track changes in pasturage by moving and splitting herds quickly; in this way, they are able to support more animals on the same amount and quality of land than a commercial rancher can. Also, compared to commercial ranchers, subsistence herders use many more of the products from their animals, such as skins and bone, instead of buying similar products in the marketplace.[8]

The relative mobility of pastoral families and how they access land and water strongly influence their interactions with the surrounding savanna and wildlife. In Africa, I have observed five types of movement strategies (Figure 2), each of which causes people to interact with, and even think about, wildlife differently. These five strategies vary in how long families live in a settlement, how far they move, how they own the land, and how much they depend on growing livestock or crops (and other sources of income) to support themselves.

First are *nomadic herders* ("pure" pastoralists), who move most of their families and herds frequently, live almost entirely off their livestock, maintain only loose connections to markets and the state, and often own land as a group rather than individually.[9] These families move frequently to reach greener pastures (and for other reasons) and often have no specific dry- or wet-season pastures. In Africa, these peoples inhabit the driest lands with the most uncertain rainfall; they include tribes in northern Mauritania, northern Kenya (such as the Turkana, Rendille, and Gabra [the Gabra are shown with the Rendille in Map 5]), and Namibia (the Himba).[10]

Second are *seasonally moving herders*, or transhumant pastoralists, who move livestock and part of the family between reasonably well defined wet- and dry-season pastures. They may have other ways to make a living—a bit of cultivation, jobs in a nearby town—and land is owned mostly by the group. The classic transhumants are the Fulani or Fulbe of west Africa, some of whom still move long distances with their cattle once a year (called transhumance), grazing during the wet season in the ephemeral grasslands of the northern Sahel and moving as much as 500 km (about 300 mi) in the dry season to the wetter woodlands and harvested

Type of herder (Examples)	Stylized movement pattern

Nomadic herders *("pure" pastoralists; Turkana, Rendille, Gabra)*	
Seasonally moving herders *(Transhumant pastoralists; Maasai, Samburu, Somali)*	
Settled herders with nomadic herds *(Agro-pastoralists; Borana, Bahima, Barabaig, Nuer)*	
Sedentary herder-farmers *(Mixed farmers; Kamba)*	
Fenced ranchers *(Laikipia, Machakos, Manyara ranches)*	

FIGURE 2.

Movement patterns of different types of herding families in Africa. Open dots are temporary settlements; gray are seasonal; black are permanent. Solid arrows show examples of daily movement paths; dotted areas are monthly or seasonal moves; boxes represent private plots of land. (Sources: Baxter 1975, Fratkin et al. 1994, Niamir-Fuller 1999c, Blench 2000, Fratkin and Roth 2005, Homewood 2008.)

farmlands of the south. In east Africa, these people include some of the Maasai of Kenya and Tanzania, most Samburu, and cattle herders in southern Somalia.[11]

The other three types of African herders have at least one permanent settlement. *Settled families with nomadic herds* fall into two subgroups, those that do not grow crops and those that do (the latter being called agro-pastoralists). In dry savannas, most of the family lives permanently near a market center to access markets, schools, and health care but sends livestock out with a herder—either a family member or hired herder—to camps far from town. Rainfall is generally too low for families to cultivate crops, but they may own land. This type includes herders in the driest parts of the African continent. In wetter savannas, permanently settled members of the family grow crops and sometimes own their land. Today in east Africa, these peoples include the Nyangatoum (grouped with the

Toposa on Map 6) and Borana of Ethiopia; the Nuer, Dinka, and Toposa of South Sudan; the Jie of Uganda; the Barabaig (near Iraqw on Map 5) of Tanzania; and some Samburu, Borana, and Maasai of Kenya.

Sedentary herder-farmers, the fourth major group, have permanent settlements, herd animals in pastures around their settlements, grow crops, and usually own their land. This group includes most people who still herd in southern Africa (Tswana, Zulu, Swazi, Herero) and former herders elsewhere, such as the Kamba of Kenya. A special case is the fifth group, *fenced ranchers,* which includes government employees or private citizens who manage large tracts of fenced land and are well connected to meat and milk markets. This strategy is rare in Africa compared to the other four types, where the vast majority of the grazing land is unfenced and owned by groups in common. The fenced ranches are important from a wildlife perspective because some of them in eastern and southern Africa are now wildlife ranches or ranches combining wildlife and livestock. These five groups are quite fluid, and today many families are settling down and shifting from a more mobile to a more sedentary lifestyle.[12]

DIVERSITY AND OPPORTUNITY IN THE SAVANNAS OF EAST AFRICA

Savannas are more abundant in east Africa than elsewhere on the continent; a full 50 percent of this region is savanna, while deserts cover another 23 percent and forests 4 percent. This is one of the only places in the world where drylands exist at the equator. Traveling in the east African highlands, where most roads are, gives one the impression that the region is crowded with farms and towns with little open space. As Map 7 shows, however, only a quarter of east Africa is like this; a full three-quarters of the land in this region is sparsely populated, and most of this land is savanna.[13]

Both savannas in east Africa and the pastoral peoples that inhabit them differ from those in the rest of Africa in fundamental ways. Many east African savannas have two wet seasons and two dry seasons each year rather than just one of each. This pattern is rare both in the rest of Africa and around the world. Bimodal rainfall, as the first pattern is called, occurs only near the equator because the belt of rainfall, called the Intertropical Convergence Zone (ITCZ), passes twice here as it moves north and south following the sun's heating of the land (the ITCZ reaches places more distant from the equator only once a year). This is something most Kenyans and many Ethiopians and Somalis take for granted; their rhythm of life revolves around these double seasons. The appearance of bimodal rainfall in east Africa about 2,500 years ago may have allowed pastoral people to focus primarily on livestock as a way to produce food rather than combining crops and livestock, as is common in monomodal rainfall areas (see Chapter 5).[14]

Agro-climatologist Peter Jones created a map of bimodal and monomodal rainfall in Asia, Latin America, and Africa by defining the end of a rainy season when it was followed by at least one month with less than 60 mm (2.4 in) of rainfall (Map 8). Geographer Russell Kruska and I measured the extent of these rainfall patterns on Peter's map to determine how widespread the double rainfall pattern is in tropical Asia, Latin America, and Africa. It turns out that it is extremely rare: less than 5 percent of these regions has double-season rainfall. Not only that, but 70 percent of the double rainfall seasons are only in Africa, with 29 percent in Latin America and only 2 percent in tropical Asia. In Africa, moreover, 83 percent of all double seasons fall in east Africa, with a few patches along coastal West Africa, Cameroon, Angola, and South Africa. Within east Africa, more than two-thirds of Kenya has two seasons of rainfall, while only about a quarter of Uganda, Somalia, and Ethiopia have two rainy seasons. In the other seven east African countries—Burundi, Djibouti, Eritrea, Rwanda, South Sudan, Sudan, and Tanzania—only a few patches, some 5 percent of their land area, are characterized by double season rainfall. Savannas with bimodal rainfall are most common along the equator; they tend to be interspersed among other savannas that receive only one season of rainfall.[15]

The juxtaposition of tall escarpments and deep valleys accentuates this diversity of rainfall seasonality in much of east Africa. In many places, you can walk from cool, well-watered highlands to hot, dry lowlands within an hour. On this walk, you would traverse great differences in the potential to grow food, for both people and wildlife.

In thinking about this, Kruska and I had a question: if you stopped at any point in Africa and walked for an hour, would it be flat or hilly? Put another way, what kind of landscape does a pastoral family or a herd of wildebeest encounter in different parts of Africa? This is important not only because hilly places are steep, but also because rainfall tends to increase with a rise in elevation. To answer this question, we used the latest radar elevation data from the space shuttle, with 30 million 1-by-1-km grid cells covering the entire continent. Then we created a map (Map 9) that shows how hilly or flat (variability of topography) Africa is within 5 km of any given point (grid cell). The result is striking: the greatest differences in elevation over short distances are in east Africa, near the coastline in southern Africa, throughout most of Madagascar, in the highlands of Cameroun/Nigeria

MAP 8.
Single (monomodal) and double (bimodal) rainy season patterns
in Africa. (Map used with permission of Peter G. Jones
and adapted by Russell L. Kruska.)

and Liberia/Sierra Leone, in mountainous areas of the Sahara, and along the coasts
of the Red Sea and northern Africa. Even though much of east Africa is hilly, there
are also large stretches of flat land in eastern Kenya and in most of South Sudan,
Sudan, and Somalia. Large areas of interior southern and central Africa, most of
western Africa, and much of the Sahara have little topographic variation.[16]

This map of Africa's hilliness matches my impressions when first traversing the
length of Africa more than twenty-five years ago. Most savannas in western and
central (and some in southern) Africa are sedate, graceful, and, from an east African

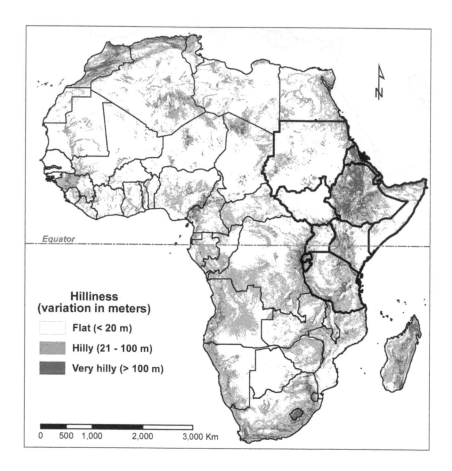

MAP 9.

Average change in elevation (m) within 5 km (one hour's walk)
of any point in Africa. (Map created from Jarvis et al. 2008 by
Russell L. Kruska.)

perspective, relatively monotonous: woodlands with seemingly identical trees
with occasional small hills stretching into shimmering heat in the distance. Those
walking in these subtle savannas delight in small variations in the surrounding
landscape, such as the variety and profusion of plants around a termite mound in
a Zimbabwean miombo forest or the tangled islets of trees on claypan outcrops
in the dry woodlands of Burkina Faso. In much of east Africa, by contrast, the
large topographic variations found over even short distances give those who live
in these savannas—particularly those who move, like pastoralists and wildlife—

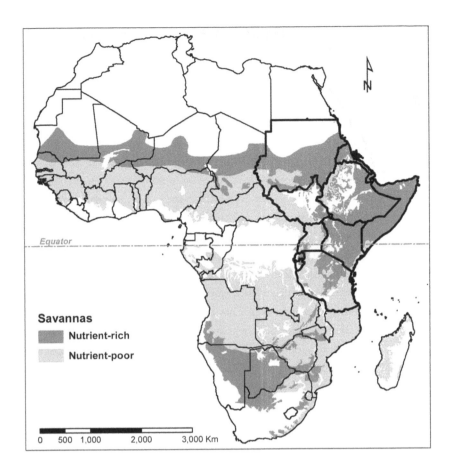

MAP 10.
Nutrient-rich and nutrient-poor savannas in Africa as determined by soils. (Map used with permission of Mahesh Sankaran
[2005] and adapted by Russell L. Kruska.)

critical opportunities to access food and water. Most scientists think that our hominin ancestors and wildlife have taken advantage of these diverse landscapes for
millions of years (see Chapter 4).

Many eastern and southern African savannas are distinct from those elsewhere
because they have extensive soils rich in nutrients. Nutrient-rich (eutrophic) savannas roughly correspond to those with fine-leaved trees—acacias, mostly—and
nutrient-poor (dystrophic) savannas to those with broad-leaved trees, as shown on
Map 10 (created and kindly lent by Mahesh Sankaran, a savanna ecologist). This

map shows that 70 percent of east Africa's savannas have rich soils, as compared to only 30 percent of savannas elsewhere. Outside east Africa, the other blocks of savannas with rich soils are along the Sahel on the southern fringe of the Sahara, in dry Botswana and Namibia, and in northwestern South Africa.[17]

That east African savannas have relatively rich soils has far-reaching implications. Soil fertility affects everything that is possible for plants, wildlife, and people in these landscapes, a story told elegantly by Richard Bell, a savanna ecologist. Bell analyzed the nutrients in grasses from different soils in a floodplain in Zambia and found that grasses that grew on soils richer in essential nutrients often contained more protein. He posited that richer grass attracts larger and more diverse herds of grazers, particularly the many small and medium-sized grazers such as gazelles and wildebeest that require high-quality grass to survive. In general, poor soils support poor-quality grass, and only a few, disproportionately large animals, such as elephants and buffaloes, are able to eat and digest enough of this forage to obtain the nutrients they need. Eutrophic savannas can support about two to three times the biomass of grazers than can dystrophic savannas. In addition, soil nutrients likely cycle more quickly from soil to plant to animal and back again in the rich savannas. Bell also found that richer soils, regardless of rainfall, support more people and more livestock. When Paul Okwi and his colleagues mapped levels of poverty in Kenya and overlaid poverty on soil fertility, they found that poorer people live on poorer soils.[18]

East Africa is also home to more species of medium to large-sized mammals than the rest of Africa, as Kruska and I learned, to our surprise, when we overlaid distribution maps of the 281 most common species these animals (excluding elephants and rhinos) in Africa, excluding Madagascar. We then counted how many species occurred in different places across Africa. East Africa, we found, has more medium to large mammal species than any other region of Africa. And within east Africa, there are more mammal species in the Rift Valley of southern Kenya and northern Tanzania and in the Ruwenzori Mountains to the west than anywhere else (Map 11). Great variety in climate and topography, and the richness of the soils, no doubt contribute to the abundance of species.[19]

East Africa also stands out as home to more than half of the continent's herding peoples, who use savannas for grazing. This is remarkable, given that the region accounts for only 21 percent of the continent's land area. Part of the reason may be that east Africa has about 20 percent more savanna than the rest of Africa.

Of the four major language groups (phyla) in Africa, pastoral and agro-pastoral peoples of east Africa fall almost entirely into two: Afro-Asiatic, includ-

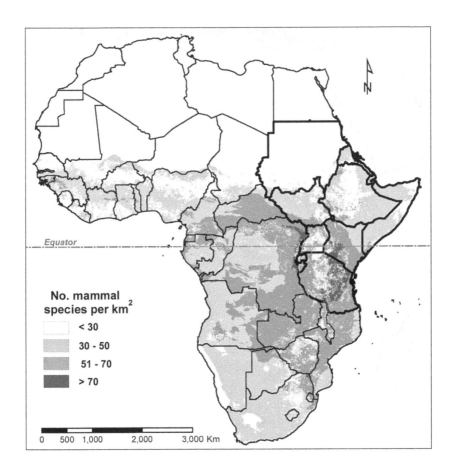

MAP II.
Density (number/km^2) of medium to large mammal species in
Africa. (Map adapted from Boitani et al. 1998 by Russell L.
Kruska.)

ing the Beja, Cushites, Semites, and Arabs; and Nilo-Saharan, including Nu-
bians, Pre-Nilotes, and Nilotes (see Map 3, based on linguist Joseph Greenberg's
1963 classification of the languages of different ethnic groups). These pastoral and
agro-pastoral peoples can be further split into more than seventy linguistic, social, cultural, and political groups that include the Rendille, Ariaal, Borana,
Gabra, Orma, Samburu, Njemps, Kipsigis, Pokot, Nandi, Tugen, Il-Chamus,
Teso, Turkana, and Kamba of Kenya; the Maasai of Kenya and Tanzania; the
Barabaig, Tatoga, and Parakuyu of Tanzania; the Dinka, Nuer, Shilluk, Toposa,

Murle, and Nyangatoum of South Sudan; the Beja, Kababish, and Fulani of Sudan; the Jie, Karamojong, and Bahima of Uganda; the Afar, Borana, Oromo, Mursi, and Daasanach of Ethiopia; the Tutsi of Rwanda and Burundi; and the Somali of Somalia. See Maps 4–6 for anthropologist George Murdock's historical distribution of ethnic groups distinguished by Greenberg's language phyla.

Besides pastoralists, the rest of the groups in the region were historically hunter-gatherers or farmers. Hunter-gatherers are now hard to find, living in small pockets around the region—so small that when Murdock drew his ethnic group map in 1959 he included only four hunter-gatherer groups out of the 220 ethnic groups he showed across east Africa. Some of these hunter-gatherers are in a third language phylum in the region, Khoisan, like the click-speaking Hadza (Kindiga on Map 5) of Tanzania. Most farming peoples belong to the fourth language group, the Niger-Congo (Bantu), but some fall into the Afro-Asiatic and Nilo-Saharan groups, with the click-speaking Sandawe farmers in the Khoisan phylum.[20]

A final reason the savannas of east Africa are special is that they are probably the "savannas of our birth." It is here that some apes probably first climbed down from their arboreal homes and ventured out onto the land. Here we find the oldest evidence of ancient hominins walking upright, their hands freed for carrying, throwing, and digging. They probably crafted the first tools on these plains; hunted and scavenged wildlife here; first loved and first killed each other here. They may have purposely lit fires here for the first time, and spoken their first words here too. As far as we know, it is from here, millennium after millennium, that our forebears repeatedly left Africa and populated other parts of the world. In this sense, every person can claim east Africa as their ancestral, hominin, home.[21]

WHAT MAKES A SAVANNA A SAVANNA?

Why are savannas where they are, how do they work, and what makes them so resilient? At first glance, it seems that African savannas occupy places too dry for a forest and too wet for a desert (Map 7). Tropical forest and desert plants have in common a relatively constant (wet or dry) climate, while savanna plants must survive a great variability in growing conditions. Tropical forest plants would have difficulty surviving a savanna dry season, while desert plants would have difficulty adapting to the frequent wetting and drying that occurs in savannas. A savanna may be lush and green one season, parched and brown the next. Because wet seasons can be short and dry seasons long, savanna plants must respond quickly to rainfall, as must the people and wildlife who depend upon them.

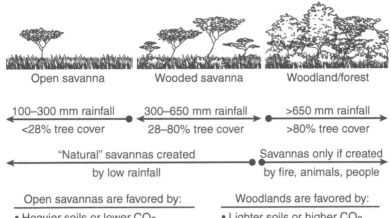

Open savanna | Wooded savanna | Woodland/forest

100–300 mm rainfall | 300–650 mm rainfall | >650 mm rainfall
<28% tree cover | 28–80% tree cover | >80% tree cover

"Natural" savannas created | Savannas only if created
by low rainfall | by fire, animals, people

Open savannas are favored by:

- Heavier soils or lower CO_2
- More elephants, more fire
- Wood-cutting for construction, fuel
- Heavy browsing by wild browsers, goats or camels

Woodlands are favored by:

- Lighter soils or higher CO_2
- Fewer elephants or fires
- Less wood-cutting
- Heavy grazing by wild grazers, cows or sheep

FIGURE 3.
Forces that determine the balance of trees and grass in savannas.
(Sources: Scholes and Archer 1997, Bond and Midgley 2003,
Sankaran 2005.)

In essence, the amount and seasonality of rainfall explain why savannas are where they are today. Yet knowledge of rainfall alone is not enough to predict if a savanna will be largely woodland, shrubland, or open grassland. As mentioned above, in wetter savannas (with greater than 650 mm of annual rainfall) with no woodcutting, fire, or browsing animals, especially elephants, trees often grow more vigorously than grasses and take over, creating woodland or forest. Below 650 mm of annual rainfall, there is not enough moisture to support a forest, except in areas with abundant groundwater, so trees and grass will coexist, though on the edge of deserts (with less than 100 mm rainfall) there may be no trees at all. In both wet and dry savannas, fires, soils, animals, and people play a large role in shifting the balance of trees and grass (Figure 3).[22]

How heavy the soil is often governs how woody or grassy a savanna is. Heavy, clayey soils support more grass; lighter, sandy soils, more trees. In the Maasai Mara region of Kenya, for example, the flat plains and bottomlands, with their heavier soils, are often more grassy than the nearby hills, which tend to be covered with dense shrubs and some trees. Grasses may succeed in heavy soils because these soils absorb water slowly, allowing shallow grass roots to capture most of the water before it can percolate down to tree roots. Indeed, when ecologists removed the grass from a heavy soil in savannas in South Africa and Kenya, the trees grew more quickly, while doing the same thing in the lighter soils of a nearby woodland had no effect on tree growth.[23]

Fire is probably the most potent force affecting the balance of trees and grass in savannas, particularly in wetter savannas. It is also extremely common: over half of the savannas in Africa burn every year. Savannas are perfect places for fires: unlike deserts, they are wet enough so that masses of flammable grass accumulate in the wet season; unlike rain forests, they are dry enough so that this fuel dries out and can burn if lit. And fire is more common the wetter savannas get. This regular burning can create the classic savanna landscape of a few trees in a sea of grass. By destroying the tops of young shrubs and trees, fire prevents them from growing tall enough to avoid the reach of most flames. Half a century ago in Zambia, Colin Trapnell experimentally recreated what hunter-gatherers and herders have done in Africa for millennia: he burned a woodland often enough (every year) to create an open grassland after a few years. It is quite possible that early humans helped spread savannas in Africa by burning (see Chapter 4).[24]

Like fire, elephants can have a huge impact on trees in savannas. Along with humans, elephants are one of the most powerful ecosystem engineers on the planet. In the past, when elephants gathered in one place for a significant period they might push down all the trees, turning a shady woodland or forest into a sunny grassland dotted with broken tree trunks. Today elephants are rarely at a sufficient density to do this; more commonly, they make existing savannas more open by knocking down single trees or patches of trees. In the mid-1980s in Kenya's Amboseli area, ecologist David Western and his long-time Maasai ecologist colleague, David Maitumo, erected a single-strand wire fence, 2 meters (6.6 ft) tall, and electrified it to keep out browsing elephants but not other smaller browsers that eat woody plants, such as impalas. After four years, the seedlings inside the fence had grown into a small forest 3 m (9.8 ft) tall, while outside the fence the trees were still only knee high in an open grassland. Western and Maitumo think that elephants were largely responsible for keeping the unfenced area in grassland.[25]

Elephants kill mature trees, too. If they break many branches or remove bark around a tree trunk, the tree is four times more likely to be dead in five years than one untouched by an elephant. Working in Kenya's Tsavo National Park, ecologist Willem van Wijngaarden calculated that it takes twenty to twenty-five years for elephants, at a high density of about 1 per km^2, to turn a closed woodland into open grassland. In his doctoral work, ecologist John Waithaka studied elephant damage across Kenya. He found that they knock down trees in ecosystems ranging from wet forests to dry woodlands, particularly when people have removed their habitats by farming and cut off their traditional travel routes so they can no longer migrate freely.[26]

Sometimes elephants and fire work in concert to rapidly turn woodland into grassland. Ecologist Holly Dublin was puzzled by the almost complete loss of trees in the Mara ecosystem from the 1950s to the 1980s. To find out the cause, she experimented with the effects of fire, elephants, wildebeest, and antelopes by burning parts of the savanna and fencing other parts off in different combinations. She found that elephants alone did not explain the tree loss; only when elephant damage was combined with fire or browsing by other animals did woodlands decline quickly.[27]

Even by themselves, smaller browsing animals can have a surprisingly large effect on trees. Joy Belsky, a savanna ecologist working in the Serengeti, showed that small browsers such as the impala, a graceful antelope that prefers feeding in wooded areas, can keep grasslands open by continually nibbling down new tree shoots. She fenced off small plots of Serengeti grassland and found that outside the fences, where impalas and other browsers fed freely, trees were less than half the size of those in the fenced-off areas. At nearby Lake Manyara, major die-offs of impalas during an outbreak of anthrax in the 1980s allowed trees previously stunted by their browsing to grow to maturity in large numbers. Even rodents can magnify the tree-killing effects of larger wildlife in central Kenya.[28]

People open up savannas as well, by burning and cutting wood. About nine out of ten fires in African savannas today are started by people. The east African Maasai, as they moved southward hundreds of years ago, used fire to open up grazing lands for their cattle. Many farmers today burn their fields to create pasture for their livestock or to clear them before planting, and these fires can escape and burn surrounding savannas. Hunter-gatherers use fire to drive animals toward an ambush. Honey hunters also accidentally light fires when they are smoking out bees. Tree fellers open up woodlands by intensively harvesting trees for construction, firewood, or charcoal.[29]

Sometimes, however, people do the opposite: encourage trees to grow where open savanna once stood. James Fairhead and Melissa Leach, anthropologists working where forest meets savanna in west Africa, describe how Guinean villagers have created forest from open savanna. In the past, farmers settled on rocky hilltops, which were easy to defend. Around the base of these hills, the farmers "lent a helping hand" to the forest by planting and fertilizing young trees and preventing fires, thus creating dense forest islands within a broad savanna landscape. Historically the trees provided a fortification around the village against raiding tribes. Today's villagers recognize the value of the forest islands created by their ancestors and maintain them to supply forest products and create a moist environment to grow valued tree crops like bananas and oil palm.[30]

Across the continent to the east, Turkana herders in northern Kenya also create miniature forests in and around their old settlements. In the 1980s, our Turkana research team kept noticing circular patches of *Acacia tortilis,* one of the classic umbrella-shaped acacia trees, throughout the area. We wanted to know how they got there and how they affect this ecosystem, and this became the subject of my PhD research. Our Turkana informants told an intriguing story of these tree patches. In the dry season, they said, the seed pods of this acacia are one of the few good sources of food for their livestock. So herders guard trees around their homesteads, and when they need the pods they hook a long stick high in the canopy and shake them down for the goats, sheep, and calves waiting below. Each pod holds about six to ten seeds, which feel like little hard pebbles. These pebble-like seeds are tough enough to pass through the ruminant stomach of a goat intact, although the rumen's digestive juices weaken the seed's shell. Sift through a handful of goat pellets in the dry season, which I did by the thousands, and you will often find several intact seeds. Because Turkana herders enclose their livestock in thorn corrals at night to protect them from attack by people, hyenas, lions, and leopards, half of the dung produced by a goat ends up on top of the soil in this corral and the other half gets scattered around the savanna as the goat grazes during the day. My Turkana team and I counted 25,000 seeds in the average 10 m–diameter goat corral after one month of use, but only 250 seeds in the same area far from settlements.[31]

This is more than a curiosity because Turkana are one of the most mobile peoples on Earth, repeatedly leaving their homesteads and corrals for new places. When it rains, thousands of the acacia seeds in the abandoned Turkana sites germinate and soon young acacia seedlings carpet the old corrals. Although many of these trees die over time, more acacias growing within the old corrals survive than those growing outside the Turkana homesteads. Dung in corrals forms a moist and nutritious mulch that improves survival of young trees when they are small and vulnerable to Turkana's long dry seasons.

HOW PLANTS AND ANIMALS SURVIVE,
AND THRIVE, IN AFRICAN SAVANNAS

Grasses and some trees are uniquely adapted to savanna environments, which are characterized by frequent fire, drought, and herds of grazing animals. Unlike other plants, grasses are tough to digest, grow from the base (from intercalary meristems, or growing points), grow prostrate (low-lying) against the soil surface, and often have more roots beneath the ground than shoots above. George Stebbins, an evolutionary ecologist, suggested that grasses developed these unique habits in order to survive grazing, or even to avoid being grazed (see Chapter 4). But ecologists William Bond and Michael Coughenour have suggested that grasses are just as likely to have evolved these habits in order to survive fire or drought as to survive grazing. Many savanna trees developed some of the same resilience: tough bark that resists fire, seeds that survive burns, and an ability to resprout (or coppice) after fire. And as we saw above, seeds of some trees germinate more quickly after passing through the gut of a grazing animal.[32]

It is remarkable how plant-eating animals (herbivores) manage to survive on a diet of grass, a difficult food. You and I, with our "simple" stomachs (like those of pigs), would starve rapidly on such a diet; even if we ate until our stomachs were brimming, the grass, replete with silica, cellulose, hemi-cellulose, and lignin, would pass out of us undigested. Although horselike grazers, such as zebras and donkeys, have digestive systems like our own, with a simple stomach and small and large intestines, they also have an enlarged caecum for some extra digestion (hence their status as *caecal digesters*). They eat a lot of grass but absorb the parts easiest to digest, and excrete the rest.

Ruminants are the consummate grass digesters, with their four-chambered stomach filled with microbial symbionts. They eat less grass than a zebra but digest it more completely. There are three kinds of ruminants. *Concentrate selectors,* or browsers, mostly avoid grass, preferring the leaves of trees, shrubs, and leafy herbs, which are generally easier to digest than grass but harder to find in large quantities. Then again, some species, as well as older parts of most woody plants, contain chemical compounds (tannins and others) that make them difficult to digest, so browsers need to be especially picky. Browsers include dik diks, duikers, bushbucks, kudus, gerenuks, giraffes, and camels (Figure 4). *Roughage feeders,* or grazers, eat large quantities of grass, digesting it slowly and more or less completely in their rumens. This group includes oryxes, topis, reedbucks, kongonis, wildebeest, oribis, waterbucks, buffaloes, sheep, and cattle. Then there are *mixed*

FIGURE 4.
Giraffes and an eland browsing together in a bushed savanna,
Kapiti, Kenya. (Photo by Cathleen J. Wilson.)

feeders, which eat grass in the wet season when it is green and easy to digest, then switch to shrub and tree leaves in the dry season. This group includes impalas, Grant's gazelles, elands, steenboks, and goats. In general, large animals can digest tougher and drier grass than small animals because large animals have lower energy requirements compared to how quickly their rumens can extract energy. Thus, (large) buffaloes can survive on tall brown grass while (small) Thomson's gazelles need more digestible green grass (Figure 5). With this knowledge, one can predict where to find different types of animals in the woodland and grassland patches of a typical African savanna.[33]

Once grass passes through the guts of animals and is deposited on the ground in the form of dung, it does not last long in an African savanna. Thirty-five years ago, biologists Jonathan Anderson and Malcolm Coe watched a 1.5 kg (3.3 lb) ball of elephant dung disappear in only two hours as 16,000 different dung beetles visited the dung and carried pieces of it away. (I remember well my amazement at

FIGURE 5.
A topi and Thomson's gazelles grazing together in an open
savanna, Mara, Kenya. (Photo by Cathleen J. Wilson.)

the speed and buzzing sound of shiny blue dung beetles arriving immediately after
my first morning break in an African savanna.) Although dung beetles exist in
most African ecosystems, the greatest diversity and number of these creatures
occur in African savannas. There are more and larger dung beetles where there
are more species of mammals. And it is in savannas that you will most commonly
see dung beetles scrupulously patting and rounding a piece of dung and then carefully (or sometimes frantically) rolling it away (Figure 6).

If you follow one beetle and its ball (do!), you will find that it may roll the ball
a long distance before stopping at a grass clump or shrub and burrowing beneath.
The beetle then takes the dung ball down to its nest, sometimes more than a meter
deep, and lays an egg in the ball (though some leave the ball right on the surface of
the ground). Remarkably, by building nests of dung to care for their young—
sometimes for as little as three days—dung beetles play a critical role in returning
nutrients to the soils of African savannas. While these "rollers" will rivet your
attention, if you push the original pat of dung aside with your toe you will see the
other main group of savanna dung beetles, the "tunnelers," who dig below the
dung pat and bury dung right there. There are also a few "dwellers" that build

FIGURE 6.
Dung beetle rolling dung, Kapiti, Kenya. (Photo by Cathleen
J. Wilson.)

their nests inside the dung pat, but these are rare in savannas. If you go out at night and move a dung pat, you will likely find larger dung beetles than during the day, mostly tunnelers. Dung beetles are typically found on wild herbivore or cow dung, but some species also roll dung from birds, carnivores (lion, leopard), tortoises, toads, or even from millipede carcasses and rotten fruit. Watch for birds and baboons that find dung beetles good food.[34]

Large mammals are so visible to the human eye that we accord them undue weight as consumers of plants in African savannas. Termites may be just as important. Ecologist Souleymane Konaté, for example, working in Lamto, a wet savanna in Ivory Coast, found that termites consume 11 percent of all of the plant matter produced above ground each year, which is similar to savannas elsewhere. Along with elephants and people, termites are critical "engineers" in African savannas, determining what savannas look like and how they work. Although termites usually consume less living grass than large mammals, they consume most of the dead grass and wood. Entomologist Paul Ferrar estimated how many termites are beneath a single square meter of soil in a dry savanna in South Africa and found that the soil may contain as many as fifty to four thousand termites, although termites are notoriously difficult to count. But in the same amount of soil there is probably not a single earthworm, because they are mostly absent in all but the wettest tropical savannas. And only in the tropics are there termites that grow fungus (found only in Africa) or eat soil.[35]

Termite mounds are important as well. Some termite mounds are ancient, lasting centuries as termites repeatedly colonize the same mounds. Building mounds mixes up the soil, bringing soil from a depth to the surface that often is lighter in texture than the surficial soil. Termites also concentrate nutrients by bringing wood and grass to their mounds from surrounding areas. African farmers have long taken advantage of this fact by planting their food crops on mounds or spreading termite mound soils across their fields. In addition, termite mounds are hotspots of biodiversity in savanna ecosystems. If you walk through a miombo forest in southern Tanzania, for example, you will find more different species of trees and shrubs growing on termite mounds than in the surrounding forest. At Mpala in central Kenya, pouched mice and other small mammals prefer to dig their burrows in termite mounds rather than in the surrounding mound-free savanna. And the spacing of termite mounds matters at Mpala: their uniform distribution makes for greater productivity in this savanna than a random distribution would.[36]

Most of life is on top of the ground in a forest, whereas in African savannas most is belowground. Scientists who painstakingly dig up whole grass plants from different savannas around the world find that more than half of each plant grows underground. This is particularly exaggerated in wet savannas, where a full 80 percent of grass biomass is in the soil. This means that, despite the abundance and diversity of large mammals in savannas, more animals are eating savanna plant food below the ground than above. In a southern African savanna, for example, ecologists Bob Scholes and Brian Walker weighed the larvae of a single species in the soil, a type of root-feeding beetle (a coleopteran). Remarkably, these buried beetle larvae weighed three times as much as all of the large mammals grazing in the same area above the ground. As Scholes and Walker suggest, if the appetite of these beetles matches their abundance, we are missing most of the grazing occurring in savannas when we focus on giraffes, zebras, and other surface-feeding herbivores. Scholes and Walker also found that over 66–99 percent of the nutrients critical to life—carbon, nitrogen, and phosphorus—occur underground in this same southern African broad-leafed savanna, in the organic matter of the soil, with the remainder above ground in the plant biomass.[37]

Charles Darwin, recounting his voyage on HMS *Beagle*, was astonished by the productivity of the savannas of southern Africa: "There can be no doubt of its being sterile country . . . covered by a poor and scanty vegetation," he wrote. "Now if we look to the animals inhabiting these wide plains, we shall find their number extraordinarily great, and their bulk immense. . . . I confess it is truly surprising how such a number of animals can find support in a country producing so little food." His observation seems obvious, but is rarely explained by biologists. The Serengeti-Mara ecosystem, for example, has twenty-eight species of hoofed grazers and browsers (ungulates) and ten species of large carnivores, which all together may add up to 35 metric tons of wildlife per square kilometer, most supported ultimately by grass. The answer to Darwin's conundrum is this: savannas can support so many animals because grass that grazers consume regrows quickly and is relatively easy for a grazer to digest. A forest, in contrast, while supporting ten times more plant mass than a savanna, is only 10 percent as productive. And much more of that plant mass is edible in savannas than in forests. Leaves, which are often the best-quality food for herbivores, make up just 2 percent of the forest plants in Ivory Coast but 15–60 percent of the vegetation in nearby savannas (and not all tree leaves are easy to reach).[38]

Not only do African savannas support a great number of individual animals, but they also support a great number (and diversity) of species. Why? When it comes to eating, large savanna mammals are experts in specialization. In the early 1960s, wildlife biologists Lee and Martha Talbot, working in the Athi-Kaputiei, Loita, central Kajiado, Mara, and Serengeti ecosystems of Kenya and Tanzania, shot ninety-two wildebeest, Thomson's gazelles, Grant's gazelles, topis, and impalas, scooped out their stomach contents, and measured what proportion of different kinds of plants each animal ate. They found that each species ate different types of plants (grass, leafy herbs, shrubs, trees), different species of each plant type (wildebeest prefer *Themeda* grass; topis, *Cynodon* grass), and different parts of plants (wildebeest are leaf-eating specialists, topis prefer grass stems). Chris Field, a biologist working in Uganda in the late 1960s, had similar findings.[39]

At about the same time, ecologist Hugh Lamprey and his team of game scouts walked along three 8,000 yd–long transects (7,315 m) nearly every day for four years in what is now Tarangire National Park, northern Tanzania, and recorded six thousand observations of where and what wildlife ate. They found even more evidence than the Talbots that animals reduce competition with each other for food: different wildlife species not only ate different types, species, and parts of plants, but they also ate in different places at the same time or at different times in the same place. They found that giraffes, rhinos, and dik diks may be in the same place at the same time, but they eat plant parts growing at different heights off the ground. Buffaloes and wildebeest both eat mostly grass leaves and stems, but buffaloes prefer grasses in wooded areas, whereas wildebeest prefer them out in the open savanna. Lamprey and his team also found that some animals switch places by season: during the dry season, Grant's gazelles and impalas, which can survive with little water, move out into the waterless, dry savanna, while zebras and wildebeest move nearer to water.[40]

Grazers often stick together for safety. Ecologists working in the Serengeti-Mara ecosystem have puzzled over this for decades. In the early 1980s, ecologist A.R.E. (Tony) Sinclair, who has studied wildlife in the Serengeti for forty years, set out to find out if the wildebeest that (famously) migrate in vast numbers from the Serengeti and north to Kenya's Maasai Mara National Reserve each year attract or repel other species of wildlife. Even though they eat different plants, as Lamprey emphasized in Tarangire, Sinclair found that zebras, topis, waterbucks, warthogs, and gazelles all grazed close to the migrating herds of wildebeest once they arrived in the Mara Reserve, preferring to feed within 300 m (1,000 ft) of the wildebeest in groups that ecologists Michael and Judy Rainy call multiple-species associations. Why? Sinclair thinks animals (especially small gazelles) may graze together with wildebeest to protect themselves from predators. Ecologist Samuel McNaughton thinks they also group together on short, nutritious "grazing lawns" to access the best food. Species too large to fear predators (elephants, rhinos, giraffes) rarely graze in groups. All these explanations may be correct in some savannas and not in others (see further discussion in Chapter 6).[41]

How do these large animals affect smaller animals? It may be that the large mammals rob opportunities for other species to flourish. Ecologist Truman Young set up an experiment in the 1990s at Mpala Research Centre, central Kenya, using different types of fencing to find out what happens to plants, soils, and smaller animals when livestock, elephants, and other large wildlife are removed from a savanna. Felicia Keesing, a mammalian ecologist, used Young's experiment to find out how large mammal grazing affected pouched mice. She was astonished to learn that in less than a year, in the absence of large mammals, the number of pouched mice doubled.

Why did this happen? Keesing speculated at first that grass cover regrew where there were no large mammals, providing the mice with more protective cover from their snake, bird, mongoose, and jackal predators. But she soon found that this was not so; mouse survival in the face of predation did not in fact change with the removal of large mammals. Rather, it appears that large mammals compete with small mammals for food. With the large animals removed, the small animals ate a full half of the extra food available. This finding suggests that removing wildlife will not benefit livestock (or vice versa) as much as one might guess. Keesing hypothesized that if wildlife continue to decline in number in east Africa, small mammalian populations will explode, the numbers of small predators will increase, and the incidence of human and livestock diseases will rise. Since then, ecologist Douglas McCauley, working with Young, Keesing, and others, found that snakes and rodent fleas increase in number when wildlife decline. (I return to the "animals affecting animals" story in Chapter 8, on the Serengeti-Mara ecosystem.)[42]

Perhaps the biggest tool that animals have to adapt to savanna environments is movement. But what triggers them to start moving? Concepts developed by social scientists to help explain why people move may help us think about wildlife migration. People move because they are "pushed" away from some places or "pulled" to new ones; sometimes the push and pull are created by the same factors, sometimes not. In most cases, animals travel to the wettest part of the savanna as dryness sets in and return to the driest part of the savanna when the rains start again. In the dry season, scarce food or salty water can push some wildlife species away from the drier part of a savanna. In the Serengeti, for example, ecologist Emmanuel Gereta and hydrologist Eric Wolanski can predict within a week when the wildebeest will leave the dry, short-grass savanna for the northern, wet savannas by the amount of salt in the water of the Serengeti's Lake Magadi. At the same time, abundant food and fresh water pull animals to the wetter part of the savanna.[43]

Why don't wildlife just stay in the wettest part of the savanna year round? There are several reasons: muddy soils underfoot, poor-quality grass on which to feed, and the inability to see predators in the tall grass. Dry-savanna grasses are more nutritious because their short stature contains few of the hard plant tissues required to support tall stature, which are hard to digest. They also possess more of the nutrients critical for growth, pregnancy, and lactation of grazers than do wet-savanna grasses.[44]

An exception to this migratory pattern occurs in the (unusual) cases where abundant water exists in the driest part of a savanna, Kenya's Amboseli being a case in point. Here, animals congregate around the park's dry-savanna swamps during the dry season, grazing on the abundant, but tough, swamp grass. They move to better grazing lands with shorter, more nutritious grass in the wetter part of the savanna when those grasses begin to regrow after the rains start.[45]

Why doesn't all this wildlife become super-abundant in savannas? The answer is straightforward: predation and seasonal shortages of food. But which of these threats is more important? Ecologists Tony Sinclair, Simon Mduma, and Justin Brashares think they are both important, with differences according to body size. When it comes to predators, they found, it is dangerous to be small. A small oribi, for example, must avoid seven lethal predators; a medium-sized wildebeest, five; and a large giraffe, only one: the lion. Adult elephants and hippos are virtually predator-proof. Every oribi ultimately dies in the claws and jaws of a predator; 75 percent of zebras die this way, but only 5 percent of giraffes. When poachers killed many of the large predators in the Serengeti in the 1980s, the populations of small grazers tripled, while giraffe numbers remained unchanged. As for lack of food, a dramatic natural experiment was carried out in 1962 when rinderpest disease was controlled. Wildebeest populations increased almost ten times in the next fifteen years, until they ran out of food. Since then the populations have remained relatively stable, just below the limit of their food availability. Beyond food and predators, disease, drought, and floods also play a role in regulating populations, and in modern times, people have become significant killers of savanna wildlife as well, as we will see in Chapter 7.[46]

Given that wildlife can run out of food, can they actually overgraze the savanna, as many suggest livestock do? I have observed that wildlife can overgraze savannas, but their impact is limited. *Overgrazing* is a tricky term—difficult to define and value-laden in its interpretation (see Chapter 7). In this book, I define overgrazing as the amount of grazing that initiates a long-term change (i.e., lasting tens to hundreds of years) in the productivity or composition of soils or vegetation and from which recovery is likely to take a similarly long time once the grazing abates. As I said, wildlife do occasionally overgraze savannas. I have observed, for example, soil loss on tracks made by migratory wildebeest crossing rivers that, even if the wildebeest were to shift to another crossing, would take many years to recover. But these impacts are quite minor.

It is often humans who cause wildlife to overgraze their habitats, such as when elephant populations become compressed into reserves. The first time I saw heavy elephant damage was in Chizarira National Park, in Zimbabwe: kilometer after kilometer of mopane trees pushed over or broken 2–3 m (about 7–10 ft) from the ground. David Cumming, a wildlife ecologist from Zimbabwe, and his colleagues nicely demonstrated that such intensive elephant damage inside parks depletes the diversity of birds and ants. In central Kenya, Darcy Ogada and colleagues saw a 30 percent increase in bird species and more woody plants and ground-living arthropods when they removed elephants and giraffes. Such overgrazing is far from insignificant: extensive elephant damage threatens the conservation of biodiversity in some of Africa's parks and reserves. All of these examples involve impacts on parts of the savanna that are very slow to recover, such as soils and long-lived trees.[47]

Finally, some wild grazers may be "speeding up" the supply of nitrogen that savanna plants need for growth. If this occurs over a wide area, it may mean that grazers are making nutrient-rich (eutrophic) savannas richer, forming a feedback loop that improves the grazers' own productivity. As we have seen, nutrient-rich savannas tend to cycle nutrients faster than do nutrient-poor savannas. Grazers such as wildebeest (and bison in North America) can accelerate this effect, converting nitrogen from a form not readily available to plants to one that plants can easily use, thus increasing the amount of nutrients the plants can take in for growth and reproduction. In Yellowstone Park, for example, it has been shown that elk and bison actually double the amount of nitrogen available for plant growth. In addition, their urine adds nitrogen to the dead plant matter lying on the surface of the soil, which causes these materials to decompose faster, thus adding even more nitrogen to the soil. Grasses grazed by bison or wildebeest have more nitrogen in their leaves than nearby ungrazed grasses, and these grazers often prefer to eat the richer plant regrowth in places grazed previously, creating a positive feedback loop.[48]

CHAPTER THREE

· Pastoral People, Livestock, and Wildlife

Ole Nkare steps through the thorn fence that surrounds his corral, bending to pick up and examine a newborn kid. He likes what he sees: many healthy goats that will probably grow quickly and fetch a good price at the market. He glances south toward the peak of Mt. Kilimanjaro, 65 km (40 mi) away, across a landscape almost unchanged since his grandmother was a girl: an open savanna with no fences, where wildlife and livestock often graze side by side. Here, in Olgulului Group Ranch, he and his family move their cattle, sheep, and goats freely as the seasons demand, through a relatively porous, open landscape.

Last night at a group ranch meeting, Ole Nkare sat patiently listening as his neighbors elaborated the many advantages they see of owning their land for the first time rather than sharing and using the land as a group, as they have traditionally done. Most want a land title because they are afraid that other people, not from this place, will push them off their ancestral lands. They can see how crowded the nearby highlands are with farmers and know that many more of them want to own land. They have also watched the government take land away to create parks for wildlife and their own wealthy kinsmen mark off large pieces of land as their own.

Then it was Ole Nkare's turn to speak, and his neighbors turned to give him all their attention. Ole Nkare told of a visit he made just last week to his age-mate Ole Dukuya at his home in the Kaputiei Plains 135 km (84 mi) to the northwest, a region nestled up against Nairobi National Park and, immediately beyond, the busy (and growing) metropolis of Nairobi. He described the tall buildings, such as a foam mattress factory and

Kenya's only ice skating rink, that now sit on top of the pasture where Ole Dukuya's grand-
father once herded his cattle among thousands of migrating wildebeest. Ole Dukuya re-
called well how, as a boy, he could graze his sheep and goats in almost any direction from
his family's boma *(homestead), through an open landscape with few boundaries and*
abundant wildlife. Now, he lives in a place where each family has its own parcel of land, so
he must ask permission to move his sheep off of his father's land and onto his neighbors'.
And to reach the water point nearby, he and his herd must walk around hard boundaries
like fencelines and buildings, traveling ever farther to reach the same point because of the
increasingly circuitous route. He also sees only a few wildebeest now, not the large herds he
remembers from his childhood. Ole Nkare finished his story by saying that neither he nor
Ole Dukuya expects their children to herd cattle, as they do; more likely, the next genera-
tion will become teachers, doctors, or engineers.

THE PEOPLE AND LIVESTOCK OF THE EAST AFRICAN SAVANNAS, AND HOW THEY LIVE

Most of the rest of this book will focus on the oldest food producers in east Africa, pastoral herders, because of one rather extraordinary fact: the remnants of the great Pleistocene herds of wildlife are most abundant and diverse in the savannas of east Africa, where they have lived side by side with the region's pastoral people for millennia (see Maps 3 and 11). These pastoralists replaced hunter-gatherers over the last five thousand years and started interacting with in-migrating farmers about three thousand years ago (see Chapters 4 and 5). Over the last century, colonial settlers and governments took some of the savannas used by pastoralists and created commercial ranches and protected areas. Today, about 9 percent of east African savannas are within parks and reserves, but in Kenya, for example, about half the nation's wildlife still live outside these preserves on pastoral lands. A much smaller fraction of savanna land is still on large, fenced, privately owned commercial ranches, which often support abundant wildlife populations. Outside parks and commercial ranches, however, most wildlife live where herders continue to move their animals seasonally, where most land is still used in common (although land is privatizing rapidly in some places), where people raise livestock for household needs more than for sale in the market, where fences remain rare, and where people see their landscapes through a "livestock lens."[1]

This book focuses on these large pastoral savannas. That wildlife are abundant here is not a coincidence; where wildlife still exist in these landscapes, it is because pastoral people choose not to exterminate them. The existence of wildlife cannot be separated from land use and culture. And, as I propose in the next chapter, the converse may be equally true: human evolution (and perhaps culture) cannot be separated from the existence of wildlife. (Farmers interact with wildlife too, of course, but nearly all of those interactions in east Africa are negative, a topic covered in Chapter 7.)

Over the last seventy years, starting with Edward Evans-Pritchard in 1940, a group of anthropologists, geographers, historians, sociologists, biologists, and economists has given us a detailed picture of the history, culture, biology, geography, and economies of pastoral peoples in east Africa. Pastoralists themselves are just starting to tell their own stories and publish their own research. I focus narrowly here on pastoral peoples' interaction with wildlife by looking at how they use wildlife directly, how they use the land, and how the wider world affects these interactions. Although wildlife studies by biologists are common, they rarely mention pastoralists (with certain exceptions), and few pastoral studies by social scientists make more than the briefest mention of wildlife (also with certain exceptions). This means that the story is incomplete: biologists focus on wildlife, social scientists focus on people, and pastoralists are not yet well published, so there is little balanced coverage of the middle ground where people and wildlife meet.[2]

Before covering the middle ground, it is important to look at how herders manage to live and even thrive in savannas. What survival challenges do they face? How do they respond to those challenges (i.e., how do pastoral people use their lands and ecosystems for food, water, and other uses)? And how are these challenges and responses changing in a dynamic world? (In Chapters 8–11, I cover these three topics using examples from across Maasailand in Kenya and Tanzania.)

Curiously, west African farmers are able to grow maize crops in places that receive almost half as much annual rainfall (400 vs. 700 mm, 16 vs. 28 in) as many places used by farmers in east Africa. Ecologist James Ellis and anthropologist Kathleen Galvin provide one reason for this anomaly: there is too little continuous rainfall to grow a crop to maturity where there is bimodal (two-season) rainfall—that is, where precipitation is spread out over the year in two seasons rather than falling all in one. But there is an advantage to the spread-out rainfall: it keeps grass growing and animals fed almost year round, which in turn keeps livestock producing milk for longer periods. This makes pastoralism more viable than crop farming in much of east Africa's savannas. Archeologist Fiona Marshall thinks that the appearance of bimodal rainfall in eastern Africa about 3,000 years ago allowed pastoral people to focus on livestock as a way to produce food, rather than combining crops and livestock as is common in places with a single rainy season like western and southern Africa (see Chapter 5). In my view, this spread-out rainfall also explains why extensive wildlife and pastoral systems still exist in east Africa: if the same amount of rain fell here in one season alone, these savannas would be covered with maize and bean farms. With climate change, if east Africa becomes hotter and thus drier, this picture may change (see Chapter 12).[3]

Pastoralists are known to live and die by their livestock. Herders are also deeply connected socially, psychologically, and politically to each other and to savannas through their livestock. Livestock provide the essentials of life: food (milk, meat, fat, blood), clothing and shelter (hides), medicines, utensils, and adornments. Herding families use livestock as bank accounts to be invested in, exchanged, and marketed; as a way to create and strengthen social ties; as religious symbols and emblems of divinity; as a form of prestige and a symbol of well-being. Small children learn early about livestock: it is common to see, early in the morning, boys as young as seven setting off alone from the busy pastoral homestead with a small gourd of milk and a herding stick in hand. Most men sing and dance to their favorite bulls. Barabaig herders (near Iraqw on Map 5) sing loudest for animals of one color with one horn pointing forward and one backward. In Rwanda, political power has long been built through cattle, as it has with the Bahima herders of Uganda. For most herders, payment of required bridewealth from the groom's family to the bride's is in animals, some of which must be borrowed from relatives because typically the payment is too large for one herder to meet alone. A cow given away cements a friendship and ensures a future obligation, while a livestock payment can make amends for misdeeds. Among the Gabra of Kenya (grouped with the Rendille on Map 5), a woman dribbling a female sheep with milk on the head and along the back three times a week may bring auspicious times to her family. Kaj Århem, a cultural anthropologist working in Ngorongoro, in northern Tanzania, concludes that "to the Maasai, . . . cattle give meaning to life; they mean life itself" (Figure 7). A central Somali camel herder agrees, though he prefers different animals: "A camel is a man, a goat is half a man, and a cow is not a man at all."[4]

Bleeding animals to use their blood as food deserves special mention, partly because it is best known in east Africa. In general, non-Muslim herders (most Nilo-Saharan and some Afro-Asiatic) bleed livestock to drink it, while Muslim herders forbid this practice. Turkana herders, for example, bleed all types of animals, from large to small: camels, cattle, sheep, and goats. For the large animals, they shoot a "blocked" arrow from a short distance to open the jugular vein in the neck of a cow

FIGURE 7.
These Maasai cattle are the lifeblood of a family that welcomed
us into their home during our research in the Mara. (Photo by
Cathleen J. Wilson.)

or camel; for the small animals, they open a small vein above the eye. I have seen
cattle complain, but only briefly, when first struck by the arrow. Herders draw blood
from an animal only every two months or so, so as not to weaken it. Turkana will
bleed as many as three liters from a single camel but only a third of a liter from a
goat.[5]

Herding in a savanna, whether the family lives in a settlement or moves with
the herds, presents some basic, inescapable challenges (Figure 8). Each cultural
group differs in how it solves these challenges, but the challenges themselves are
generally the same. First and foremost, families must maintain a reliable source of
food by keeping the animals they depend on alive, producing milk, and reproduc-
ing themselves consistently and reliably year after year. Pastoral families buy
other types of food, such as grains, and gather wild plants and animals, sometimes
including large mammals and fish, particularly in dry times when there is little
milk. As anthropologist Kathleen Galvin found out, food shortages are common
among the Turkana; indeed, by Western standards Turkana families may be under nutritional stress all the time. The spread-out, bimodal rainfall helps keep
livestock growing and milk flowing in parts of east Africa, by keeping some green

grass growing for about half the months of the year. But other parts of the region, such as Sudan and most of Tanzania, experience one season of rainfall, with a dry season that lasts eight or nine months, often without a break.

Droughts are another common challenge. In Somalia, for example, local droughts occur every three to four years, and wider regional droughts every eight years or so. Floods, often associated with El Niño in east Africa, are rarer than droughts, but are still a formidable challenge. People also face the challenges of finding reliable water for themselves and their livestock, keeping themselves and their livestock as healthy as possible, and protecting themselves from human and non-human predators.

Anthropologist Peter Little simplifies the basic challenges facing pastoralists into what he calls the "evil three": drought, disease, and insecurity. As Figure 8 illustrates, these basic challenges also threaten wildlife, which employ some similar biological and social strategies to survive. If, on the other hand, forage, water, health, and safety are taken care of, both people and animals are likely to be success-ful in the fifth key ingredient for life: reproduction. Beyond these basic needs, a wide range of other things are important to pastoral families—education, political power, prestige—but these are secondary to the basic needs (and also can help satisfy the basic needs).[6]

Over time, pastoral people have developed a wide range of strategies to survive these basic challenges. A key to pastoral survival is to be in the right place at the right time, where water and pasture are abundant. Rain does not fall uniformly; one day it falls on one part of the savanna, the next day on another. Finding water and pasture is particularly challenging in drier savannas, where, as rainfall de-clines, its location and timing become increasingly unpredictable. As the weeks go by, savannas become a mosaic of green grass and leaves mixed with brown grass and leaves, with the best food and water resources located in different places. Thus herders have no choice but to move, more often in the drier savannas, where the rainfall is particularly patchy. As we saw in Chapter 2 (Figure 2), some pasto-ral groups move entire families with livestock (nomadic herders); others move the herds seasonally with part of the family (seasonal herders); others split the family into different homesteads with different herds in each; and still others hire other herders to move their animals for them.[7]

Herders move to find better pasture and water, but they move for other reasons, too. In Turkana, in northern Kenya, anthropologist Terrence McCabe followed families as they moved, asking them why they move. He then drew maps of where they went. About half the time, they moved to find forage and water, as one would

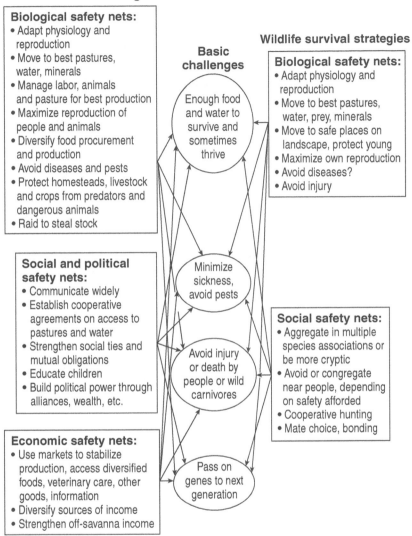

Pastoral survival strategies

Biological safety nets:
- Adapt physiology and reproduction
- Move to best pastures, water, minerals
- Manage labor, animals and pasture for best production
- Maximize reproduction of people and animals
- Diversify food procurement and production
- Avoid diseases and pests
- Protect homesteads, livestock and crops from predators and dangerous animals
- Raid to steal stock

Social and political safety nets:
- Communicate widely
- Establish cooperative agreements on access to pastures and water
- Strengthen social ties and mutual obligations
- Educate children
- Build political power through alliances, wealth, etc.

Economic safety nets:
- Use markets to stabilize production, access diversified foods, veterinary care, other goods, information
- Diversify sources of income
- Strengthen off-savanna income

Basic challenges

Enough food and water to survive and sometimes thrive

Minimize sickness, avoid pests

Avoid injury or death by people or wild carnivores

Pass on genes to next generation

Wildlife survival strategies

Biological safety nets:
- Adapt physiology and reproduction
- Move to best pastures, water, prey, minerals
- Move to safe places on landscape, protect young
- Maximize own reproduction
- Avoid diseases?
- Avoid injury

Social safety nets:
- Aggregate in multiple species associations or be more cryptic
- Avoid or congregate near people, depending on safety afforded
- Cooperative hunting
- Mate choice, bonding

FIGURE 8.
People and wildlife face some of the same basic challenges when trying to survive in savannas. (Sources: Gulliver 1975, Little 2003.)

expect. But the other half of the time, they moved away from dangerous places where they feared neighboring tribes would raid their camps to steal their livestock. They also moved to be closer to relatives. Some herders shifted camp to avoid harmful insects and disease or to avoid associates who were uncooperative, careless, held grudges, or were otherwise unpleasant. And how often do herders move? Some never move (cultivating and settled herders, fenced ranchers), others once a year (many seasonal/transhumant herders), and others several times a year (many nomadic herders). In the early 1980s, Turkana families were perhaps the most mobile people on Earth, moving their homes or campsites more than once a month.[8]

When you see herders grazing animals away from their home or a water source in a savanna, they are almost always herding one type of livestock (the exception being mixed sheep and goat, or "shoat," herds). That does not mean the family holds only one type of animal, just that different types are herded in different places. Around the world, pastoral families herd alpacas, llamas, the two-humped Bactrian camel and one-humped dromedary, water buffaloes, taurine and zebu cattle, donkeys, goats, sheep, yaks, horses, and reindeer, as well as ducks and geese. Historically, pigs were herded in the Middle East and Europe. In east Africa, herders usually herd three, and sometimes up to five, types of livestock: sheep, goats, zebu (or humped) cattle, camels, and donkeys. They also keep domestic dogs and sometimes chickens.[9]

Each of these domesticated animals eats different vegetation, can walk different distances to reach food, needs water on a different schedule, and represents a different economic and social value to a pastoral family. For these and other reasons, each has a contrasting, though often complementary, purpose in a household. And family needs for different species will depend on what pastoral lifestyle they lead. Sheep and goats, or shoats, for example, are not herded far from the homestead (2–4 km, or 1.2–2.5 mi) and need water every two to three days. Sheep eat short grass, while goats will eat both grass and woody plants, so having both in a herd allows pastoral families to harvest a wide range of forage in their savanna environment. Shoats reproduce rapidly (20–40 percent reproduction rate per year); goats produce milk, sheep a little; and both are a ready source of hides, meat, and cash when sold in a market.

Most pastoral sheep in east Africa are "hair" sheep that produce no wool. In many places, pastoral families prize sheep for their fat tails. (These fat tails help distinguish, from a distance, sheep from goats in a mixed herd.) Once in northern Kenya, while measuring my study acacia trees along a dry riverbed, I watched as two young Turkana men cut the fat-bloated tail from a struggling sheep with a rounded knife that one of the men wore as a wide bracelet on his wrist. He then sewed up the wound with several long, straight acacia thorns and threw the tail on a small fire the other man had built. The fat from the tail soon began to sizzle and pop in the fire, and then the two men ate it, quickly and with relish.[10]

Cattle can produce abundant milk as well as meat, manure, and other products, but they reproduce much more slowly than shoats (3 percent per year), and herders must move them to better pastures to survive droughts. Two main species of cattle occur in east Africa, a humped zebu (*Bos indicus*) and a humpless taurine animal (*Bos taurus*). The humped cattle can tolerate lower quality forage, less drinking water, and more heat stress than their humpless cousins. Unlike sheep and goats, cattle can harvest food a long way from the homestead (up to 30 km, or about 19 mi, but usually less than 15 km, or 9 mi, round trip), but they must be watered every other day and they prefer to eat grass. Some herders travel in different directions on alternate days, accessing water on one day and allowing their livestock to forage on another. Pastoral families use cattle hides for sleeping mats, leather sandals, house construction, ropes, and milk containers. As described in Chapter 2, unlike most commercial ranchers, who herd cattle to produce beef, east African pastoralists rarely slaughter and eat their cattle, instead keeping them mainly for their milk and sometimes blood. In dry savannas, herders must move cattle to wetter pastures to keep them alive, particularly in the dry season and during droughts. Compared to cattle in the cool highlands, the coats of cattle in hot lowland savannas are lighter in color to reflect the sun's radiation and thus help lower heat stress in the animals.[11]

The consummate dry-savanna animal in east Africa is the single-humped camel, or dromedary. Here, camels are the backbone of pastoral survival: they can tolerate and survive drought more easily than cattle, going four to five days between waterings, which also allows herders to access pastures very far (more than 30 km, or 19 mi) from the homestead. In the dry season or during a drought, when there is no grass, camels survive well on shrubs and trees. Although camels produce less milk than cattle per milking, they produce it more steadily; over the course of a year in a dry place like northern Kenya, a camel will produce six times more milk than a cow. In Turkana, our research team found that 85 percent of the energy that flowed from the sun through plants to livestock and people came from just one shrub, *Indigofera spinosa,* which is eaten by camels and delivered to people through camel milk, which is delicious as well as nourishing.[12]

What might be considered the "best" pasture depends on the needs of the livestock using it. Herding families often split their herds according to species, age, and lactation status. They do this partly to keep some livestock near their homes, but also so they can send different herds to different parts of the landscape to find localized patches of green forage or forage of different types. They might keep the milking herd near the homestead, for example, because milking animals with young cannot walk far and need more nutrition than the rest of the herd, and because the family needs to use some of this milk. Male and "dry" (nonmilking) females can then be sent off to find pastures at a distance. Northern Somali herd owners stay behind with the sheep and goats near the wells and send the young men with the camels to pastures up to two weeks' walk away. Borana herders in northern Kenya, meanwhile, split up herds into five groups: lactating cows, dry cows, lactating camels, dry camels, and shoats. Eldest brothers then herd the most prestigious stock—cows and camels—while younger brothers take charge of the sheep and goats. Families also split herds to create social obligations and to widen the geographic spread of their livestock, lending or giving them to other families.[13]

Herders take careful advantage of "key resources," places with sufficient moisture to be green year round such as swamps, riverine forests, and highland pastures, to keep animals alive during dry seasons and droughts. Access to key resources "often determines whether or not herders survive harsh years without massive livestock losses." Anthropologists Katherine Homewood and Jeffrey Lewis followed herds near Lake Baringo and in the nearby Tugen Hills of Kenya during the 1983–1984 drought. As the drought deepened, the livestock near the lake concentrated more and more on the wetlands that ring Lake Baringo. Goats that foraged here throughout the drought fared just as well as they did in a year with normal rainfall, but the goats up in the Tugen Hills, away from the rich forage and water at the lake, suffered heavier losses than they did in normal years. Droughts also force herders to go to dangerous places they usually avoid, such as those with tsetse flies (which transmit the livestock disease trypanosomosis, related to human sleeping sickness), livestock raiders, or wild predators.[14]

Nonpastoralists often accuse pastoralists of irrationally keeping too many livestock. But keeping many animals is a proven survival strategy in Africa's savannas. Those with large herds are left with relatively more animals after drought than those with small herds. And herds composed primarily of females regrow the fastest after a devastating drought or disease outbreak. Theoretically, families could sell their herd at the onset of a drought, live off part of the profits during the drought, and repurchase animals with their remaining profits after the drought. But this is impractical in most pastoral lands, for it is impossible to predict when a drought will occur or how long it will last. In any case, not only is access to livestock markets difficult in many places, but the value to herders of their stock amounts to much more than cash on the hoof.[15]

The second of the evil three is disease. Most of the diseases of livestock also involve wildlife and thus will be covered in the next section and in Chapters 7 and 8.

Herding in east Africa can be dangerous because of both human and nonhuman predators, and represents the third major challenge for pastoralists. One early morning in Turkana in the late 1980s, I heard Lopeyon and Achukwa, two of my Turkana colleagues, in side-splitting laughter. I leaned out of my cot and lifted the mosquito net to see them pointing at hyena tracks traversing the narrow space between our parked Land Rover and the cot of our camp manager, Mohammed. The animal had obviously passed by while Mohammed was fast asleep. Hyenas can be deadly to humans, so Mohammed was in real danger sleeping in a cot in the open air, but he did not expect one to pass by because we hadn't encountered one for two months. (Turkana often laugh at hardship, especially when it involves a dangerous near miss.) Suddenly they all fell silent as Lopeyon squatted to examine a faint imprint in the soft sand of the riverbed. He spoke and pointed to the nearby hills; in the still morning air I picked up the word *Ngoroko*, a reference to a group of people who often raid Turkana settlements, stealing livestock and sometimes killing people. The footprint had been made several days before, so we were in no immediate danger. Still, it was sobering to see how quickly my colleagues switched from mirth to seriousness. Ngoroko are also Turkana in fact, and this one had walked many kilometers to reach southern Turkana, where we were camped. For Turkana people of the south, the word *Ngoroko* spells danger, injury, and possibly death.

Fear of raiding is constant for many pastoral families of the region, affecting where they choose to live and graze their animals. Terrence McCabe, an anthropologist on our research team, discovered how deeply raiding affects Turkana families while interviewing a Turkana woman. When he asked what livestock she preferred, a seemingly simple question, she ranked cattle the lowest, even though her family depended on cow milk for survival. McCabe asked her why. She put it this way: "Cattle bring us to our enemies."[16]

In dry times, the only grass available to keep cattle alive in southern Turkana is along the border with Pokot country to the south. Turkana view Pokot as their traditional enemies (the Pokot have the same view of Turkana), even more dangerous than the Ngoroko. Raiding here was particularly deadly in the early 1990s, when over 10,000 Turkana were killed in just a three-year period, this time by armed Pokot herders allied with powerful government officials from the south.

Raiding throughout northern Kenya, southern Ethiopia, northeast Uganda, and Somalia is still common today, though it is becoming less common in Maasailand than it was a few decades ago, according to my pastoral colleagues.[17]

Livestock raiding is an old and widespread practice. In the early 1900s in what is now South Sudan, colonial administrators created empty areas between Nuer and Dinka herders to avoid conflict. In southern Ethiopia, also in the 1900s, the Daasanach (also Reshiat, Map 4) and Nyangatoum (grouped with Toposa on Map 6) occasionally broke out into what anthropologist David Turton calls "all-out war." In the early 1960s, for example, the Daasanach killed 400–500 Nyangatoum, a tenth of their entire population. In northwestern Tanzania, Kuria herders (not mapped, south of Gusii, Map 5) no longer raid to replenish home herds or to amass bride wealth; they now raid to steal cattle to sell for cash. As one Kuria man recently told anthropologist Michael Fleisher, "Thieves use guns to steal cattle and kill people pointlessly. They don't care about old people or young people. They kill everyone they meet in their houses to take the cows away." It is not uncommon for families to lose everything in a raid, becoming destitute overnight.[18]

Why raid? Herders raid to steal livestock and grain and to kill their enemies. Over time, raiding may allow one pastoral group to expand into new territory with new resources at the expense of other groups, as we will see in Chapter 5. At the root of many raids is the felt need by one group of herders to extract revenge on other groups that in previous raids killed members of their community. More recently, some raiders have obtained the support of powerful politicians or "entrepreneurs," the latter seeking cattle to feed armies at war or to sell for profit. Colonial governments also sometimes inadvertently "bred ill-feelings and led to increased hostility between local peoples" in northern Kenya in the early 1900s, as historian Neal Sobania describes, by imposing new political boundaries and limits on pastoral movement, fracturing the formerly fluid and extensive social ties among herding groups.[19]

Pastoral families also have to protect themselves and their livestock from attack by lions, leopards, hyenas, cheetahs, and wild dogs. Night is the most dangerous time; among many families who live among predators, some members sleep with one eye virtually open and sandals on, ready to defend their livestock from an attack. They also have to worry about lions that approach a homestead (or *boma*) upwind, purposely sending their scent into the livestock corral to spook the cattle so that they burst out of the thorn fence surrounding the homestead into the claws and jaws of the waiting lions. Lions and leopards will also silently penetrate or jump the high thorn fence to kill or drag out a cow, sheep, or goat. In central

Kenya, ecologist Mordecai Ogada studied predator attacks on livestock across one communal ranch and nine commercial ranches. He found that lions take more livestock than any other predator inside *boma*s, but leopards, cheetahs, and hyenas take more sheep and goats during the day while herders are grazing and watering their animals outside the *boma*. This is especially true if the herder is a young boy rather than a more imposing adult. Leopards are particularly hard to guard against because they are so quiet, can scale even very tall fences, and are undeterred by the presence of people or dogs. A *boma* with lots of people and dogs is less likely to be attacked by lions than one with few household members or dogs. A *boma* solidly built will deter hyenas. In the end, although top-level predation is a constant concern for east African pastoralists, it accounts for a relatively small loss (at least as measured in numbers). Ecologist Laurence Frank, for example, found that on Laikipia ranches in central Kenya (north of Nanyuki), predators take less than 1 percent of the cattle and 2–2.5 percent of the sheep and goats each year, about half the rate of loss due to disease (see below).[20]

Big, territorial grazers, like elephants and buffaloes, are also a problem. Ecologist Dickson Kaelo found that, in the pastoral lands north of the Maasai Mara National Reserve of Kenya, elephants kill people and livestock when they are surprised, and particularly if they have young. Elephants in the Mara most often kill lone men who are walking home drunk late at night. One lucky man survived an elephant encounter on the way home from a ceremonial gathering, describing it this way: "Suddenly I heard a loud explosion, it was like a helicopter was about to land on my head; in a split of a second I heard a bang and my body went numb." He passed out and woke up hours later to find his abdomen wall punctured and his intestines hanging out. He lay, waiting almost ten hours for the sun to rise, hoping no hyenas would find him, and then struggled home to get help. Outside Nairobi and Tsavo national parks, a full third of the farmers and herders interviewed by tourism scholar John Akama knew someone who had been injured by wildlife. In places near parks where crop farming is widespread, such as outside Tarangire in Tanzania, villagers complain most strongly about crop destruction by elephants and buffaloes and fear for their personal well-being when they encounter these territorial grazers (for more, see Chapter 7).[21]

In facing these multiple challenges, most pastoralists agree that their most important survival strategy lies in the complex web of social connections and mutual obligations that form their social safety net, the glue of herding societies. Safety nets, usually created or strengthened by an exchange of livestock, are everything to pastoralists, particularly those who live in the driest and riskiest environments.

As anthropologist Paul Baxter explains, "East African pastoralists seek to convert stock which is surplus to their subsistence needs into social relationships, through acquiring affines (in laws), or by tying clients to them with gifts and loans, thereby increasing their range of social involvement."[22]

These pastoral safety nets take many forms. Ties are created not only by birth (with kin) and marriage (affines), but also by friendship or partnership, age set, clan, neighborhood, section, and tribe. Historian Neal Sobania describes the wide diversity of bond partnerships that occurred in the mid-1900s between unrelated men in the Daasanach (or Reshiat) society of southern Ethiopia, for example. Two warriors may participate in a "bond of smearing" by cutting open the stomach of a slaughtered ox and smearing the partially digested grasses on each other's chests, eating the meat, then exchanging headrests. Or an older man may establish a "bond of holding" with a young boy who is going through circumcision by providing the initiate with the materials needed during the ceremony and wrapping his arms around the initiate during the "cut." The strongest bond is that of name-giving, when a man gives his name to a newborn boy who is not his son. In the past, Daasanach also created these partnerships with other ethnic groups, like the Samburu and Rendille.[23]

The giving of livestock to strengthen bonds is especially important in these societies, and is done by both men and women. This sharing creates fluid ownership of livestock, as Baxter described in the mid-1970s: "It is exceptional for animals to be individually owned; most beasts are the focus of a number of claims, so that each needs to be perceived as being *a mobile bundle of rights.*" This remains largely the case today as well.[24]

The use of social ties and flexible land ownership to secure access to water and pasture is particularly important during dry seasons and droughts. As Peter Little puts it, "Reciprocal grazing rights are a way that herders, even those in areas of conflict, adjust to climatic volatility." Most land and water in the savannas of east Africa are held in common by many people; private ownership is rare. This does not mean that access is a free-for-all, as is commonly assumed in discussions of the "tragedy of the commons." Rather, tribal sections, clans, factions, councils, or whole tribes develop customary rules that regulate who can use what, when, where, and how. Visitors must negotiate permission to use common pasture or water, which is often granted. Conventional wisdom holds that privatization of land is the inevitable consequence of land scarcity, but in many cases common property may be more efficient, less costly, and just as sustainable as private ownership (if not more so, in my opinion). Still, privatization continues, creating what range scientist

María Fernández-Giménez calls the "pastoral paradox": although small land holdings cannot support a family's livestock, herders nevertheless feel compelled to secure individual rights to land. Maryam Niamir-Fuller, a socioecologist, identifies a critical problem with changing land rights in pastoral lands when she writes, "Biases inherent in the 'modern' judicial and administrative systems against pastoral mobility actually may increase resource competition and conflict."[25]

One at a time, the "evil three"—drought, disease, insecurity—can make life for pastoral families difficult. But when they come all at once, devastation can result. Sometimes, however, desperation can be the spur for innovation. Peter Little describes one such incident. In April and May of 1996, the good, long rains did not come on time in northeastern Kenya. Local Somali cattle herders looked at their options. They heard from travelers that the rain was good south of them in Kenya and to the east across the border along the coast of southern Somalia. But during the rains, tsetse flies infest the lands to the south, bringing deadly trypanosomosis, and in Somalia they feared that bandits armed with AK-47s would ambush them, stealing their cattle and possibly killing them as they knew had been done to clan members. What to do with bad choices? They contacted Somali elders in southern Somalia who had influence over the bandits and negotiated safe passage for themselves and their livestock to the coastal grasslands. In exchange, the elders required the herders to travel unarmed. By acting carefully, the herders succeeded in avoiding all of the "evil three."[26]

PEOPLE AND WILDLIFE IN THE SAVANNAS OF EAST AFRICA: CONFLICT, COEXISTENCE—OR CONFLICTED COEXISTENCE?

The rest of this book focuses on how people and wildlife interact with each other in east African savannas and how the wider world changes these interactions. The ways in which wildlife, and savannas more generally, respond to people in east Africa depend partly on the type of savanna (climate, soils, vegetation), the species of wildlife, and how intensively people use the savanna and wildlife. Most of this story will focus on pastoral people who mainly herd livestock for subsistence, since grazing is the most widespread use of the Earth's savannas today, and east Africa is no exception. However, it is useful to start more broadly, by examining the influence of human populations and land use (herding, hunting, farming, and wildlife conservation) on wildlife. While the rest of this book then focuses on nomadic, seasonally moving, and settled herders, it is important to acknowledge that where human populations grow, some pastoralists will settle and become farmers; where climate dries as predicted, some farmers may move or become pastoralists.

The way people use the land to produce food strongly determines how they "see" wildlife and whether they feel, on balance, that wildlife exert a positive, neutral, or negative influence on their lifestyle. In general, the way people use land determines how direct their contact with wildlife is, what people and wildlife compete over (or how they facilitate each other), and how much people allow or curtail the access of wildlife to critical natural resources.

The least intensive and most wildlife-compatible human use of land is hunting and gathering (as long as the offtake is sustainable; Figure 9a). This lifestyle and land use, while quite rare in east Africa today, depends almost entirely on a healthy and abundant supply of wild foods in the form of wild plants and animals. Hunter-gatherers do modify savannas but usually lightly, for example by burning vegetation and collecting wood. Where wildlife hunting is a community's main subsistence strategy and customary rules for judicious hunting of wildlife are in place, the hunting pressure is likely to be moderate, allowing wildlife populations in these savannas to remain high. Where, however, hunting pressure on a given species is higher than the ability of those animals to reproduce, they can all but disappear. Ecologist Mohammed Said led our team in analyzing changes in wildlife over time in Kenya and found, for example, that few wildlife remain in the dry savannas of Kenya's Kamba country, lands that were once rich in wildlife. We surmised that this decline is largely due to the rising human population and spread of crop farming, but also to the still-thriving hunting tradition among the Kamba people.[27]

At the other end of the spectrum, the most intensive land use and the one least compatible with wildlife is farming, which usually causes significant wildlife loss (Figure 9c). Farmers tend to create hard barriers or boundaries on their land to keep wildlife out of their fields, water sources, and pastures. Farmers in savannas typically view the wildlife around them as a nuisance, incurring more costs than benefits for them. The only wild animals that farmers tolerate on their farms are usually those that are small in size and do little damage to crops. In western Tanzania in the 1800s, for example, wildlife were so abundant that farmers had to band together in compact villages to protect their crops from being consumed.[28]

Occupying the middle ground between these two extremes is pastoralism, where living with wildlife has obvious and significant costs but also some less

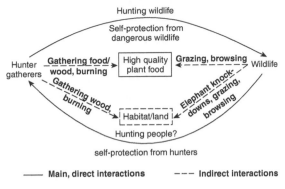

Hunting wildlife

Self-protection from
dangerous wildlife

Hunter gatherers — Gathering food/ wood, burning → High quality plant food ← Grazing, browsing — Wildlife

Gathering wood, burning

Elephant knock-downs, grazing, browsing

Habitat/land

Hunting people?

self-protection from hunters

—— **Main, direct interactions**   --- **Indirect interactions**

a. Hunter-gatherers and wildlife. Soft-boundary landscapes.

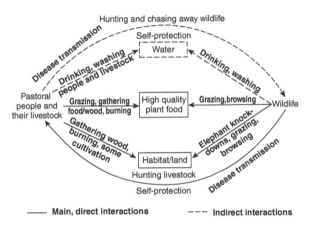

Hunting and chasing away wildlife

Self-protection

Disease transmission

Drinking, washing people and livestock

Water

Drinking, washing

Pastoral people and their livestock — Grazing, gathering food/wood, burning → High quality plant food ← Grazing, browsing → Wildlife

Gathering wood, burning, some cultivation

Elephant knock-downs, grazing, browsing

Habitat/land

Hunting livestock

Self-protection

Disease transmission

—— **Main, direct interactions**   --- **Indirect interactions**

b. Herders and wildlife. Soft- to mixed-boundary landscapes.

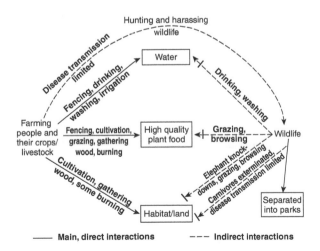

Hunting and harassing wildlife

Disease transmission limited

Fencing, drinking, washing, irrigation

Water

Drinking, washing

Farming people and their crops/ livestock — Fencing, cultivation, grazing, gathering wood, burning → High quality plant food ← Grazing, browsing --- Wildlife

Cultivation, gathering wood, some burning

Elephant knock-downs, grazing, browsing

Carnivores exterminated, disease transmission limited

Habitat/land

Separated into parks

—— **Main, direct interactions**   --- **Indirect interactions**

c. Farmers and wildlife-only parks. Mixed- to hard-boundary landscapes.

FIGURE 9.

How people interact with wildlife depends in part on how
they use the land.

recognized benefits (Figure 9b). The balance of these costs and benefits determines whether herders and wildlife generally tolerate each other or come into conflict—or both (Figure 10). In wet areas where crops are a major part of the pastoral lifestyle, and near markets and towns, herders, like farmers, have incentives to create hard boundaries. Generally, however, herders—especially the nomadic herders, seasonally moving herds, and settled herders with nomadic herds of Figure 2—rely on open landscapes to move their livestock to distant pastures and build fewer social and physical boundaries than farmers, but more than hunter-gatherers. Even in these open savannas, of course, pastoral people and wildlife do sometimes come in direct and strong conflict, when predators kill or injure livestock or herders kill or injure wildlife. Many (though not all) herders use wildlife as food and as raw materials to make shields, medicines, belts, ropes, and containers, and when times get tough, even those herders who do not customarily hunt will turn to wildlife as another source of food. According to ecologist David Western, for example, older Maasai in Kenya's Amboseli ecosystem still call wildlife their "second cattle," recalling Africa's great rinderpest panzootic of the 1890s when the Maasai lost all their livestock. Similarly, lions and hyenas hunt livestock when other wildlife prey are scarce, making cattle in essence a carnivore's "second wildebeest."[29]

But most of the interactions between pastoralists and wildlife are more subtle and sometimes even positive. Herders and wildlife create, modify, and help preserve the grazing habitat and plant diversity that both depend on. To some degree, the presence of wildlife and pastoralists prevents the land from being used by farmers and others in ways that are not compatible with wildlife. This means that wildlife conservation on pastoral community land may be able to achieve the elusive goals of conserving wildlife, increasing income for pastoral households, and promoting the human and land rights of pastoral people.

Although pastoral people burn savannas, removing trees and shrubs that wild browsers thrive on (at a cost, therefore, to browsers), burning can also create more grassland for wild grazers (a benefit to grazers) and sometimes enriches the soil. And this facilitation benefit can go both ways. When elephants knock down whole woodlands, they open up grazing lands for cattle and sheep (but not for camels and goats, which thrive on woody plants). Wildlife and herders can also improve the amount and nutrient richness of forage and protect each other from predators (see Chapters 6 and 8). Zoologist Rudolph Bigalke found that wildlife in South Africa grew faster where they grazed with cattle than where they did not, but the opposite was true for cattle: they grew slower when grazing with

Costs of wildlife to herders	Benefits of wildlife to herders
Injury or death	"Second cattle" when times get tough
Disease	Hunting and tourism revenue
Competition for water,	Cultural and educational value
forage, or habitat	Habitat preservation
	Nutrient flows
	Predator warning

Which way will the balance tip?

Conflict

Coexistence

Costs of herders to wildlife	Benefits of herders to wildlife
Injury or death	"Second wildebeest" when times get tough
Disease	Moderate burning
Competition for water,	Habitat preservation
forage, or habitat	Nutrient flows
Too much burning	Protection

FIGURE 10.
Both herders and wildlife can bring costs and benefits to each other. (Source: Emerton 2001.)

wildlife. Wildlife benefit herders (and many others) by attracting revenue from hunting and tourism, while herders may benefit wildlife by creating new water points and diversifying savanna habitats. Biologist Patricia Moehlman, for example, describes how Afar herders share forage and water with the endangered African wild ass in the Danakil Desert of Eritrea. (For more depth on benefits, see Chapter 6.)[30]

The cohabitation of wildlife and pastoralists in savannas comes with significant conflicts and costs for both (see Figure 10 and Chapter 7). People and their domestic dogs hunt, injure, and kill wildlife, and wildlife hunt, injure, and kill people and livestock. As described earlier in this chapter, leopards, lions, cheetahs, and hyenas eat domestic stock, both in the open savanna and inside people's homesteads at night. Naturally, pastoral people and wildlife both engage in self-protection, sometimes harming or killing each other in the process.

People, livestock, domestic dogs, and wildlife also pass diseases back and forth. Viral rinderpest, related to measles and mumps in humans, was introduced to Africa with cattle in the 1890s and became a major pandemic across the continent, sometimes killing most of the susceptible ruminant livestock and wildlife. Rinderpest was officially eradicated in 2011. Transmitted by the nasal secretions of newly born wildebeest, malignant catarrhal fever is a viral disease that is deadly for cattle. The parasites that carry tsetse-transmitted trypanosomosis and tick-transmitted East Coast fever (theileriosis) need the blood of either wildlife or livestock to complete their life cycles, and can be fatal, more so to livestock than to wildlife. Cape buffaloes are the host for three closely related viral variations (serotypes) of foot and mouth disease, which cause outbreaks in cattle; three other serotypes of this disease came with cattle to Africa. Domestic dogs pass deadly diseases to wild predators, such as canine distemper to lions and rabies to wild dogs and jackals. Wildlife have ticks and other parasites that can transmit diseases to nearby livestock.[31]

Pastoral people and livestock also compete with wildlife for resources such as plant foods, water, and minerals (see Chapter 7). In a review of competition, ecologist Herbert Prins found little evidence of *direct* competition between livestock and wildlife but much evidence of *diffuse* competition. In other words, livestock generally compete with several types of wildlife at the same time, and if livestock are removed, wildlife numbers will grow. Presumably, removal of wildlife would have the same effect on livestock. In another study, ecologist Wilfred Odadi and his colleagues found that wildlife compete with cattle for forage only in the dry season; in the wet season, wildlife grazing in fact improved forage quality and weight gain for cattle. Those herders who also grow crops find that elephants and other wildlife can destroy whole swaths of their fields in a single night. Herders in Longido, in northern Tanzania, recently gave up growing crops because elephant damage was so pervasive.[32]

Herders exclude wildlife from pools, springs, and dams during the day, but wildlife often use these same water points at night. In northern Kenya, because livestock use water during the day, Grevy's zebra (*Equus grevyi*) shifted their visits to the water points to the night, even though the predation risk is high then. In Kenya's Mara region, where wildlife are plentiful, women and children often fear injury by predators or large herbivores when they go to fetch wood and water. Predators, elephants, and buffaloes may also limit herders' access to water. Heavy grazing by livestock around water points changes the types and abundance of plant species, can shift plants from those palatable to grazers to those that are not, and concentrates soil nutrients, which in turn affects wildlife. Wildlife presumably produce the same effects around water points, but the evidence is sparse and the impacts are probably not widespread. (See Chapter 7 for a summary of current debates on overgrazing and desertification.)[33]

Like land use, the number of people living in a savanna affects wildlife. Generally, savannas with higher human populations also have more livestock and more land given over to farming. And as might be expected, the more people in a savanna, the fewer wild animals (with some exceptions, discussed in chapter 6). Although this is widely believed, surprisingly little research exists to support this assumption, so we don't really know how many people a savanna can support without strongly diminishing wildlife populations. What we do know is as follows. In farmland, elephant numbers drop markedly in Zimbabwe in areas with more than 15 to 20 people per km^2 (about 6–8 people per mi^2), as measured by ecologists Richard Hoare and Johan du Toit, while in nutrient-rich savannas in Kenya elephants persist in areas with up to 80 people per km^2 (about 31 people per mi^2). In pastoral areas with no farming in the Mara of Kenya, our team found that wildlife numbers begin to decline when human populations reach 8–10 people per km^2 (about 3–4 people per mi^2) and that about 90 percent of the wildlife disappear when there are more than 75 people per km^2 (about 29 people per mi^2) (see Figure 21, Chapter 6). The pattern of decline, however, can differ greatly from one species to another. The positive side of this story, from a human-centric point of view, is that as human populations rise, conflicts with wildlife decrease and shift to smaller, less dangerous species, simply because the larger species disappear.[34]

We are only just discovering that the impact of people reaches deep into parks and reserves. Zoologists Rosie Woodroffe and Joshua Ginsburg suggest that, in particular, people negatively impact carnivore populations in reserves around the world. In Ghana, savanna reserves that have lost the greatest number of mammal species over time are those with the most people living around them. And these problems are only growing: as biologist George Wittemyer and his colleagues have found, human populations are growing faster near parks than away from parks.[35]

Why are the densities of wildlife and people related in savannas? More people in a savanna in some cases can mean more hunting pressure and thus more wildlife deaths. Many pastoral families keep dogs, and as the human pastoral population rises, so does that of dogs, which, as we have seen, can transmit diseases deadly to wild carnivores. Perhaps most important, as more people settle in savannas, they become less tolerant of wildlife and start to purposefully exclude wildlife from their land and water points. Farmers chase elephants, baboons, bush pigs, warthogs, and other wildlife from their fields to protect their crops from destruction, while people in pastoral settlements not only actively chase carnivores away from their livestock at night, but also "passively exclude" wildlife with their daytime noise. The few studies available suggest that wildlife start to disappear when farms and settlements cover 25–50 percent of savanna landscapes (see Chapter 7).[36]

In savannas where there is little farming, herders can strongly affect wildlife by building boundaries like fences or large settlements, making movement of wildlife (and herders with livestock) ever more difficult. In Map 12, Russell Kruska and I overlay maps of land use and human population density in the savannas of east Africa to show where conflicts and coexistence likely occur, assuming that wildlife densities respond to human densities at rates similar to those found in the studies mentioned above. We also assumed that savannas that now contain more than 25 percent cropland or that have been built up into towns or cities will have few wildlife or little open savanna left. In what we call *hard-boundary savannas* (Figure 11), with population densities greater than 75 people per km^2 (about 29 people per mi^2), we assume that wildlife populations are generally low, though just how low depends on the local species. In these savannas, movement of people, livestock, wildlife, and even water and other natural resources is constrained by numerous more or less impermeable boundaries, often defined by private ownership of land, fences, and strong rules regulating land use. People in these savannas are substituting their former dependence on flows of natural resources with dependence on flows of market resources. These savannas occur especially in very wet areas, near hills and highlands, or near markets and towns. Hard-boundary savannas butt up against the city of Khartoum on all sides in southeast Sudan; exist long the edges of highlands in Eritrea, Ethiopia, and Kenya; surround Lake Victoria in Tanzania, Kenya, and Uganda; and make up all the savannas in Rwanda and Burundi.[37]

We then distinguished *mixed-boundary savannas*, with 15–75 people per km^2 (about 6–29 people per mi^2) (Figure 12). People in these savannas generally engage in some cultivation or settlement activities but use most of the land to raise livestock. Here, the number of wildlife ranges from low to high, depending on how willing people are to live with wildlife. Although there can be significant costs from living with wildlife like those described above, people in these mixed-boundary savannas can also enjoy significant benefits from doing so—if the mechanisms for sharing the profits from wildlife tourism or hunting, for example, are in

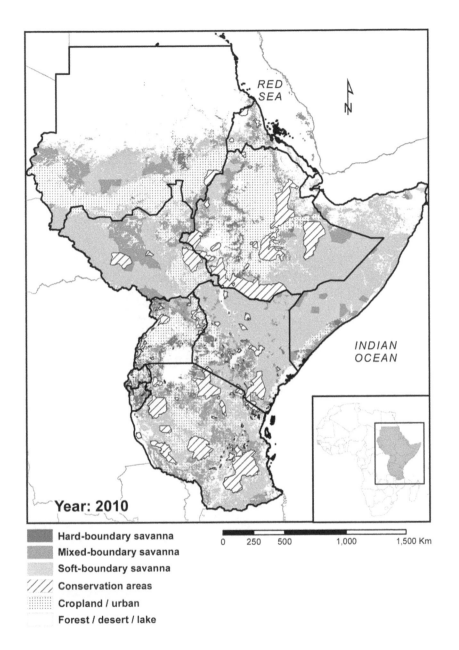

RED
SEA

INDIAN
OCEAN

Year: 2010

Hard-boundary savanna
Mixed-boundary savanna
Soft-boundary savanna
///// Conservation areas
Cropland / urban
Forest / desert / lake

0   250   500   1,000   1,500 Km

MAP 12.

Savannas with hard, mixed, and soft boundaries in east Africa in
the year 2010. (Map adapted from GLC 2003 and UMP 2004 by
Russell L. Kruska.)

FIGURE 11.
Hard boundary between Nairobi National Park (left) and
Nairobi city (right). (Photo by Cathleen J. Wilson.)

FIGURE 12.
Mixed-boundary savanna, with open savanna, fenced maize
fields, and an industrial flower farm in the distance, Kitengela,
Kenya. (Photo by Cathleen J. Wilson.)

FIGURE 13.
Soft-boundary savanna, with pastoral land in the foreground
and Nairobi National Park in the background beyond the river.
(Photo by Cathleen J. Wilson.)

place. People, livestock, and wildlife are all in transition in mixed-boundary savannas, where the boundary edges are fast hardening and the wildlife fast disappearing. Herder mobility can be greatly limited. These savannas are found throughout east Africa where there is moderate rainfall and thus the prospect of growing some crops, and also near towns.

Below 15 people per km^2 (about 6 people per mi^2), we find *soft-boundary savannas* (Figure 13), where interactions between people and wildlife are both positive and negative and people can harvest great benefits from wildlife if conditions are right. Here, people, livestock, and wildlife are relatively free to move if and when needed. Almost all this land is held in common by communities or in trust for them by the government, with very few private landholdings. If people do not purposely kill wildlife in the soft-boundary savanna, wildlife populations can be high. These savannas generally occur where rainfall is moderate to low. Soft-boundary savannas exist in many countries of east Africa, especially in South Sudan, northern and eastern Kenya, most of Somalia, and in large patches of Tanzania. These savannas are less common in Uganda, Rwanda, Burundi, Djibouti, and Eritrea.

The last type of savanna is reserved only for wildlife. These are inside wildlife parks and conservation areas or on commercial wildlife ranches, where little to no settlement is allowed.

What is remarkable and new about Map 12 is that it shows how rare hard-boundary savannas are (5 percent of all savannas) and how abundant the soft-boundary savannas are (66 percent of all savannas). By overlaying Maps 7 and 12, moreover, we found that 42 percent of soft-boundary savannas receive enough rainfall to be farmed, while the rest are probably too dry for farming. All of these vast areas represent opportunities for people, livestock, and wildlife to continue to mix and move, if wildlife populations remain abundant. Mixed-boundary savannas in transition occupy 19 percent of all savannas, where land use is rapidly changing and opportunities for wildlife fast declining.

CHAPTER FOUR

Moving Continents, Varying
Climate, and Abundant Wildlife

Drivers of Human Evolution?

The most remarkable demonstration of the ancient relations between people and wildlife in east African savannas exists in the fossil footprints of three small hominins, surrounded by other animal footprints, in the now solid ash that, 3.7 million years ago, spewed from a volcano in what today is northern Tanzania, a place known as Laetoli. At that time, three individuals of Australopithecus afarensis, *a species that may be ancestral to that of all people alive today, walked on the volcanic ash, leaving behind foot imprints that are the first physical evidence of hominins walking on two feet, a habit they had developed some time earlier. What strikes me most about the trail of Laetoli footprints is not so much their humanness (which is overwhelming) but the familiarity of everything else found in the ash layer: imprints of acacia-like leaves and of hundreds of other animals, including guinea fowl, equids, and giraffids. I find these fossil prints to be unmistakable, and profoundly evocative, evidence that, nearly 4 million years ago, hominins and wildlife existed in a primordial east African savanna woodland side by side, if not necessarily in "harmony."*[1]

HOW WE GOT HERE

The last two chapters described how rich soils, double-wet seasons, and varied topography created conditions in east African savannas that are rarely found elsewhere. In this chapter we will delve into the paleosciences to find out when savannas first formed and how their very existence—particularly that of savanna woodlands—may help to explain our own evolution. The opportunity (and problem) of exploiting highly productive savanna mammals for food may have allowed those hominins with greater social organization to survive better than those without. The first ancient humans of the genus *Homo* became increasingly skilled at "ecosystem engineering," using information and technology to capture ever more savanna resources as a means of satisfying their growing needs.

Timelines help to explain why savannas evolved the way they did and may also help to suggest how they might change in the future. To simplify things, I have divided the development of savannas, people, and wildlife in east Africa into four periods or stages, starting with the birth of the planet about 4,600 million years ago (Ma) and ending just before people began producing food, about 12,000 years BP (before present, or 10,000 BC). The first period, "deep time," started long before anything like humans were on Earth, when essential building blocks fell into place for the creation of life. This period includes the formation of the solar system and Earth, the first cells, the first land plants, the first amphibians and reptiles, and the age of the dinosaurs. It also saw the launch of critical ecological dynamics such as the cycling of water, energy, and nutrients; the first plant-consuming animals (herbivores); the first fires; the first proto-savannas; and the age of the dinosaurs. At the end of this period, Africa separated from Asia, and remained so for 70 million years.

The second, post-dinosaur, stage ushered in the evolution of grasses, many modern mammal species, and modern savannas. About 38 million years ago, the southern continents separated from Antarctica, initiating a global cooling and drying spell that eventually led to the spread of grasslands. Africa rejoined Asia some 30 million years ago, ancient elephants and apes moved out of Africa, and the ancestors of savanna grazers and browsers that evolved in Asia walked in. At the end of this second stage, about 6–7 Ma, proto-chimps and proto-humans separated into two evolutionary lines, coinciding with another strong global cooling and drying cycle.

The third stage began as australopithecine hominins started walking upright in savanna woodlands. One likely ancestor of humans, *Homo erectus* (sometimes called *Homo ergaster* in Africa), appeared in Africa during this stage, and we also see the first evidence of hominins in Asia.

In the final stage, humans came of age as ecosystem engineers, learning to scavenge, gather, and then hunt. Our own species, *Homo sapiens,* evolved in eastern or southern Africa about 200,000 years ago, eventually moving out of Africa and replacing other hominins to become the only *Homo* on Earth.

STAGE 1

DEEP TIME

*4,600–65.5 Ma: Pre-Cambrian, Paleozoic,
and Mesozoic Eras*

Earth time started as the solar system and Earth formed about 4.6 billion years ago (or 4,600 Ma; see Figure 14). Then there followed a whole sequence of Earth "firsts" in the Pre-Cambrian and Paleozoic eras: the first complex organic molecules (4,000 Ma), first life with single cells (3,500 Ma, found in black chert rock in Africa), oldest rocks (basement complex) in east Africa (3,250 Ma), beginning of the movement of the Earth's plates (3,000 Ma), first cells with a nucleus (1,400 Ma), first life with multiple cells (800 Ma), and first vertebrates (fish, 500 Ma). Up to this point, all life was in the sea. Then the first land plants appeared, as did scorpions and millipedes (420 Ma), followed by the first large, treelike plants (409 Ma). Air-breathing land animals appeared with the first amphibians (370 Ma) and the first reptiles (300 Ma) and, much later, the first birds (160 Ma). Near the end of this very long stretch of time the Earth's crust folded and warped, creating the large basins of Africa. Here, I compress the vastness of deep time—which includes three of the four eras of all of Earth's geologic time (the Pre-Cambrian, Paleozoic, and Mesozoic, 4600–65.5 Ma) and encompasses 98.6 percent of all the time that the Earth has existed—into a few paragraphs to focus this book on the ephemerality of our own species and on more recent savannas and land mammals.[2]

A listing of firsts does not do justice to the complexity of biological interactions that began in deep time. Any basic text on ecosystems first describes their structure and the relationships of their elements: food webs, trophic levels, nutrient cycling, predators and prey, and the like. This was the time when all these interactions, the foundation of our changing life on Earth today, took form, and when we see the "origin, assembly and modernization of terrestrial ecosystems," as paleoecologists William DiMichele and Robert Hook put it. For example, this was when thick-walled cells first allowed the earliest land plants to survive repeated wetting and drying, a trait essential to plant survival in savannas today. By about 390 Ma (Devonian period), food webs were weakly connected: early plants, which had neither roots nor leaves, were not eaten by early animals (like millipedes) directly, but only once they became part of the soil. Africa was located at the South Pole at this time and was part of the supercontinent Pangea. At first, plants grew in new, open ground, but quickly the best growing spots were filled and plants began competing for space, light, nutrients, and water. In the late Devonian (360 Ma), *Archaeopteris*

Era	Ma	Events
Precambrian	15,000	-- Origin of the universe
Precambrian	4,600	-- Origin of the solar system and earth
Precambrian	4,000	-- First complex organic molecules
Precambrian	3,500	-- First single-celled life, found in Africa (3,500), oldest rocks in Africa (3,500, present-day South Africa), first basement complex rocks in east Africa (3250), oldest known sedimentary rock (3,000)
Precambrian	1,400	-- First nucleated cells
Precambrian	800	-- First multi-celled life, life only in sea
Precambrian		-- Probably bacterial mats on land surface (600)
Other	500	-- First organisms with backbones (primitive fish)
Other	0	-- Evolution of higher plants, animals, humans

(magnified after 500 Ma)

Era	Ma	Events
Paleozoic	500	-- First organisms with backbones (vertebrates, 500)
Paleozoic	450	-- First land plants, scorpions and millipedes (409–439)
Paleozoic	400	-- First animals eat plant matter in soil (363–409), first
Paleozoic	350	amphibians (370), plants produce enough fuel for fires (350–400)
Paleozoic	300	-- First reptiles (300), oxygen at 25% level (280–310), first modern insects (245–290), first aboveground herbivory (about 250–300)
Mesozoic	250	-- First true mammals and dinosaurs (220)
Mesozoic	200	-- Present-day Senegal next to present-day Florida (200)
Mesozoic		-- Ancestors of dung beetles appear (160–208)
Mesozoic	150	-- First birds (160), major savanna tree families (150)
Mesozoic	100	-- Africa separates from Asia (100)
Mesozoic		-- Dinosaurs decline (66), first primates (65)
Cenozoic	50	-- First grasses, archaic African mammals (60)
Cenozoic	0	-- Global cooling, drying (38), bovids move into Africa from Asia (30), first monkeys (23), proto-humans (8), oldest tools (2.5), first use of fire (0.5–1), *Homo sapiens* (0.1), first herding (0.01)

FIGURE 14.

Earth time from the origin of the universe to the present
(Ma = million years ago). (Sources: Maglio and Cooke 1978,
Johanson and Edey 1981, Waters and Odero 1986, DiMichele
and Hook 1993, Reader 1997, Sues and Reisz 1998, Klein 1999.)

trees evolved, with trunks up to 1 m (3 ft) in diameter, and formed Earth's first shady forests. By the Mississippian epoch (early Carboniferous period, 350 Ma), all major evolutionary lineages of plants had evolved, and a forest, viewed from afar, probably looked much like one today, even though none of the same plant species are alive today.[3]

If savannas are defined as a mix of trees and grasses (Chapter 2), then savannas did not exist before grasses evolved (about 45–60 Ma). But I think *proto-savannas*, with dynamics much like today's savannas, likely existed long before there were grasses. The first of Earth's proto-savannas may have been a *scorpion and millipede proto-savanna* that existed during the Mississippian epoch, starting about 359 Ma. At this time, Europe and North America were in the tropics (the equator cut across Ontario to northwestern Mexico, for example) and Africa was still completely south of the equator (still part of Pangea). The climate, at least in these paleo-tropics, was seasonal (monsoonal), with at least one wet season and one dry season per year, as we see in many African savannas today. Land plants, amphibians, scorpions, and millipedes were well established, having first appeared well before this time. The main trees were lycopods (polelike trees with lateral or terminal cones) and early gymnosperms (ancestral to today's conifers).

Models of paleoclimates during this time suggest that there was more oxygen in the atmosphere (25–35 percent) than today (21 percent), which means that fires, which burn better with more oxygen, could ignite more easily. With lots of dry plant biomass for fuel, wildfires were widespread, as attested by the abundant fossil charcoal (or fusain) in paleosoils of the period. Fuel might have been particularly abundant at this time because basidiomycete fungi (like mushrooms or shelf fungi), organisms that decompose lignin in wood, probably were not widespread yet, and thus wood probably lasted a long time before decomposing. Geologist Howard Falcon-Lang examined fossil charcoals and found that well-drained uplands in eastern Canada and Britain burned every 3 to 35 years, a bit less often than grassy savannas do today (every 1 to 10 years). He suggests that the ancient communities of gymnosperm plants in these uplands were savannalike, analogous to those in east Africa at the present time.[4]

The Paleozoic era ushered in an ecological invention that is crucial to the story of savannas: the beginning of tetrapod herbivory, or the direct eating of plants for food by insects, amphibians, reptiles, and, eventually, mammals. This step, which directly connected plants and animals in the food web for the first time, made one major element of ancient ecosystems dependent upon and affected by another. Thus began a long and continuing set of changes in plants, to allow them to survive being eaten by animals, and in animals, to allow them to eat and digest these plants.[5]

Another likely proto-savanna developed during the age of the dinosaurs, in the late Jurassic period, about 160 Ma. Africa was still part of the global supercontinent, Pangea (although the continent was breaking up), but it was no longer at the South Pole and, instead, spanned the equator. In Pangea, the convex part of what is now the west African coast fit like a puzzle piece up against the concave part of the east coast of Florida and Georgia. Climate in the African tropics was seasonal and dry and the land was a mix of forest patches within widespread semiarid vegetation. The trees in the open landscapes of what could be called *dinosaur proto-savannas* included conifers, cycads, and ginkgos instead of the acacias of today, and the ground plants were ferns, horsetails (sphenophytes), and mosses (bryophytes) instead of grasses. Africa, like other regions, was dominated by small and large reptilian dinosaurs.[6]

One such dinosaur proto-savanna existed at present-day Tendaguru, about 70 km (43 mi) northwest of Lindi on the southeastern coast of Tanzania (see locations on Map 2). Tendaguru was home to the largest dinosaur that ever lived, the plant-eating sauropod *Brachiosaurus*, which was twenty times larger than today's African elephant. Here was also an abundance of other dinosaurs such as stegosaurs and small ornithopods. These dinosaurs probably had to migrate long distances to reach seasonally available vegetation, foreshadowing the great mammalian migrations in savannas millions of years later. Dinosaurs probably ate such large quantities of plant food that there was less fuel to burn in fires. Troops of ancient dung beetles may have followed the giant reptiles and buried their dung, beginning an ancient recycling of nutrients that is still important in today's savannas (though today's dung beetles recycle mostly mammalian dung). It is not clear how long these proto-savannas were in existence, but parts of Africa near the equator may have been dry and seasonal for a huge expanse of time, possibly 170 million years, from the Triassic to mid-Tertiary (200–30 Ma).[7]

At this time, mammals were still small and inconspicuous. The oldest fossil examples of true mammals are from about 220 Ma (Triassic) and include *Megazostrodon*, a tiny shrewlike creature found in South Africa. This early mammal could fit in the palm of your hand, measuring just 10 cm (4 in) long and weighing only 20–30 g (about an ounce). *Megazostrodon*, which likely ate insects and was active at night, played an ecological role different from that of any dinosaur in the Mesozoic landscape. Some scholars hypothesize that the key adaptation unique to mammals was a slightly higher body temperature, allowing them to be more active in the cooler night-time air than cold-blooded dinosaurs. However, no conclusive evidence exists to prove that early mammals were more warm-blooded than dinosaurs.[8]

Despite plant life having existed for millions of years up to this point, no flowers had yet appeared. The evolution of flowering plants (angiosperms) about 130 Ma in the early Cretaceous ushered in a great expansion in the numbers and types of plants on Earth, and hence of new foods—nectar, fruits, nuts, and berries. Insects, which fed on these new foods, greatly diversified about this time, with the evolution of bees, butterflies, and ants. These insects in turn became a favorite food of many early mammals. Today, without flowering plants, there would be many fewer options for food, medicines, and fiber. The history of the peopling of the Earth is replete with stories about aboriginal humans domesticating or adopting some high-energy flowering plant, such as potatoes, yams, maize, and rice, which then allowed human populations to grow. And the flowering plants eventually became most of the grasses and trees in modern savannas, a foundation of the food webs that sustain all savanna life today.[9]

And then, about 66 million years ago, the dinosaurs all but disappeared, leaving mammals to slowly rebuild a new ecological structure. Scholars still debate what caused the precipitous decline of dinosaurs. Physicist Luis Alvarez and colleagues took the scientific world by storm in 1980 by suggesting that a giant asteroid hit Earth about the time of the dinosaur extinctions, kicking up sufficient global dust to suppress plant photosynthesis and eventually dinosaur populations. Ten years later planetary scientist Alan Hildebrand and colleagues found a likely location for the impact at the Chicxulub Crater in the Yucatán Peninsula of Mexico. Since then, the scholarly debate has centered on the age of the Mexican crater and of another potential crater in India. Other scholars, meanwhile, suggest dinosaurs were wiped out by volcanism, climate change, or fungal disease. Still others think that there was no sudden extinction at all. Whatever the cause of the dinosaur extinctions, mammals found great opportunity in the gap the reptiles left, eventually playing a major role in all ecosystems on the planet.[10]

STAGE 2

THE EVOLUTION OF WILDLIFE AND SAVANNAS

*65.5–5.3 Ma: Cenozoic Era, Paleocene through
Miocene Epochs*

After the dinosaur era, the next stage included the evolution of grasses, many modern mammal species, and modern savannas (Figure 15). It ended about 5–7 million years ago at about the beginning of the Pliocene epoch, just before the first evidence that hominins could walk upright. At the beginning of this stage, Africa was still detached from other continents and remained so for about 70 million years, during the extinction of the dinosaurs. This long period of isolation led to a unique archaic fauna in Africa. The ancient lines that led to two of today's major ecosystem engineers, elephants and their relatives (proboscideans) and humans and their relatives (anthropoid apes), evolved in Africa at this time. So did a wide array of hyraxes, archaic meat-eaters (creodonts, not ancestors of true carnivores), and other animals. Some creodonts, such as *Hyainailouros sulzeri,* with a skull twice the size of that of a modern lion, resembled a dog. During this isolation, it is also extraordinary to note the many animal types that, though common today in Africa, had not yet migrated onto the continent: there were no "true" or fissiped carnivores (ancestors to cats, hyenas, dogs), no hooved ungulates (ancestors to antelopes, rhinos, giraffes, pigs, zebras, asses), and no rodents or bats.[11]

When and why did the first modern savannas—woodlands or grasslands—appear in Africa? To answer this question, paleobotanist Bonnie Jacobs summarized a wide range of studies on ancient pollen, fossil plant parts, soil carbonate, and fossil mammalian teeth from Africa. At the end of the age of dinosaurs (Cretaceous period), she found no evidence that the grasses and trees of today's tropical African savannas existed yet. Rather, the main trees were conifers and palms. About 70 Ma, the climate in tropical Africa was likely "ever-wet," which favored the expansion of forests. The first African grass pollen appeared in the fossil record in Nigeria (and elsewhere) about 55–60 Ma (though molecular clock studies suggest that grasses originated about 80 Ma outside of Africa). Even so, grasses were apparently quite rare at this time in Africa, forming only 4 percent of the ancient pollen record, while pollen samples from savannas and woodlands today contain 20–60 percent grass pollen.[12]

The first savanna trees may well have evolved in the interior of Africa in the early Eocene epoch (55 Ma), as indicated by fossilized wood of *Combretoxylon,* an early member of an important tree family, the Combretaceae. By the middle Eo-

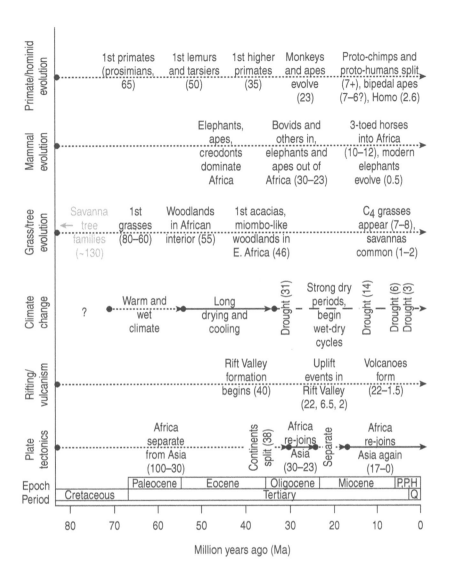

FIGURE 15.

Timeline of the emergence of savannas in east Africa, 80 million years ago to present. (Sources: Coppens et al. 1978, Maglio 1978, Waters and Odero 1986, Cerling 1992, Cerling et al. 1997, Jacobs et al. 1999, Klein 1999, Herendeen and Jacobs 2000, Retallack et al. 2002, Kingdon 2003, Jacobs 2004, Turner and Anton 2004, Segalen et al. 2007.)

cene (46 Ma), there is a savanna woodland that is structurally similar to present-day miombo woodland in the fossil record of Singida, north-central Tanzania. Here, Jacobs and her colleague Patrick Herendeen found the first evidence of the genus *Acacia*, the most recognized modern African savanna tree. Their models suggest that rainfall was much like it is today, though with a less severe dry season (660 mm or 26 in), a finding that contradicts other climate reconstructions for the continent, which suggest that all of Africa was wet at this time.[13]

Starting about 45 Ma, grasses and grazing mammals probably evolved in tandem. Today, as we saw in Chapter 2, grasses are consummately adapted to being eaten, and grazers are consummately adapted to digesting grasses. In 1980, evolutionary ecologist George Stebbins elaborated on a hundred-year-old idea to explain why grasses and grazers appeared almost simultaneously in the fossil record. Early grass leaves contained opaline bodies made of silica, a compound that promotes grass growth yet is highly abrasive on ancient grazers' teeth. At the same time in the fossil record, archeologists found the first grazers with "hypsodont," or high-crowned, ever-growing teeth, which are tough enough to withstand the abrasion of grass tissue containing abundant opaline bodies, allowing these grazers to survive on a diet of grass rich in silica. Putting these two facts together, Stebbins reasoned that grasses and grazers coevolved, meaning that a change in one initiated a change in the other. His idea is supported by the fact that more heavily grazed grasses in the Serengeti today have more silica in their tissues than lightly grazed grasses. More recently in India, paleobotanist Vandana Prasad and colleagues found evidence of both silica-rich grasses (in fossilized dinosaur dung, or coprolites) layered with hypsodont teeth of ancient mammals, which suggests that Stebbins may have been right, though this coevolution happened much earlier than he thought, even before the dinosaurs went extinct (late Cretaceous). Other scholars disagree, however, suggesting that insect grazing may have been involved in grass evolution, or that there was no coevolution of grasses and grazers at all.[14]

About 38 million years ago in the late Eocene, the southern continents separated from Antarctica, initiating a global cooling and drying spell that eventually led to the spread of grasslands. Grasses and grazers further evolved, and moving tectonic plates and volcanism created the topographic diversity and rich soils that are so important to savannas of east Africa today. Easily visible from space, Africa's Great Rift Valley (see Map 1) started to form about 40 Ma as a massive doming and then splitting of the Earth's crust, with three major episodes of uplift 22, 6.5, and 2 million years ago. Over time, the Rift became a 6,400 km–long (about

4,000 mi) lake-strewn scar in the Earth's surface, running from the Red Sea in the north to beyond Lake Malawi in the south. In the Rift, the oldest volcanoes tend to be on the western side and the youngest on the eastern. For example, in Kenya, Mt. Elgon, in the west, formed 15–22 Ma; across the Rift to the east, the Aberdares and Ngong formed 3–6.5 Ma, Mt. Kenya 2–3.5 Ma, and Mt. Kiliman-jaro (in Tanzania) about 1–1.5 Ma. Ol Doinyo Lengai, in northern Tanzania, is still active and has periodic major eruptions today, with smaller lava flows and smoke emissions several times a year. As these new volcanoes and the Rift formed, they created a close juxtaposition of highlands and lowlands that provided plants, mammals, and early hominins new opportunities to access water and food (see Map 9). In some places, lava, ash, and colluvium flowed from these volcanoes and new, rich soils began to form on top of the ancient basement rocks of old Africa (see Map 10).[15]

As the Rift Valley formed, Africa reconnected with Asia (about 20–30 Ma), and Africa's savannas changed forever. The ancestors of most of the large mam-mals so prominent in African savannas today walked or swam to Africa at this time. Moving into Africa were the continent's first of many types of hooved animals (ungulates): ancient suids (pigs), giraffids (giraffes), bovids (bushbucks, buffaloes, wildebeest, hartebeest, reedbucks, dik diks, gazelles, goats), and rhinos—although they did not appear all at the same time or diversify at the same speed. Many of the migrants had hypsodont teeth, probably developed as they grazed on early grasses in the Asian steppes. The migrant bovids probably also had a specialized digestive ruminant system with a four-chambered stomach that was specially adapted to digest cellulose in grasses. These teeth and digestive systems were new to the Afri-can continent and probably allowed the migrants to consume grass in large quan-tities with subsequent impacts on African vegetation. (Today, one out of three of the genera and families of mammals on Earth are ungulates, outnumbering even the diverse rodents.) The three-toed *Hipparion,* the first horse ancestor in Africa, arrived much later, about 10–12 Ma; *Equus,* the genus of today's zebras and asses, arrived even later, at 2 Ma. Giraffes came from Asia about 8 Ma. Hippos likely evolved in Africa.[16]

Members of some African groups, such as the ancestors of elephants and an-thropoid apes, moved out of Africa as the ungulates moved in, and other groups in Africa went extinct in the Miocene: the carnivorous creodonts, piglike anthra-cotheres, rhinolike embrithopods, elephant-related gomphotheres, semiaquatic *Moeritherium,* and the bizarre *Barytherium,* which, looking somewhat like an Asian elephant, had eight short tusks, four pointed vertically downward from its upper jaw and four pointing forward horizontally from its lower jaw. The hyraxes, which were the most abundant medium-sized browsers in Africa before this time, were largely replaced by the incoming hooved mammals.[17]

Ancestors of modern "true" carnivores, along with amphicyonid bear-dogs and saber-tooth machairodontine cats, came from Asia about 20 Ma, and some creodonts moved from Africa to Asia at the same time. The saber-tooth cats were part of the African fauna for millions of years, disappearing only around 1.5–2 Ma and leaving only today's cats extant. The genus of today's lions, *Panthera,* may have originated in Africa. Gradually, over some 15–20 million years, more true carnivores migrated in from Asia, including the ancestors of living canids (dogs), felids (cats), mustelids (zorillas, honey badgers), mongeese, genets, civets, and hyenas. These migrating Asian carnivores slowly replaced creodonts as the dominant meat eaters in Africa, until finally the creodonts went extinct at the end of the Miocene. The canids evolved in North America and arrived in Africa relatively late, probably about 3–4 Ma.[18]

This rich fauna existed for millions of years in the widespread savanna woodlands of Africa. Then, about 10 million years ago, a new type of grass appeared, adapted to lower rainfall and less CO_2, and started to replace existing grasses. Previously, all grasses (and trees and shrubs) were C_3 plants, which create plant tissue (by capturing or assimilating carbon) more easily from CO_2 in a cooler, wetter, shady, CO_2-rich world. The new kind of grass, a C_4 plant, captured carbon better in an oxygen-rich (and CO_2-poor), warmer, drier, sunny world. After 7.5 Ma, CO_2 rose and fell, but likely stayed well below 335 parts per million (ppm) most of the time, which favored the new C_4 grasses. As the climate gradually became drier, C_4 grasses gradually spread. Dry climates and more grasses probably meant more frequent and extensive wildfires. Grasses also spread at the expense of trees because trees (which are C_3's) grow too slowly to recover from frequent fires in a low-CO_2 environment. Geologist Gregory Retallack suggests that the spread of grasslands may even have accelerated drying and cooling trends because the newly abundant grass roots buried large amounts of carbon in the soil, thus lowering CO_2 further. Today, rising CO_2 levels may be causing the opposite trend: a shift from more open grassy savannas to denser woody savannas (see Chapters 7 and 12).[19]

As woodland savannas gradually became more dry and open, the ancestors of humans split into different lines of apes. About 12 Ma, our ancestors diverged from the orangutans, with whom we share more than 97 percent of our genetic material. About 7–8 Ma, our line split from the gorillas, with whom we share 98.4

percent of our DNA, and finally, 6–7 Ma, from the chimps, from whom we differ genetically by only 1.23 percent. While this may seem to be a minor difference, recent research suggests that all life on Earth shares similar genes, but other genetic material (like "junk" DNA) may play an important role in controlling how those genes are expressed, turning scorpions into scorpions and humans into humans. Even so, chimps and humans are more genetically alike than are zebras and horses, or African elephants and Indian elephants. When our line split from chimps, there were perhaps 50,000–100,000 of our common chimp-hominin ancestors on Earth.[20]

STAGE 3

THE EVOLUTION OF US

5.3–1.8 Ma: Pliocene Epoch
and First Half of Pleistocene Epoch

The split between the African ape and human lineages occurred during a drying phase known as the Messinian Salinity Crisis at the end of the Miocene epoch (5–7 Ma), when global drying reduced the Mediterranean to a salty lake. The subsequent Pliocene epoch (5.2–2.6 Ma), a wetter period, was when hominins started becoming more human, at first walking on two legs and then, much later, using tools. Just what the human ancestral tree looks like is unclear, since the ancestral picture gets more complicated and contested with every new hominin fossil or genetic discovery. Recent fossils from Afar in Ethiopia, for example, suggest a straight-line descent from the australopithecines (small-brained hominins, likely ancestors of *Homo*) to *Homo habilis* about 2.5 Ma, to *Homo erectus* 1.7–1.8 Ma, and then on to *Homo sapiens* 150,000–200,000 years ago. The resulting ancestral tree would be tall and slender. Paleontologist Bernard Wood, however, suggests a "bushy" model with many branches, representing "a series of successive adaptive radiations—evolutionary diversification in response to new or changed circumstances—in which anatomical features are 'mixed and matched' in ways that we are only beginning to comprehend." Support for the bushy model comes from recent discoveries that suggest there were many types of hominins, whose ancestry remains unclear (see Figure 16).[21]

   The oldest hominin now known, about 7 million years old, is *Sahelanthropus tchadensis*, from Chad in north-central Africa, found by paleontologist Michel Brunet and colleagues. Another, slightly younger hominin is *Orrorin tugenensis*, found in the Tugen Hills in Kenya by paleontologist Brigitte Senut and colleagues, about 6 million years old. *Orrorin* may have been bipedal. At issue in determining the "oldest" is whether the new fossils have traits that place them closer to modern humans (*Homo sapiens*) or closer to chimpanzees (*Pan troglodytes*). One scholar even suggests that, genetically, today's chimps and humans are close enough that they should share the genus *Homo*. The diversity of hominins is far greater than anyone imagined only a few decades ago, and the picture will likely become even more complicated in the future, as tangled mixtures of different hominin lineages are found in various parts of the world.[22]

Several aspects of our ancestry are more certain, however. We do know that "robust" and "gracile" australopithecines (robust forms are often called *Paranthropus*) walked side by side at the same time in eastern and southern Africa. Hominins became bipedal many millions of years (6–4 Ma) before they first used tools, at the beginning of the Pleistocene (2.5 Ma). While our direct ancestors developed marked "encephalization," or larger brains per unit of body size, it is possible that one of the hominin dead-ends, *Homo rudolfensis,* also had largish brains. Be that as it may, big steps in brain size evolution occurred with the appearance of the first species of *Homo, H. habilis,* at 2.5 Ma and then *H. erectus* (sometimes called *H. ergaster,* with *erectus* evolving later) about 1.8 Ma. *Homo erectus* may have been the first species of hominin to walk out of Africa into what is now Europe and Asia, although new evidence suggests that *H. erectus* could have evolved in western Asia and migrated into Africa. Other recent discoveries in Indonesia suggest that the ancestors of a species known as *H. floresiensis* may have been an even earlier migrant out of Africa, either just before or after *H. erectus* first appeared (about 2.0–1.5 Ma). As for modern humans, evidence from widely different corners of the scientific world, including blood typing frequency, mitochondrial DNA, Y-chromosome DNA, "junk" DNA sequences, and linguistics, all point to eastern or southern Africa as the ancestral home of our species, *Homo sapiens*. Remarkably, it seems that all humans on Earth today are related to a relatively small group of *H. sapiens* who walked out of Africa about 50,000 years ago.[23]

What were savannas like when hominins started to walk on two legs? *Ardipithecus ramidus* lived in a more open savanna, but its tree-climbing anatomy suggested it frequently used trees about 4.4 Ma. *Australopithecus,* who lived about a million years later, probably lived in a more wooded environment at first. Perhaps the best-known australopithecine is Dinkenesh, from Hadar in eastern Ethiopia. Dinkenesh (Amhara for "Thou art wonderful" or "She is wonderful"), or Lucy as she's known in English, lived about 3 Ma and is of the species *A. afarensis*. She was just over 1 m (3.3 ft) tall and weighed about 30 kg (66 lb), had longish arms,

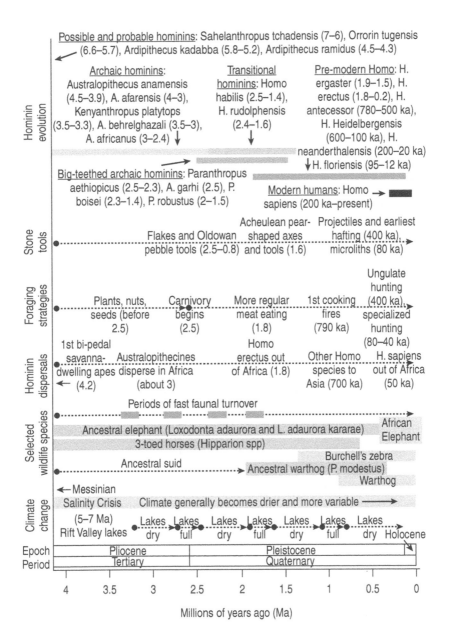

FIGURE 16.
Timeline of human evolution, 4 million years ago to present
(ka = thousands of years). (Sources: Schick and Toth 1993, Klein
1999, Bobe and Behrensmeyer 2004, Trauth et al. 2007, Bradley
2008, Wood and Lonergan 2008, Foley and Gamble 2009.)

shortish legs, and probably also body hair. At most, her species had a brain about 37 percent the volume of a modern human brain (about 500 cc [2.1 cups] vs. 1,350 cc [5.7 cups]), more or less the size of a modern gorilla brain. The males may have been half again as large as Dinkenesh (45 kg, or 99 lb).[24]

Dinkenesh, like other australopithecines, probably spent much of her time in trees but also walked on the ground. It was wetter at that time than now, and Hadar was mostly a forested and woodland landscape, with occasional open grasslands, particularly around swamps—a savanna somewhat woodier than the landscape where australopithecine footprints were found farther south in Laetoli, Tanzania. For Dinkenesh, it was probably a dangerous woodland with almost twice as many species of carnivores as today, including saber-toothed cats and ancient hunting hyenas, as well as modern hyenas and cats. Although about a quarter of the species of large mammals alive at the time still exist today, many species have become extinct, including three-toed horses (*Hipparion*), various pigs (suids), horselike three-clawed chalicotheres, and elephant-like deinotheres (with tusks in their lower jaws). Whereas we know that Dinkenesh climbed and walked, there is no evidence to suggest that she used stone tools.[25]

Why did hominins evolve in the first place? Evolution is a random process, and some hominins survived changes in their environment better than others, more or less by chance. It is becoming clear that a widely varying climate at distinct times in the past permitted some hominins to survive, reproduce, and pass on their genes and others to fail. Scholars have three main hypotheses linking climate to human evolution. For some time, it has been thought that the expansion of savannas in particular, caused by a drying climate, allowed those hominins that could adapt to these new environments to survive (the "savanna hypothesis"). Paleontologist Elisabeth Vrba, building on this idea, suggests that it was not just general drying that set the stage for evolution, but a series of discrete, rapid drying episodes, which in turn caused sudden extinctions (and new appearances) of grazers (bovids), rodents, and hominins (the "turnover pulse" hypothesis; for a visual of these fast turnover periods, see the bottom of Figure 16).[26]

Paleontologist Rick Potts sees a third, finer distinction: major evolutionary steps, he maintains, occurred not only during abrupt drying episodes but also when climates swung more *widely* from wet to dry and back (the "variability selection" hypothesis). These swings prompted "resilient and novel responses to new conditions" by all life, responses that allowed the survival of more flexible and adaptable lineages. For example, while many mammalian species that became extinct relied on low-quality forage, common in wetter savannas, the new, more successful species were smaller and could switch from grass to tree forage as climate and vegetation changed. This variability may have favored groups of hominins that invented certain kinds of technologies, such as stone tools, to exploit more parts of savanna ecosystems, such as underground tubers, termite mounds, and meat. It could also be that *Homo* survived instead of other hominin lines (like *Paranthropus*) because *Homo* was more mobile, moving to find seasonally available food sources, and able to subsist on a wider variety of foods.[27]

Over the last decade, a range of scientists from oceanographers to paleontologists to ecologists have built an intriguing story linking climate and human evolution, using evidence from Rift Valley lake levels, dust deposited in deep-sea sediments during dry periods, the extent of polar ice, and the types of large mammalian and hominin fossils. From these diverse data, it is clear that the climate has dried and its variability doubled or even tripled over the last 5 million years and savannas expanded. Over this long stretch of time there were six relatively short periods when climate swung back and forth between wet and dry, Rift Valley lakes filled and emptied, savanna life went through significant changes, and twelve of fifteen hominin species first appeared in the fossil record (80 percent of hominins appeared during one-third of the time period).[28]

The first three (of the six) short bursts of climatic variability were before 3 Ma, and this is when various species of australopithecine appeared in the fossil record. The latest three of these periods correspond to major global climatic transitions (such as the onset of northern hemisphere glaciation at 2.8–2.6 Ma). The fourth burst of variability occurred about 2.8–2.6 Ma, which coincided with the first appearance of *Paranthropus* and *Homo* (*habilis*) (and the disappearance of two australopithecines), the first widespread use of Oldowan stone tools by hominins, and major extinctions and first appearances of several mammalian species. During the fifth burst, from 1.9 Ma to 1.7 Ma, mammalian species changed again and hominins started using larger, Acheulian tools. It was at this time that *Homo erectus* first appeared in the fossil record (our first long-legged, large-brained [907 cc, 3.8 cup] ancestor). The most recent and sixth episode of variability occurred at 1.0–0.7 Ma, corresponding with the extinction of *Paranthropus* and *Homo ergaster*. Oddly, there does not seem to be a similar episode of variability corresponding with the appearance of our own species, *Homo sapiens*.

What did climate drying and variability mean for hominins trying to find food in savannas? Anthropologist Thomas Plummer suggests that early hominins subsisted on the abundant nuts, fruit, and leaves in widespread woodlands and forests. As climate dried and openings appeared in these woody landscapes, this rich food source was gradually replaced by grasses, which hominins could not consume directly. At the same time, however, animals—another food source—likely became much more abundant in the open savannas. As described in Chapter 2, even though a forest appears to have more plant food than a grassland, grassland often has the same amount of edible leaf biomass, or more. The new open savannas, if similar to modern savannas, would have supported two to three times more herbivores and carnivores than the woodlands or forests they replaced. Thus, as Plummer suggests, climate drying likely caused a major shift in food availability for both hominins and grazers: although hominins no longer had easy-to-digest plant food readily available, many grazers, with their specialized teeth and digestive systems, could easily subsist on the newly abundant savanna grasses. The hominins that thrived were those that could take advantage of the new savanna foods: underground plants and, if they could catch them (dead or alive), animals.[29]

About this time, hominin brain size grew substantially. The big change in brain size came with *Homo erectus,* whose brain was almost twice as large as that of *H. habilis* and the robust australopithecines. Female *H. erectus* not only had larger brains but also larger bodies than *H. habilis* (female body size grew by 50 percent, whereas males were only 18 percent larger than *H. habilis*). Because brains are energetically expensive to maintain, *H. erectus* had to consume higher-quality foods than its forebears. The logical sources of that high-energy and -nutrient food in open savannas were underground plant parts (tubers and rhizomes, also called underground storage organs, or USOs), termites, and meat, whether from scavenged carcasses or live animals that had been hunted. Some scholars doubt that hominins scavenged carrion because chimps, bonobos (a type of ape), and modern hunter-gatherers all avoid carrion even when it is quite fresh. Like grazers, USOs are much more abundant in drier, open savanna than in wet forest, perhaps 10–400 times more so. Modern hunter-gatherers eat large amounts of USOs, particularly when other food is scarce. Although the digestibility of tubers improves with cooking, two-thirds of tubers in some savannas are edible raw, and there exists some (although contested) evidence for the use of cooking fires by 1–1.6 Ma (see below). While *H. habilis* likely was the first tool user, it was *H. erectus* who regularly used stone tools to butcher meat and cut plants, as shown by anthropologists Kathy Schick and Nick Toth. Capturing and consuming a carcass or a live animal may have also required more complicated social (and living) organization for detailed planning, coordination, cooperation, and food sharing. Thus, to fully understand our own evolution, we cannot separate it from the evolution of savannas and of the plants and large mammalian fauna of east Africa.[30]

As the climate dried and savannas spread, the Rift Valley became a richer and more diverse place for other reasons. Remember that the main episode of recent rifting and volcanic activity in east Africa was before and during early hominin evolution. Uplift and rifting created hills adjacent to valleys that today support a stunning array of vegetation types and animal habitats (see Map 9). New volcanic deposits flowed out over the old African shield and slowly broke down into rich volcanic soils. Rich soils support eutrophic, nutrient-rich savannas (Map 10) and two to three times more large mammalian biomass than the underlying older, nutrient-poor soils. Part of the reason this happens is that when there are plenty of nutrients, they can cycle faster from soil to plant to grazer and back into the soil. In turn, at least today, grazing by animals can actually accelerate this effect, sometimes causing grasses to produce more than when they are not grazed.[31]

As we have discussed, when hominins began to hunt regularly and to burn savannas, they became ecosystem engineers. But it is not clear when hominins first started to hunt; they may have scavenged for a very long time. A longstanding archeological question is, Who got to meat first—and did hominins share in this resource? Working in the Serengeti and Ngorongoro, Tanzania, anthropologist Robert Blumenschine tried to answer this question by studying how much meat was left and for how long on 264 carcasses killed by lions, hyenas, and cheetahs. If large quantities were left for long periods, then this could have been a productive source of food for early hominins. Blumenschine found that meat and marrow left on larger kills (elephants, giraffes) by carnivores remained for up to 150 hours, but that left on smaller kills (Thomson's and Grant's gazelles) remained for up to only 4 hours. Hyenas consumed almost all of the carcasses they fed on, but lions left behind most of the bone marrow and head contents (brain), particularly on large animals. There was 50 percent more meat to be had in the dry season than in the wet, and more also when migratory herds were present, creating an immediate source of prey and thus carcasses. From this, Blumenschine concluded that early hominins could find bone marrow on most kills and meat on larger kills in the dry season near streams and rivers, where hyenas have more difficulty finding carcasses. In later work, he concluded that cats (felids) got to carcasses first, hominins second, and hyenas third, the latter scavenging what the hominins left behind. Other scholars found evidence suggesting that hominins got to carcasses first or had "first access" by chasing carnivores away from kills or by hunting the animals themselves.[32]

STAGE 4

1.8 Ma to 12,000 BP: Second Half of Pleistocene Epoch

Mastery of fire was a key step in hominin evolution, giving them greater control over their environment and the resources available to them. The first hominin cooking fires may be as much as 1.5 million years old. Strong evidence comes from Europe, where evidence of fire inside hearths dates back 200,000–400,000 years. Archeologist Naama Goren-Inbar and colleagues found burned seeds, wood, and flint in Israel dating to about 790,000 years ago. Older possible evidence of fire includes charred wildebeest, zebra, warthog, baboon, and hominin bones from 1–1.5 million years ago, found along with unburnt hominin bones and bone tools by archeologists Charles Brain and Andrew Sillen in a cave at Swartkrans, South Africa. Brain and Sillen experimentally burned hartebeest bones at different temperatures and found that the ones in the cave matched ones heated to campfire temperature. Such charring could be caused by cooking, though it could also could be the side effect of fire-building for warmth or protection from predators. Archeologist John Gowlett and colleagues found reddened soil that looks like soil heated under a hearth at Chesowanja, Kenya, as did archeologists Randall Bellomo and Jack Harris farther north at Koobi Fora, dated to 1.4 million years ago. Thus, hominins likely started using fire 0.8–1.5 million years ago, long before modern humans (*Homo sapiens*) evolved some 200,000 years ago.[33]

That hominins built fires for warmth or cooking does not mean they also lit fires to burn savannas. Although there is ample evidence that savannas burned throughout the evolution of hominins, particularly as the climate dried during glaciations, we don't know if those fires were set by lightning strikes or by hominin hands. Fires would have helped them capture insects and small mammals running from fire, as storks do today when they stand, alert and immobile, in grass at the edge of a burning fire. Early hunters likely discovered the value of roasted animal flesh and plants by consuming cooked meat or tubers after a fire had passed. Today's hunter-gatherers set fires in woodland savannas and forests to drive their prey toward other concealed hunters or nets. Okiek people in Kenya burn the more open savanna in patches to attract and concentrate grazers to the short green flush of grass that regrows within weeks of a fire. Many cultures burn thick grass so they can better see predators, snakes, or scorpions underfoot. It may not be possible to know when hominin hunters and scavengers discovered how to burn savannas. Whenever it was, this discovery allowed hominins to manipulate (or engineer) large parts of savanna landscapes for the first time.[34]

It is also unknown when early humans developed the technology (or social organization) to kill *large* mammals. The world's oldest known spears, some 400,000 years old and from Germany, presumably allowed early *Homo* to hunt and kill large mammals there. Scholars differ on when *Homo* in Africa had the technology to kill large (and dangerous) mammals efficiently. It was not until about 100,000–200,000 years ago that points and "hafting," the mounting of stones on handles or shafts, appear in the African archeological record. Perhaps some of these were later used with a bow. But did *Homo* use these points to kill large mammals? Paleoanthropologists Sally McBrearty and Alison Brooks show that hunters killed the long-horned buffalo and other animals about 80,000 years ago in southern Africa. (They go on to argue that most cultural innovations seen outside Africa originated in Africa.) But they do not say that these early *Homo sapiens* killed large mammals often. On the other hand, archeologist Stanley Ambrose proposes that *Homo sapiens* became "superpredators" earlier, as much as 70,000 years ago, by constructing effective networks of information sharing to strategically plan hunts. Curtis Marean, an anthropologist working at Lukenya Hill outside Nairobi, has found repeated mass killings of a now extinct animal, similar in body form to a hartebeest or topi but about 25 percent smaller, not more than 20,000 years ago. By this time, *Homo sapiens* was using spear throwers, and a little later the bow and arrow.[35]

Toward the end of the Pleistocene (50,000–10,000 years ago) the Earth experienced a global "extinction spasm," when over half the genera of mammals disappeared. In the Americas, Europe, and Australia, nearly all of the genera of largest mammals (megafauna, greater than 1000 kg, or 2,200 lb), three-quarters of the medium-sized mammal genera (100–1000 kg, or 220–2,200 lb), and 40 percent of the small-mammal genera (5–100 kg, or 11–220 lb) became extinct. This is when the world saw the last mastodons, woolly rhinos, and other giants. Many hypotheses have been put forth to explain these extinctions, with few strong data to allow us to embrace or exclude any of them definitively. Climate change at the end of the Pleistocene may have been one cause of these great mammalian die-offs in some parts of the world. Other hypotheses for the extinctions include "hyper-diseases" or invasive species, possibly introduced by aboriginal humans; widespread habitat change caused by burning; and protracted or rapid overhunting or "overkill" by humans. If Ambrose is right about humans becoming superpredators 70,000 years ago, overkill may well have been a significant factor. But the evidence supporting the overkill hypothesis is not conclusive, and some evidence even points against it, including in North America and Australia, where the evidence for overkill that does exist is strongest. This debate will likely continue, but it seems pretty clear that humans had some part in these mass extinctions. Whatever the causes, ecologist Norman Owen-Smith points out that such a huge loss of megafauna, which are major ecosystem engineers, must have had large spillover effects on vegetation structure, fuel biomass (and thus fires), and populations of smaller animals.[36]

The late-Pleistocene extinctions were less dramatic in Africa, which lost only 18 percent of its genera of megafauna. East Africa, for example, lost the long-horned buffalo at this time, but no other mega-species of note. Why was Africa mostly spared? It's possible it wasn't, that in fact the extinction event happened here much earlier than on other continents. Between 130,000 and 200,000 years ago, in the mid-Pleistocene, we know that Africa lost a range of large mammals, including carnivores and cercopithecoid monkeys. Archeologist Todd Surovell and his colleagues even argue that hominins helped push early proboscideans (*Elephas, Deinotherium*) to extinction in Africa earlier still, based on the coincidence of their extinction with evidence of subsistence exploitation by *Homo erectus* in Upper Bed I of Oldupai Gorge, dated to about 1.8 Ma. The evidence for early-hominin-caused extinctions in Africa, while tantalizing, is scanty. Some scholars think that as hominins slowly evolved in Africa into more proficient hunters, mammals adjusted their behavior to avoid hominin predators. This stands in sharp contrast to the first-time entry of a fully proficient hunter into the Americas in the Pleistocene. It could also suggest that the existing wildlife in Africa may be more resistant to extinction at the hands of humans than on any other continent.[37]

CHAPTER FIVE

Ecosystem Engineers Come
of Age

Animals apparently healthy the night before were found
dead in the morning. Others dropped on their way to
water. Whole camps were left with no more than
half-a-dozen head and many families lost everything.

RICHARD WALLER, *historian, commenting in 1988 on
rinderpest, a disease that killed massive numbers of livestock
and wildlife across Africa in the 1890s*[1]

[Rinderpest] in some respects . . . has favoured our
enterprise. Powerful and warlike as the pastoral tribes
are, their pride has been humbled and our progress
facilitated by this awful visitation. The advent of the
white man had not else been so peaceful.

FREDERICK LUGARD, *colonist and hunter
in east Africa, 1893*[2]

Up to about 12,000 years ago, all people on Earth gathered or hunted the natural
bounty that grew wild on the land and in the sea.[3] This likely created "soft boundary" landscapes (see Map 12), a fluid mixing of people and wildlife in savannas.
During that time, people diverted minor amounts of energy and nutrients from
the food web for human needs, making for the most part only light and ephemeral
impacts on the land. Then, about 12,000 BP, people began to domesticate and
grow animals and plants for food—and the ecosystem engineering of savannas
started in earnest. In the last few millennia, moreover, foreigners, abetted by east
Africans, have scoured the region to enslave people and kill elephants for their
ivory. These practices led to massive human dislocation and elephant deaths, disease, social trauma, and the economic transformation of east Africa, including how
people use savannas. These more recent events set the stage for the often conflicting arguments we have about conservation and development in east Africa today.
This chapter focuses on the period from the start of food production to the late
twentieth century, with somewhat more detail about the history of savannas of
present-day Kenya and Tanzania, to give background to the discussion of that
region today in Chapters 8–11.

TAMING AND HERDING FOOD
ON THE HOOF

12,000–2,000 BP

People invented food production in just nine places in the world; in most of these, humans domesticated and grew plants before they domesticated and raised animals. Given the proximity of east Africa to Mesopotamia in the Fertile Crescent, one might expect that the first east Africans to produce food were farming peoples who had migrated south with their crops and livestock from this "cradle of agriculture" on what is now the Turkish-Syrian border. Instead, the first people to produce food in eastern Africa were hunter-gatherers who had become mobile herders (see timeline, Figure 17). Some of the earliest people to use domesticated animals lived in the Nabta Playa in what is now the Western Desert of southern Egypt about 9,000 years ago. The bones they left behind were mostly of gazelles and lagomorphs (rabbits and hares), but there were also a few bones of domesticated cattle in their settlements in and along a seasonal lake bed. Cattle were probably first domesticated earlier in Africa, perhaps to the west next to one of the Saharan massifs on what is now the Sudan-Chad border. Cattle thus are native to Africa, but not to Africa south of the Sahara. People then brought sheep and goats to Africa, probably from Asia, about 7,000 years ago. Egyptians probably first domesticated the donkey (*Equus asinus*) from one or more types of African wild ass (*Equus africanus*) about this time, as zooarcheologist Fiona Marshall suggests. Camels were brought from Arabia to Africa much later, about 3,000 years ago.[4]

Verdant grasslands across what is now the Sahara started to dry out and shrink in size about 8,000 years ago, likely pushing these early African herding hunter-gatherers to the south to find wetter pastures in present-day Ethiopia, South Sudan, and locations even farther south (Map 13). By 6,000 years ago, the habit of herding had spread to the hunter-gatherers along the Nile in what is now Sudan. For context, consider that Egypt's first dynasty was under construction by 5,000 BP. About 5,000–4,000 years ago, southern Cushitic (Afro-Asiatic–speaking) herders moved south from present-day Ethiopia to Kenya, bringing with them goats, cattle, and sheep and eventually spreading across what is now Kenya's highland grasslands and south into Tanzania. As they pushed south, according to archeologist Stanley Ambrose, these herders encountered, lived side by side with, and eventually absorbed click-speaking (Khoisan) hunter-gatherers who used the forest environments of what are now Kenya and Tanzania. Today's remaining Khoisan speakers, some of whom still practice hunting and gathering, are descendants of the most

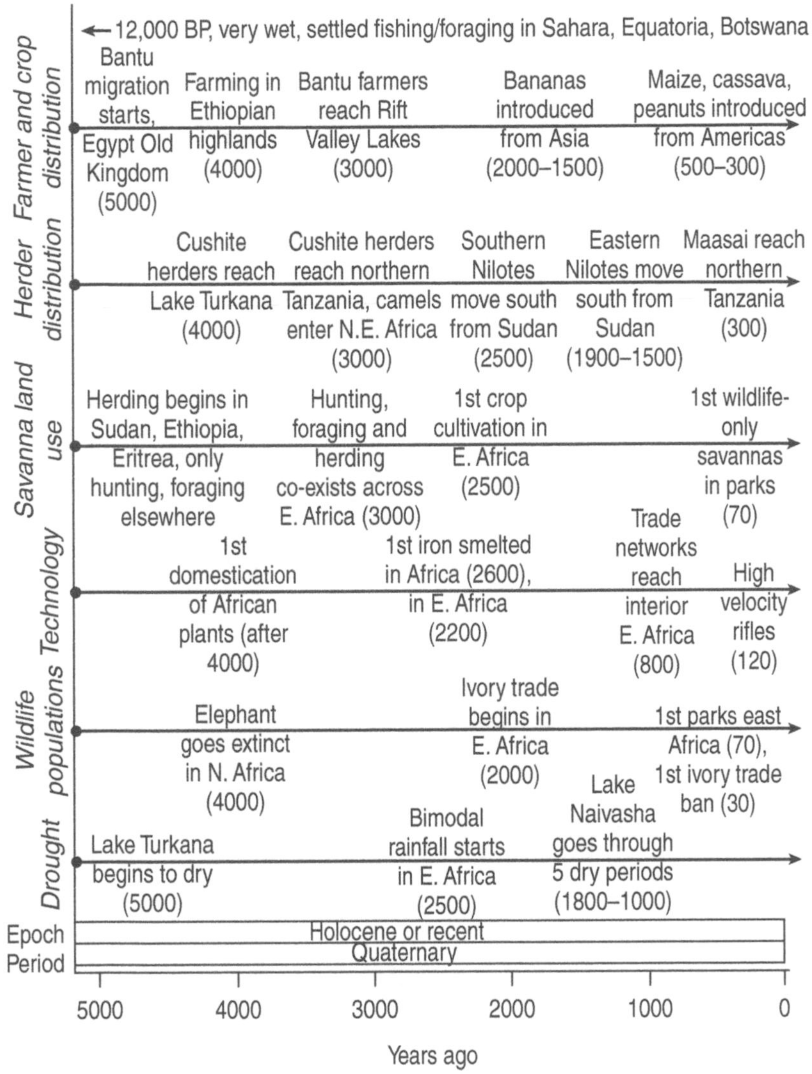

←12,000 BP, very wet, settled fishing/foraging in Sahara, Equatoria, Botswana

Farmer and crop distribution

Bantu migration starts, Egypt Old Kingdom (5000) — Farming in Ethiopian highlands (4000) — Bantu farmers reach Rift Valley Lakes (3000) — Bananas introduced from Asia (2000–1500) — Maize, cassava, peanuts introduced from Americas (500–300)

Herder distribution

Cushite herders reach Lake Turkana (4000) — Cushite herders reach northern Tanzania, camels enter N.E. Africa (3000) — Southern Nilotes move south from Sudan (2500) — Eastern Nilotes move south from Sudan (1900–1500) — Maasai reach northern Tanzania (300)

Savanna land use

Herding begins in Sudan, Ethiopia, Eritrea, only hunting, foraging elsewhere — Hunting, foraging and herding co-exists across E. Africa (3000) — 1st crop cultivation in E. Africa (2500) — 1st wildlife-only savannas in parks (70)

Technology

1st domestication of African plants (after 4000) — 1st iron smelted in Africa (2600), in E. Africa (2200) — Trade networks reach interior E. Africa (800) — High velocity rifles (120)

Wildlife populations

Elephant goes extinct in N. Africa (4000) — Ivory trade begins in E. Africa (2000) — Lake Naivasha goes through 5 dry periods (1800–1000) — 1st parks east Africa (70), 1st ivory trade ban (30)

Drought

Lake Turkana begins to dry (5000) — Bimodal rainfall starts in E. Africa (2500)

Epoch: Holocene or recent
Period: Quaternary

Years ago: 5000 4000 3000 2000 1000 0

FIGURE 17.

Timeline of early food production to the present, the last 5,000 years (BP = before present). (Sources: Richardson and Richardson 1972, Ambrose 1984a, Ambrose 1984b, Marshall 1990, Miller and van der Merwe 1994, Illiffe 1995, Newman 1995, Haaland 1995, Schmidt 1997, Hanotte et al. 2002, Marshall and Hildebrand 2002, Hakansson 2004, Field 2005, Marshall 2007, Rossel et al. 2008.)

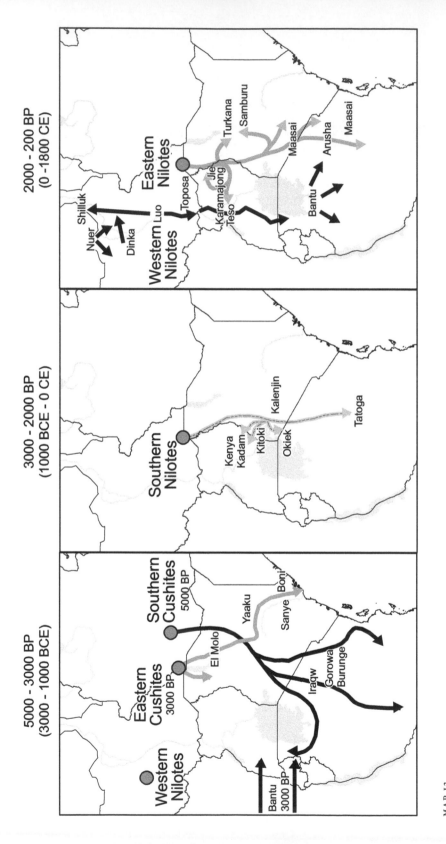

MAP 13.

Major prehistorical and historical movements of people in east Africa. (Map adapted with permission from James L. Newman [1995] and Yale University Press by Russell L. Kruska.)

ancient living peoples on Earth (see Map 5). It took more than a thousand years for the southern Cushitic herders to become fully established in present-day Kenya, probably because of changing climates and new diseases afflicting their livestock. By 3,000 BP, the southern Cushitic herders had reached what is now northern Tanzania.[5]

Farmers came into east Africa first from the north and then from the west (Map 13). As long ago as 7,000 years ago, people were living in a large settlement, herding cattle, and cultivating large quantities of wild sorghum just 20 km (about 12 mi) north of what is today Khartoum (Sudan). The great Egyptian dynasties to the north certainly built their power partly on a farming economy, starting about 5,000 years ago. By 3,000 years ago, the Afro-Asiatic or central Cushitic speakers (Agaw) lived in small farming communities and cultivated a native grass called tef, still grown today in Ethiopia, as well as finger millet in the highlands of what is now Tigray, in Ethiopia and in Eritrea. Perhaps about this time (3,500–4,000 years ago), other Afro-Asiatic–speaking Semitic farmers migrated from southern Arabia to the Ethiopian highlands, bringing wheat, barley, oxen, and the plow and mixing with the central Cushites.[6]

Farther south, in what is now Uganda, Kenya, and Tanzania, it is not clear who planted the first crops. Reconstruction of the vocabulary of the earlier southern Cushitic herders, who moved south from present-day Ethiopia to Kenya's Lake Turkana, suggests that these herders knew about finger millet and sorghum, but there is no clear archeological evidence that they cultivated these grain crops. Instead, farming and more sedentary lifestyles definitely came with Niger-Congo–speaking peoples (also known as Bantu) who moved in from the west. Dependent on cultivating yams, foraging, and fishing, these migrants started moving out of what is now west Africa's Cameroon/Nigerian borderlands about 5,000 years ago, arriving 2,000 years later on the west side of Lake Victoria. By this time, southern Cushitic herders had reached what is now northern Tanzania. Thus, when the Bantu farmers from the west later reached the Great Rift Valley, they found bimodal rainfall and herders already using the savannas there. It was only much later that groups like the Maasai arrived in this region, taking land from hunter-gatherers and herders who had lived in this area for thousands of years before them.[7]

Today, geographer James Newman describes the countries around Lake Victoria as "arguably the most complex ethnolinguistic region in the continent" because this is where all four major language groups meet: Afro-Asiatic, Khoisan, Niger-Congo, and Nilo-Saharan (see Map 3). This complexity arose as follows: about

2,500 years ago (c. 500 BC), as the Niger-Congo farmers moved into the equatorial region from the west, the first wave of Nilo-Saharan herders (also called southern Nilotes) walked south from what is now South Sudan into Kenya. They were the forebears of the Kalenjin (Pokot, Tugen, Nandi, Kipsigis on Map 5) in the western highlands of present-day Kenya and also the Tatoga (Barabaig) of northern Tanzania.[8]

By 100–500 AD, the eastern Nilotes, who now include the Bari, Kuku, Karamojong, Turkana, Maasai, Teso, and Samburu, moved south from what is now South Sudan into the Rift Valley, a region still dominated by some of these groups today. Much later, western Nilotes (today's Luo people) came from present-day South Sudan into the Lake Victoria region. One of the last places the Niger-Congo farmers settled was present-day central Tanzania. You can see this complex linguistic mix in north-central Tanzania and southwestern Kenya today, where ethnic groups representing all four major language families live nearly side by side. These include the Afro-Asiatic–speaking Iraqw, the Khoisan-speaking Kindiga (or Hadza), many Niger-Congo (Bantu)–speaking groups like the Sukuma and Sonjo, and three different branches of the Nilo-Saharan language group: the Tatoga and Kipsigis (southern Nilotes), Maasai (eastern Nilotes), and Luo (western Nilotes).

Soon after the Niger-Congo, or Bantu, farmers arrived west of Lake Victoria, they adopted a technology that allowed people to further modify landscapes to their needs: ironwork. People living in settlements in present-day Rwanda were using iron by about 2,800 years ago (800 BC), and people were smelting iron in what is now Uganda by about 2,200 years ago (200 BC) and in the Usambara Mountains of present-day Tanzania by about 1,800 years ago (200 AD). It is possible that farmers using iron hoes and other implements could produce more food faster than farmers using wooden implements (depending on the crop cultivated) and that this ability helped Bantu farmers to spread across sub-Saharan Africa. Historian James Webb further argues that Bantu people succeeded in colonizing most of sub-Saharan Africa because they developed early resistance to the fatal form of malaria caused by the parasite *Plasmodium falciparum,* common in their ancestral home in the rainforests of west Africa. These Bantu migrants were then able to survive malaria as they moved across Africa, presumably better than the peoples they encountered along their path.[9]

The Bantu adopted not only iron, but also plants and animals that allowed them to cultivate and herd in savannas with bimodal rainfall. Loan words suggest that the Bantu farmers learned sorghum cultivation and possibly cattle herding from herders speaking Nilo-Saharan languages that they met in eastern Africa. The double-wet-season rainfall that began in east Africa about 2,500 years ago likely made cultivation more risky, causing many Bantu farmers to take up herding as well.[10]

Unlike in other parts of the world, herding thus came before farming to east Africa (and, in fact, to the rest of sub-Saharan Africa, too). But why? Zooarcheologists Fiona Marshall and Elisabeth Hildebrand think that Africa's changing climates and the accompanying expansion of savannas are the key: unstable, marginal environments made early food producers more concerned about producing reliable food day to day (scheduled consumption) than producing a large quantity of food (high yields). As described in Chapter 3, farming requires several months of sufficiently high rainfall to support plant growth, whereas herders and their livestock can move to take advantage of patches of green grass and ephemeral water sources to survive dry spells. The flexibility of moving with herds was clearly a better way than farming to access food reliably in east Africa, at least at first. This lifestyle also blended well with the existing hunting and gathering lifestyles of the people of the region and the diverse assemblages of wild animals and plants they depended on.[11]

Archeologist Andrew Smith describes the historical introduction of herding in an area already occupied by hunting peoples like this: "Herders use the same wild resources as the hunter/foragers but will have the additional benefit of a sustained yield in the form of meat and milk from their domestic herds. This will give them the necessary edge in expanding their population density at the expense of hunters." Hunter-gatherers depend directly on the animals they hunt and the plants they forage for survival, while herders rely more on finding good pastures and water sources and avoiding predators and disease. Because some wildlife compete with livestock for forage and wild carnivores kill domestic animals, these new herders likely considered wildlife more of a problem than their hunting-only neighbors. Hunters, who usually do not think of themselves as "owning" wild animals, will range over wide areas to reach them in their own habitat. Herders, in contrast, do develop a strong sense of ownership and husbandry, which limits them to areas within walking distance of water for their livestock. When present-day hunter-gatherers adopted herding in the Kalahari, for example, they began to move their residences less often.[12]

The transition from hunter to herder built new, albeit soft, boundaries between herders and wildlife as people concentrated in favored locations in savannas. These settlement locations may have been favorite locations for wildlife, too. As described in the next chapter, by defending livestock from predators, herders may also have created a "safe halo" around their settlements for grazing wildlife and other prey of abundant predators.

We can only guess at how the shift from hunting and gathering to herding affected the abundance and diversity of wildlife in the savannas of east Africa. For example, it is not known if herders came into a landscape in which mammals had already been reduced by hunters to levels below what the vegetation could support. To explore this question, prehistorian David Collett examined the archeological evidence from several sites across present-day Maasailand from about 2,500 years ago. He suggests that, at the time, wildlife were not a large part of the diet of most herders but rather represented "a crisis resource," one that herders resorted to only when they had lost their livestock. "Such a strategy is unlikely to have produced any large-scale threat to wildlife," he writes, "a thesis supported by the fact that all the species of wild fauna from these sites occurred in Maasailand at the time of European contact, despite more than 2,500 years of exploitation of the area by pastoralists." On the other hand, zoologist Herbert Prins argues that modern pastoral groups have long histories of hunting and eating wild meat and he thinks that, as a consequence, savannas and their wildlife are headed down the "pastoral road to extinction" today. Both of these scholars could be right: Collett could be right about the past, and Prins may be right about the present. We will return to this issue in the next two chapters.[13]

And then came the farmers—those ecosystem engineers who settled down, cleared land, and grew (and defended) crops. To grow successful crops in a savanna with abundant wildlife requires at least two things: access to enough water to grow a crop to maturity and protection of that crop from wildlife. To grow crops to maturity, farmers must cultivate either in wet savannas with sufficient rainfall or in areas where there are key resources, where groundwater extends the crop growing season, such as in wetlands or along rivers. Wet savannas and key resources are also critical to wildlife (and other organisms) for their green forage, particularly during dry seasons or droughts.

Like farmers today, early farmers no doubt struggled to keep animals of all sizes, particularly destructive large mammals, from consuming their crops. This in turn may have encouraged them to form compact settlements. Reviewing human-elephant conflicts before the colonial period, zoologist Richard Hoare speculates that "elephants . . . were probably a major obstacle to the evolution of arable farming in precolonial Africa. . . . Within elephant range in both savannas and forests, agriculturalists could probably only prosper in large, well-defended villages." Early farmers likely had localized but strong impacts on wildlife by clearing land, thus removing habitat. There is, for example, evidence of deforestation in east Africa, presumably to create farmland, as long as 2,000 years ago. A combination of farming, livestock grazing, and iron manufacturing probably caused gully erosion starting about 900 years ago in the Irangi Hills, in what is now north-central Tanzania.[14]

SLAVE AND IVORY TRADE, PESTILENCE, AND THE RISE OF COLONIAL CONSERVATION

2,000 BP to the Nineteenth Century

The slave trade disrupted human economies all over Africa for more than a thousand years, with an estimated 17 million people taken from their homes by force and sold into slavery between about 650 to 1800. In east Africa, according to historian Paul Lovejoy, "slavery was already fundamental to the social, political, and economic order of parts of the northern savanna, Ethiopia and the East African coast for several centuries before 1600 . . . sanctioned by law and custom" in Muslim societies and driven by their demand. Eastern and southern Africa played a small role (1 percent or 63,400 of the 6 million slaves traded) in the massive Atlantic slave trade via European traders at its height in the 1700s. These forceful enslavements no doubt depopulated parts of east Africa, at least temporarily, and likely retarded economic development.[15]

Even before the slave trade started, the ivory trade reached deep into Africa. This trade may well have had a larger impact on savanna wildlife and vegetation than all the hunting, herding, and farming before the trade started. The ivory traders from southern Arabia reached the east African coast about 2,000 BP (0 BC), and used Aksum in what is now Ethiopia as the collection point for all of the ivory from the Nile basin. Thomas Hakansson, a human ecologist, estimates that significant ivory trade networks spread inland from the coast to interior savannas and forests about 800 years ago (1200 AD) and were extensive by 500–600 years ago (1400–1500 AD). Much of the ivory from the early trade flowed through overseas networks to China and India. The inland trade network was already thriving when the Portuguese arrived in Mombasa in 1498.[16]

Demand for ivory soared in the mid-1800s, and by the 1840s east Africa was the foremost source of ivory in the world. With ivory flowing out from the interior through Zanzibar and Khartoum, the impact on elephant populations was dramatic. "The reduction and regional extermination of elephants in eastern and southern Africa by the end of the 19th century," Hakansson summarizes, "can be directly related to the expansion of industrial capitalism and the consequent rise in world market prices for ivory." In other words, elephants were slaughtered in large numbers to make billiard balls (especially from female tusks, "being softer and malleable"), ivory-handled cutlery, combs, and piano keys for consumers in the United States and Europe, and jewelry for consumers in India and China. By the late 1870s, half a million porters carrying ivory and other export goods passed through Tabora in what is now central Tanzania each year.[17]

Farmers, herders, and hunters who lived in savannas participated in hunting wildlife "for the pot" and hunting elephants to profit from the ivory trade. The Kamba of Kenya, for example, were major elephant hunters from the late 1700s through the early to mid-1800s. At one point the Kamba ivory capture and trade network stretched from Lake Turkana in what is now northern Kenya to Mt. Kilimanjaro in northern Tanzania; today, some wealthy Kamba families trace the origins of their wealth to their participation in elephant hunting in the 1800s. The Orma people (now living near the Pokomo, Map 5) hunted ivory in present-day northern Kenya, while the Nyamwezi hunted in their homelands south of Lake Victoria. And the Dorobo, though a hunting culture, often shifted from hunting to pastoralism by using their profits from the ivory trade to buy cattle. Hakansson even speculates that the sixteenth-century expansion of the Maasai people in the savannas of east Africa was partly spurred by their participation in the growing ivory trade.[18]

The ivory trade and agricultural expansion devastated elephant populations—a clear case of long-term, irreversible loss of a keystone species in savannas. Before the 1800s, our best guess is that the ivory trade was small compared to the size of the elephant populations in east Africa. Between 1814 and 1890, perhaps two-thirds of Africa's elephant populations disappeared, with numbers falling from 19–20 million to about 6–7 million elephants. The ivory export from Zanzibar and Khartoum over time represented tens of thousands of animals; in 1848 alone, some 7,000 were killed to supply exports from Zanzibar. By this time, elephants were gone from the east African coast and some places inland, such as Ukambani in present-day Kenya. In 1883, elephants were so rare that when Joseph Thomson walked the hundreds of kilometers from Mombasa to the Ngong Hills (near present-day Nairobi) through the Amboseli area, which supports 1,500 elephants today, he saw no elephants until he reached the Ngong Hills. Others reported few elephants in the Serengeti in 1900.[19]

Trade in ivory went into a lull after 1914, only to pick up to colonial levels again in the 1970s with large increases in demand for ivory from Asian countries. In 1987, just before the Convention on International Trade in Endangered Species of Wild Fauna and Flora (CITES) banned the ivory trade in 1989, Africa's elephant population had dropped, relative to the previous century, from 6–7 million to about 720,000 animals, partly due to poaching, partly due to the expansion of farming. By 2002, the continent's elephant populations had further dropped to about 400,000, a reliable estimate based on animal and dung counts, though best guesses suggest that there may be another 260,000 elephants, making a total of 660,000. About 20 percent of this total—134,000 animals—is located in the ten countries of east Africa. (Note that elephant populations in only 15 percent of their current range area have been counted in a reliable and repeatable way.) Overall, today's population of African elephants stands at about 2–3 percent of the pre-1814 estimate of the continent's capacity to support elephants.[20]

The booming trade of the 1800s reorganized and stressed human societies and savannas of east African in new ways (see Figure 18 and next section). Juhani Koponen, a historian, emphasizes the role of trade in promoting "war, pestilence, and famine" in nineteenth-century east Africa, even though all of these were present in previous centuries. Traders brought firearms for the first time into what is now Tanzania in the 1800s, at first old, inaccurate muzzle-loaders. By 1888, Zanzibar was sending 100,000 firearms to the mainland each year, and over half of the population used guns as their chief weapons in 1890. The ivory and slave trades essentially gave east Africans something to fight about: "Trade contributed towards increasing violence both by providing an impulse for ambitious African leaders to strengthen their own power and by giving them a means of doing so."[21]

Disease was another significant stressor on humans, particularly in the 1800s. Spread by more extensive trade in that century, new and existing diseases became more virulent than in "social and biological memory," as Koponen puts it. Cholera first appeared on the coast of east Africa in 1821, and widespread epidemics occurred in 1836–37, 1858–59, and 1869–70. The 1870 epidemic came from Mecca to Somalia and then via Maasai cattle raiders to what is now Kenya and Tanzania. Travelers at the time described piles of rotting corpses along the trade routes, villages depopulated, and a quarter of Zanzibar's population dead. In 1879, perhaps 100,000 people died of cholera in what is now Tanzania and Kenya. Smallpox was even more virulent. Although this disease was endemic to central Africa before the 1800s, starting in 1809 it broke out in east Africa every few years or decades

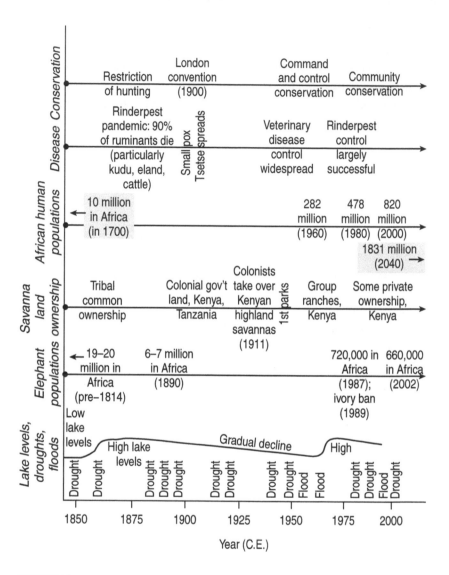

FIGURE 18.
Timeline of east African savannas from 1850 to the present
(C.E. = current era, or A.D.). (Sources: Atang and Plowright
1969, Ford 1971, Richardson and Richardson 1972, Anderson
and Grove 1987, Douglas Hamilton 1988, Waller 1988, Rutten
1992, Newman 1995, Leader-Williams 2000, Hughes 2002,
Brockington 2002, Blanc et al 2003.)

thanks to increased trade. In 1890, 30–50 percent of those infected with smallpox eventually died.[22]

Not all diseases affected humans, of course. In 1889, the Italian army unloaded some Indian cattle off a ship at the port of Massawa, in present-day Eritrea that carried the deadly viral disease known as rinderpest. These cattle triggered a panzootic epidemic that left scores of wildlife, livestock, and pastoral peoples dead (the latter from starvation) across the continent. This viral disease, which biologist John Ford described as "the biological consequence of infiltration of European culture," traveled from Massawa to the southern tip of Africa in just seven years, killing most of the domestic and wild animals it infected so quickly (as fast as three days) that entire herds appeared to drop dead en masse. Historian Richard Waller describes the speed of death from rinderpest in the quote that began this chapter. Not just cattle, but even-toed hooved wild mammals (Artiodactyls) were susceptible, including Cape buffaloes, elands, warthogs, and bushpigs, "followed," according to veterinarian Walter Plowright, "by giraffe, greater and lesser kudu, roan antelope, bushbuck and finally, wildebeest." Observers at the time estimate that 90–95 percent of the cattle and buffaloes in east Africa died, while populations of a few other species, such as bongos, roans, and greater kudus, probably never fully recovered. Meanwhile, sheep, goats, Thomson's and Grant's gazelles, hartebeest, topis, reedbucks, impalas, and waterbucks were less affected by rinderpest, and others, like elephants, hippos, zebras and rhinos, were not affected at all. A second wave of the rinderpest panzootic came in 1897 and more deaths ensued. The disease was finally controlled in the 1960s, when Plowright developed a vaccine, but not eradicated until 2011.[23]

When rinderpest struck in east Africa, many pastoral societies—the Jie, Karamojong, Mursi (with the Suri on Map 3), and Samburu—were just starting to recover from their first experience with the bacterial cattle disease CBPP (contagious bovine pleuro-pneumonia). In the highlands of present-day Ethiopia, death of cattle meant death of plow oxen and little crop cultivation, which in turn meant that desperate pastoralists from the lowlands could not depend on their farming neighbors in the highlands for food, and vice versa since the lowland herds were also devastated. The Mursi, along the Omo River in what is now southern Ethiopia, had other options for accessing food, so they gave up pastoralism entirely and took up hunting, fishing, and gathering. Other herders who depended almost solely on pastoralism, like the Maasai, deserted their pastures and joined their farming or hunting neighbors. The Karamojong, Jie, Samburu, Il Chamus, and others turned to killing elephants to trade ivory so they could rebuild their herds.

Desperation was often the rule: as one Jie informant told historian John Lamphear, "My own grandfather lived like a monkey collecting wild things in the bush." A few pastoral communities were spared because they lived in remote areas beyond the reach of rinderpest such as the lowlands of Turkana, or because, like the camel-herding Rendille and the goat-herding Labwor, they relied on animals little affected by the disease.[24]

In the early 1890s, smallpox swept across east Africa, probably killing up to half the people who became infected. Waller writes of the Maasai, whose condition was very weak due to the devastation of their herds by rinderpest: "Victims 'died in one day' or simply 'dropped in their tracks.' Murran (warriors) eating meat in the bush died together, their soup kettles still boiling on the fires." One Maasai observer said that the corpses of people and cattle were "so many and so close together that the vultures had forgotten how to fly." On top of these scourges, a vigorous invasion of jigger fleas, which can cause severe foot infections, reached east Africa from present-day Angola in the 1890s. This death and desperation destroyed the foundation of pastoral societies and pastoral reciprocity for a decade or more. Conflict among different Maasai sections broke out, for example, with starving herders raiding both agriculturalists and their Maasai neighbors alike.[25]

By the turn of the twentieth century, several of the major ecosystem engineers, including people, livestock, and elephants, were uncommonly rare in many of the savannas of east Africa, and the effects may still be seen today. Ecologists Truman Young and Keith Lindsay, for example, comment on the mature, similarly sized (and probably similarly aged) stands of *Acacia xanthophloea* (fever trees) in Amboseli, Lake Naivasha, Lake Nakuru, Ngorongoro, and Lake Manyara that all began dying at the same time in the 1970s and 1980s. Herbert Prins and Henk Vanderjeugd, zoologists, studied patches of similarly sized trees of a different, slower-growing species, *Acacia tortilis,* in Tanzania's Lake Manyara, Serengeti, and Tarangire. All these acacia trees were probably established between 1885 and 1890, just when the rinderpest and ivory trade killed many of the wildlife and livestock that would naturally have kept the acacia trees in check. At least in Amboseli, some scholars attribute the decline of these trees to heavy elephant use, but these trees may simply have died simultaneously of old age, a legacy of the rinderpest and ivory trade more than century ago.[26]

Rinderpest and other events not only allowed individual acacia trees to become established more easily, but also encouraged woodlands to spread at the expense of grassland. Historical records indicate that some woodlands spread rapidly, others slowly in the decades after the rinderpest. Biologist John Ford, for example, describes the rapid expansion of woody vegetation when cattle and human populations were decimated as rinderpest swept through Karagwe, in what is now northwest Tanzania, and Ankole, in southwestern Uganda, in the 1890s. Tsetse flies then became more abundant as their favored bushy habitat expanded, which discouraged people from returning to their former homesteads for fear that their cattle would contract tsetse-transmitted trypanosomosis (sleeping sickness in humans). Because it sometimes took decades for people, their livestock, and wildlife to repopulate these savannas, woodlands continued to spread.[27]

Ecologist Holly Dublin reports that between 1890 and 1910, the period during and after the rinderpest, travelers described the Mara and northern Serengeti in colonial Kenya and Tanganyika as an open grassland dotted with acacia trees. Twenty years later, the delayed effects of rinderpest became clear: ground and aerial photographs taken in the 1930s and 1950s "show widespread hilltop thickets and acacia woodlands throughout the northern Serengeti and Mara." This woodland spread began to be reversed in the early 1960s, when a new vaccine against rinderpest helped to control the disease in livestock. Only then, partly because the Maasai no longer feared that their cattle would die of rinderpest, did they move back into the tsetse-infested woodlands, burning them to kill the fly and thus creating the open grassy landscape of the Mara today.[28]

Even though few people remained in the savannas after the rinderpest to light fires, savannas probably burnt more often, which slowed the spread of some woodlands. The death of wildlife and livestock grazers meant there was a lot of grass fuel to burn. Biologist Lindsey Gillson found that in the Tsavo ecosystem of Kenya, a sediment record spanning the seventeenth to the twentieth century had twice as much charcoal (from wildfires) deposited in it between 1880 and 1925 than during any other period. People burned savannas to clear the bush and also as part of stock raiding to restock their herds. With abundant grass fuel, each fire was probably intense and widespread. Several authors suggest that the combined changes in people, livestock, wildlife, vegetation, and fires must have reverberated throughout the savannas, slowing rates of decomposition and nutrient cycling as plant matter accumulated and leading to a decline in ungulate-dependent predators and the spread of rodents, zebras, hippos, and other species not susceptible to rinderpest. The control of the rinderpest in the 1960s (see Chapter 8) may have helped to reverse some of these effects and created (or re-created) the massive migration of wildebeest in the Serengeti today.[29]

THE TWENTIETH CENTURY:
MARGINALIZATION OF PASTORAL
PEOPLE AND THE BEGINNING
OF "DEGRADATION"?

As the twentieth century dawned in east Africa, European powers were establishing colonies across the region. Most of the countries of east Africa were colonized—by the British (Kenya, Uganda, eastern Sudan), Germans (Tanganyika), Belgians (Rwanda and Burundi), Italians (Somalia, Eritrea), or French (Djibouti, western Sudan)—and remained so for fifty to sixty years, from the late 1800s to the mid-1900s. Ethiopia was the only country that defeated European invaders (Italians in 1896 at the battle of Adwa) and was only occupied by Italy for six years, from 1935 to 1941. At the 1884–85 Berlin conference, the British and Germans split up what would become Kenya and Tanzania into British and German East Africa. British East Africa became the Kenya colony in 1920. Germans remained the colonial authority in Tanganyika until 1921, when the British took over. This also happened in Rwanda and Burundi, with the Germans ceding control to the Belgians after World War I.[30]

The European colonial period came to a close in east Africa starting in the late 1950s (Sudan) and early 1960s (Somalia, Tanganyika, Uganda, Rwanda, Burundi, Kenya) and ending in 1977 (Djibouti). Eritrea was federated with Ethiopia in 1952 and seceded from Ethiopia in 1993. Zanzibar joined Tanganyika in 1964 to become Tanzania, and South Sudan gained independence from Sudan in 2011.

How did pastoralists and wildlife fare during colonialism? Before the rinderpest panzootic, the Maasai lived and herded their animals over much of the central and southern Rift Valley, most of present-day Narok and Kajiado districts, and the southern part of Laikipia (near Nanyuki), an area of $55,000 \text{ km}^2$ ($21,236 \text{ mi}^2$). This was land the Maasai had taken from others only a few centuries before, but now the rinderpest forced the Maasai to vacate much of this territory. European settlers, arriving by train on the newly built railway line from Mombasa to Nairobi, saw much "open" land and occupied it. Although the Maasai quickly objected to the loss of their land to the colonists and other African groups, they were only partly successful in getting it back. In 1904, they made a treaty with the British to protect their communal ownership of land they had been using for generations, only to have the treaty broken seven years later, when thousands of Maasai were forcibly moved from the better-watered pastures in the Laikipia and Rift Valley to the "Maasai Reserve" in the southern part of British East Africa (present-day Kenya). As a result, the Maasai lost 50–70 percent of their land in 1911.[31]

Maasai elders remembered the impacts of these moves nearly a century later. Historian Lotte Hughes writes: "They describe the impact of the move in 'pathological' terms, believing that the British deliberately sent them 'to that land where *ol-tikana* [East Coast fever] is' in order that they might die there. They claim that they and their herds succumbed to diseases in the Southern Reserve which were unknown or not prevalent in their northern territory, most specifically Laikipia, and that they have been blighted by sickness ever since." Indeed, both East Coast fever, a fatal disease of cattle, and malaria were little known in Laikipia compared to the southern reserve.[32]

In Tanganyika, by contrast, colonialism meant the loss of pastoral land with the exception of some protected areas. It is only since the independence of Tanzania (formed from Tanganyika and Zanzibar) in 1964 that the state, investors, farmers, (and pastoralists themselves) have converted former pastures to crop farms, mines, settlements, additional conservation areas, and other uses.[33]

Herders became marginal in colonial and postcolonial economic and political systems, as historians Richard Waller and Neal Sobania describe for present-day Kenya:

In 1890, pastoralists were at the center of regional networks of exchange. Broadly speaking, the terms of trade favored stock producers, and the dominant value systems reflected this. The herds they controlled were the universally acknowledged store of durable, investable and reproducible wealth. . . . By 1950, however, pastoralism had been relegated to the periphery of an economic and political system that was now dominated by the needs of export agriculture and in which stock had been bypassed for new avenues of accumulation. Agricultural producers could invest the products of labor directly in land and indirectly in the patron/client relationships that underpinned modern politics. . . . [Pastoral] access to markets was challenged by competitors who were favored by government—white settlers at first and then mixed [African crop and livestock] producers of central and western Kenya.[34]

Colonialism resulted not only in losses of land and wealth for herders but colonial authorities also restricted their mobility to reduce stock raiding and purportedly to slow the spread of livestock diseases. In what is now South Sudan, colonial administrators strove to reduce stock raiding by separating the Dinka and Nuer peoples, which restricted their grazing movements but also cut important social ties between the two groups. Colonial authorities enacted a similar policy in the Kenya colony, limiting different ethnic groups to what were called "tribal grazing areas." For the Rendille and Samburu, these policies not only restricted long-distance movements but also served to solidify the previously fluid boundaries between groups, which sometimes bred ill feelings, increased hostility, and cut both trading and symbolic social ties. In the Northern Frontier District of the Kenya colony, colonialism ushered in a new era of conflict over scarce resources among different groups of pastoralists, while precolonial conflicts were more political or social in nature. Colonial quarantines preventing movement of stock slowed the spread of disease, to be sure, but they also allowed white settlers to keep pastoral producers from marketing their stock, thus reducing competition. Restrictions on stock movement also made it more difficult for herders to cope with and recover from drought.[35]

The colonial authorities came with a set of values concerning the separation of people and nature (see Chapter 1) and a desire to control resources that determined both their perceptions and their prescriptions about pastoral land use. Africa was portrayed as "offering the opportunity to experience a wild and natural environment which was no longer available in the domesticated landscapes of Europe. . . . European man sought to rediscover a lost harmony with nature." Note that this perception of Africa, particularly the savannas of east Africa, as an unspoiled Eden still applies to the marketing of tourist safaris today. On top of these values, the British government sought to control access to the source of the Nile (one reason for building the railway from the east African coast to present-day Uganda), explorer-hunters sought to control access to wildlife, and European settlers sought to control access to land and water. These values and needs affected pastoral society more in some places (e.g., present-day Kenya) than others (present-day Ethiopia) because of differences in the relative influence of colonial authorities and other factors.[36]

One way to protect, and establish control over, this Eden was to develop policies based on "narratives" about the environment. As anthropologist Allan Hoben explains, "The environmental policies promoted by colonial regimes and later by donors in Africa rest on historically grounded and culturally constructed paradigms that at once describe a problem and prescribe its solution." These narratives typically portrayed Africans as the main force damaging the environment. For example, European settlers in Baringo, Kenya colony, in the 1920s constructed a "compelling master narrative of African backwardness, ignorance and incompetence in land husbandry," designed partly to justify their own settlement on what had been pastoral land. In describing the elite group of white hunters in early-twentieth-century British East Africa, historian John MacKenzie writes: "While none of the hunters doubted their own right of access to extravagant killing, they argued for conservation policies and the need to restrict the access of others." Many of these "degradation" narratives about environmental change in Africa not only were poorly informed about the past, but also predicted disaster in the future if trends continued unchecked.[37]

Some degradation narratives were not entirely misguided, however. European observers did see change in the African environment in the late nineteenth and early twentieth centuries, some of which could be classified as "degradation," that is, difficult to reverse. (For the rest of this book, I will distinguish between "decline," a change that is reversible over the short term [less than a decade]; "degradation," a change that is difficult to reverse, requiring decades or more; and "irreversible loss," where there is no prospect of recovery. See Chapters 6 and 7 for further discussion.) These observers, for example, recognized that their own actions, aided by local hunters, had caused the decline of the elephant through the ivory trade. Other wildlife species declined as well because precolonial explorers and missionaries used wildlife as a subsidy for survival, feeding themselves and their African labor largely on wild meat. In addition, African farmers and herders, once they traded for firearms and hunted with them, probably contributed to the wildlife loss. European hunting caused the quagga to go extinct in the Cape colony of South Africa in the 1880s. Europeans were particularly effective at killing wildlife after the introduction of high-velocity rifles in the late 1890s. European establishment of farms and towns was responsible for the loss of wildlife as well. Colonial observers in Machakos, Kenya colony, saw widespread erosion and tree removal on African farmsteads during the 1930s. In the Cape, when settlers realized the forests were disappearing, they developed policies to protect and sustain them for the future. In any event, what seems clear is that both colonialists and Africans caused major environmental changes in these centuries.[38]

But were these changes difficult to reverse, or were they irreversible? In the case of the quagga, extinction was certainly an irreversible loss for Earth's biodiversity. The modern loss of 97 percent of Africa's elephant populations is already long term and probably irreversible in most areas (although in very limited areas elephant numbers are on the rise). In addition to being reduced in number, many species today are limited to much smaller ranges than they were in the past, a condition that is most likely irreversible. On the other hand, the soil erosion and tree removal in Machakos was reversed as farmers adopted conservation measures over time—though it took decades to accomplish this. Even here, however, species sensitive to intensive human land use were probably lost, thus degrading the broader gene pool. We will return to this issue in Chapter 7.[39]

These colonial narratives and observations helped initiate policies and actions that evicted herders from their traditional grazing lands to create parks only for wildlife. By pitting the needs of wildlife against those of people, these policies and actions often disregarded and disadvantaged local pastoral populations.

Initially, conservation was the preserve of the wealthy—the colonial "elite"— and the focus was hunting. Between 1884 and 1900, colonial authorities started limiting the number of animals killed by each license holder in British East Africa and established early protected areas. In 1900, colonial representatives from British East Africa established two huge reserves covering about half of present-day Kajiado and large portions of the country around Mt. Kenya. From the late 1800s to early 1900s, the German and subsequent British colonial authorities created laws to protect people and crops from wildlife, restrict hunting, and create game reserves in colonial Tanganyika.[40]

Colonists established early protected areas to safeguard wildlife from overhunting by Europeans, but this goal soon shifted to protecting wildlife from African farmers and herders and to generating revenues for the state. As geographer Roderick Neumann puts it, "Correcting 'destructive' African society-environment interactions required expanding state power in rural areas through land use restrictions, hunting bans, destocking, evictions, and land alienations." In colonial Kenya, the British considered the Maasai uneconomical and irrational because they sought to accumulate livestock, and saw pastoralism and wildlife as ultimately incompatible. Later, this led to the conclusion that wildlife-only conservation areas were crucial to the survival of wildlife, particularly as European settlers began to kill off wildlife in the former highland pastures of the Maasai in Laikipia, Naivasha, and Nakuru. The colonial government hoped the game reserves would generate income for the colonial administration. It was not until the 1940s that today's set of protected areas in the colonies of Kenya, Tanzania, and Uganda were established, beginning with Nairobi National Park in 1946 and Tsavo National Park in 1947. In Ethiopia, the first national parks were not planned and established until the 1960s.[41]

Postcolonial African governments continued to evict herders to create new protected areas after independence. A case in point is Mkomazi, in northeast Tanzania, an area with a long history of use by Parakuyo and Maasai herders and Pare, Sambaa, and Kamba farmer-herders. When the colonial government created the Mkomazi Game Reserve in 1951, the first warden, David Anstey, allowed Parakuyo herders to continue to graze there because he judged their impact on the environment to be "slight." The Pare district commissioner was likewise concerned that the reserve protect wildlife "without in any way interfering with the legitimate present and future needs of the local people." Pastoral use of the reserve continued up to 1988, when officials evicted herders from the reserve, citing overgrazing and erosion. Anthropologist Dan Brockington studied the government assessment records carefully and concluded that there was insufficient evidence at a fine scale to make a clear judgment of herder impacts, but that the rapid recovery of the reserve after removal of herders and their cattle suggested they were not degrading the land in irreversible ways. Brockington concluded that limits on cattle populations are unnecessary because drought and disease will control numbers before they can damage the environment. We will return to this subject in Chapters 6 and 7.[42]

Some of the most important pastures for herders are now largely locked up inside wildlife-only conservation areas. Most of the savanna parks were created in regions where wildlife congregate in the dry season, which was also the dry-season grazing reserve for herders and their livestock. For example, Amboseli National Park sits on top of swamps that are crucial to the survival of livestock and wildlife alike in the dry season; Tarangire National Park includes the Tarangire River, which serves a similar purpose.[43]

Even though land was being locked up in protected areas, initiatives began in the 1970s to devolve (or transfer) more decision-making power and profits from wildlife (hunting fees, park gate fees, tourism income) to local pastoral communities, particularly those in savannas still rich in wildlife. This largely started on pastoral land adjacent to protected areas, where local pastoral families bore the costs of living with wildlife but received few of the benefits. The goal of these community-based conservation efforts was to improve incomes for local families and to better conserve wildlife, a classic win-win plan. Today, those profits are being distributed from government, tourism businesses, and nongovernmental organizations to pastoral community groups or to individual landowners, but with mixed success in achieving this win-win. We will return to specific community-based conservation efforts in Chapters 8–11.[44]

HOW PASTORAL SOCIETY AND
WILDLIFE ARE CHANGING TODAY

For hundreds and even thousands of years, it has not been uncommon for herding families to move in and out of a nomadic lifestyle, taking advantage of opportunities to live safer and more healthy lives as they arose. Like herders elsewhere, pastoral families in east Africa today are generally less nomadic than they were a century ago, or even a decade ago, in part due to governmental programs that are replacing communal property rights with individual land ownership. Giving up some of their traditional practices to take advantage of new opportunities, some are settling down and becoming herder-farmers and taking up other professions. Pastoral people want the same societal goods and services that people in places with higher rainfall want: good health care for their families and their livestock, good schools for their children, good markets for selling their livestock, and new technologies such as mobile phones and faster transportation. New market opportunities, such as the ability to sell premium high-cost vegetables or camel milk or to engage in ecotourism, allow some herders to expand family incomes. And many pastoralists are moving off the land to pursue employment in rural towns and larger urban centers. Education, job opportunities, new economic and land policies, up-to-date technologies, and better infrastructure encourage pastoral people to diversify their income sources and become more closely linked to their national economies and world markets.[45]

Many herders are losing their best land, wetlands and wet savannas, as farmers from outside the savanna buy or move onto the most productive pastoral lands and some pastoralists themselves settle and start farming. This only happens in the wettest pastoral lands, for farming is rarely successful in the vast drylands (see rainfall-determined savannas, Map 7). The Borana Plateau of southern Ethiopia, for example, had six times more people in the late 1990s than in the mid-1980s. Many of the newcomers were farmers moving from crowded farmland elsewhere to grazing lands on the plateau. New migrants tend to claim the best watered land to grow their crops, which is also the pasture "safety net" for herders and their livestock during dry seasons and drought. In some areas, the values and norms of agriculturalists, whether incoming farmers or former pastoralists, now determine how people use the land in wetter savannas.[46]

In many places, the herding lifestyle is simply less economically viable than it used to be. The chief reason is that there are so many more mouths to feed from the same, or a shrinking, land base. It is a simple calculus: if the number of people is rising faster than the number of livestock, pastoral families either become poorer or turn to other ways of making a living. In Kenya, ecologist Mohammed Said and his colleagues showed that this is occurring principally in wet savannas, where there are more and more people but the same number of livestock. In southern Ethiopia, in contrast, there are more and more Borana cattle, though each animal produces less and less. Why are cattle populations rising there? Leaders point to better veterinary care; also, with few markets to sell animals, herders hold on to livestock to preserve family assets. Yet Borana herders are losing the best land to farmers as well. Pastoral scholars Solomon Desta and Layne Coppock predict that better education and improved economic opportunities in rural Borana will help families find other sources of income.[47]

In some countries, national policy now allows pastoral families to own and sell land privately for the first time. While this may lead to secure ownership, it can also hurt the less powerful. Wealthy private landowners exclude others from pasture and water on their land by building fences, and convert rangeland to crop farms, businesses, quarries, and other uses. According to economist Michael Kirk, private ownership of land "secures primarily the rights of influential minorities, such as the urban elite, rather than the rural population; those of farmers over herders; and those of male heads of households rather than women." Even within pastoral society, private land often first goes to the wealthy, who also get the biggest pieces of land.[48]

In Chapters 8–11, I will focus on four case studies in Maasailand in Kenya and Tanzania, where land ownership is now rapidly changing, in contrast to many other pastoral lands in east Africa. Most of the pastoral land in east Africa is held in common by pastoral groups, as trust land administered by the government. But Maasailand in Kenya was the scene of an experiment to devolve trust land into local group ownership, in what are called "group ranches." The idea of group ranches was to allow pastoralists to secure loans to intensify livestock production with better veterinary care and more water development. Group ranches began to form in Kenya in the 1970s, with each eligible household having one member of the group ranch who participated in decisions regarding the ranch. In the early 1980s, some ranches decided to "subdivide" and grant each group ranch member his or her own plot or parcel of land with a land title. This subdivision occurred first in areas near roads and towns and is still in process in the Mara, Amboseli, and other places that are more rural. This means that Maasailand today is a mix of privately owned plots in some former group ranches, group ranches that are now subdividing, and others that are not yet subdividing (see Chapter 10 for more details).[49]

In countries with pastoral savannas in east Africa, government policy usually favors farmers over herders, a situation that is reinforced by ethnic prejudices, since most government administrators come from farming backgrounds. In northern Tanzania today, government officials consider pastoral land "open" and not fully used, justifying resettlement of farmers from crowded farmland onto grazing lands. As described above, national governments still also push pastoralists off land to make way for other uses, such as wildlife parks. This marginalization of pastoral communities to favor farming communities is not limited to east Africa; it is an age-old phenomenon common the world over.[50]

These changes have started to break down traditional ways of using the land. The sophisticated pastoral rules about who can use land and water may weaken as populations grow, connections to markets strengthen, and nonpastoralists gain control over land and other resources. With more limited land, for example, pastoralists stop reserving some land for grazing when forage is scarce in the dry season.[51]

Well-intended development efforts, such as the introduction of techniques from wetter farming lands into drylands, can inadvertently make pastoral families more vulnerable. Across Africa, many policymakers, development officials, and religious groups still advocate that pastoral families stop their nomadic movements and settle like farmers near towns to have access to health care, schools, famine relief, and other social services. Government policies for pastoral peoples are often made for restrictive purposes to collect taxes or prevent international border crossings. As anthropologist Roger Blench explains, "National governments are often hostile to pastoralists, and many countries have policies of sedentarisation that derive as much from political considerations as a concern for the welfare of those they wish to settle." Especially in drier pastoral savannas, support for more, not less, movement and opportunistic grazing is needed, but this requires finding innovative ways for health care, education, and other services to reach mobile peoples. Here, settling like farmers makes herding families more vulnerable. We will return to this subject in Chapter 12.[52]

Although settlement seems to be the logical future for pastoralists, many families are choosing to combine the advantages of both settled and mobile lifestyles. Economist John McPeak and anthropologist Peter Little describe families in northern Kenya who use both strategies: part of the family lives in a town, where the children attend school and there is ready access to markets, while the other part of the family—typically, many of the young men—live far from town in herding settlements or very crude satellite camps, moving often to find the best forage for their cattle and camels. One part of the family weaves a segment of the family's safety net by connecting to the broader society through schools, towns, and markets; the other part fulfills subsistence needs by watering and grazing livestock far from towns. (These roles are not fixed: people regularly move back and forth between herding camps and towns, for example.) Many children from pastoral families now have access to formal education, a rarity only a generation ago. Families are changing how they produce food, what they eat, what they wear, and how they aspire to live their lives.[53]

But these changes are not happening at the same speed everywhere. The first places to change are those near water, at the edge of wetter highlands, and near towns and markets. In the rangelands at the base of Mt. Kilimanjaro, for example, cultivating herders and farmers plow up precious swamplands and riverbanks to plant onions and tomatoes for local and national markets, and in the well-watered areas at the base of the Nguruman escarpment in Maasailand of south-central Kenya, farmers now produce French green beans for international export. A large-scale farming enterprise will soon produce coffee for export in the wetter part of what used to be Pianupe National Park, in Uganda. In the extensive lands away from towns and markets in parts of Somalia, Sudan, South Sudan, southern Ethiopia, and northern Kenya, by contrast, changes remain slow and opportunities limited.[54]

Open savannas are thus becoming increasingly bounded and fragmented. People are establishing fences, rules of access, and other "hard boundaries" to protect crops and property from wildlife and other people. In the still-open savannas, land use is becoming ever more polarized: at one end of the spectrum are protected areas for wildlife only, and at the other end, settlements, fenced areas, and water points used mainly by people and livestock. In the vast majority of savannas, however, land use still occupies the middle ground between these two extremes, with people, livestock, and wildlife continuing to mix.[55]

As people use savannas more intensively, wildlife are often in decline, but not everywhere. Let's refer back to the analysis by Said's team, which included wildlife. From 1978 to 1994, they found that wildlife populations grew in only one of Kenya's eighteen rangeland districts, fell significantly in seven districts, and declined slowly in three, with the remaining seven districts seeing no change. For ecosystems that are the focus of Chapters 8–11, wildlife are declining strongly in some areas (the Kaputiei Plains and Mara) but not in others (Amboseli, Serengeti, and Ngorongoro). In Tanzania, wildlife-protected areas support more wildlife than unprotected areas with farms and settlements. This is especially true for most wildlife of large size, the exception being elephants, which are just as abundant inside and outside parks in Tanzania.[56]

The wildlife story in Sudan (which is now Sudan and South Sudan) and Ethiopia has been bleak, but it may be improving. In Darfur, only five of thirty-one species censused pre-1950 could be found in aerial surveys in 1976 (the patas monkey, jackal, and three types of gazelle). Biologist R. Wilson posits that the wildlife disappeared here because of hunting in the dry north and competition with farmers and herders for land and forage in the wetter south. Carnivores were exterminated by poisoning and hunting in the 1940s and 1950s. But in what is now South Sudan, the decades-long conflict with Sudan inadvertently protected the continent's second largest migration of large mammals in Boma National Park. In Ethiopia, wildlife in Bale Mountains National Park almost disappeared after unregulated hunting following the 1991 change in government. Endemic and rare Ethiopian wolves also almost disappeared during a rabies epidemic in the early 1990s, passed to them from a growing population of domestic dogs, though wolf populations are now recovering. [57]

Rwanda and Uganda have seen massive losses of wildlife. During the 1994 war, Rwandan gunmen killed perhaps 90 percent of the wildlife in that country's large savanna park, Akagera. Uganda's army militias likewise killed much of the country's wildlife between 1970 and 1985. Wildlife ecologist Richard Lamprey and his colleagues recently surveyed Ugandan wildlife and found a partial recovery in some areas since 1985, but not in others. In northeastern Uganda, armed Karamojong pastoralists killed virtually all the wildlife in the reserves and controlled hunting areas. Some wildlife in Queen Elizabeth National Park recovered after the 1980s but then declined again, shot by army personnel. In 2004, the Ugandan government degazetted much of Pian-Upe Wildlife Reserve, a park that held Uganda's last five roan antelopes, for cotton cultivation. Rwandan refugees killed a significant amount of wildlife in Uganda's Katonga Wildlife Reserve. But Lamprey sees some bright spots. After hunters killed 80 percent of the impalas in Lake Mburo National Park, this graceful gazelle might be recovering, thanks to local communities that now benefit from fees gathered from sport hunters. In Toro-Semliki Wildlife Reserve, Lamprey credits removal of cattle grazing and prohibition of poaching for recent wildlife recovery. And Kidepo National Park in northern Karamojong still has wildlife. But these are fragile exceptions; most of the region has seen large losses of wildlife over the last few decades. [58]

Hard boundaries and segregation of pastoral people and wildlife bring unintended consequences, as we will explore in the next two chapters. The colonial and postcolonial authorities did not realize that they were widening the gap between people and wildlife when they separated people from wildlife, made central government the "owner" of wildlife, and allowed the lion's share of the benefits from wildlife to go to people outside pastoral lands. By creating a perception among herders that wildlife are "theirs" and not "ours," these actions put an age-old coexistence at risk.[59]

CHAPTER SIX

Can Pastoral People and
Livestock Enrich Savanna
Landscapes?

From a distance, the green patch looked like an emerald embedded in a yellow sea of grass, impossibly brilliant and round. Perhaps a small swamp or wetland of some sort? But it was in an odd place to be a small swamp, too far from the base of a hill or a river to make any sense. We drove slowly through the tall grass, pushing away a herd of gazelles grazing on the patch of short green grass as we stepped out of the Land Rover. Ole Sidai, his red shuka *flapping in the dry breeze, walked us slowly around the green patch, about 20 m (66 ft) across, talking as he went. The old man pointed to a lone tree on the edge of the patch where his first grandchild had learned to climb and to another spot where his favorite bull had been born. He then vividly described the structure of the settlement itself, showing us where he built his sheep and goat pen, walking us to his old cattle pens, and nudging the stones left from the firepits that once were within the huts of each of his three wives. "Welcome to our old home of eight years ago," he smiled. We then sat and dug our hands into the cool luxuriant lawn, and he told us the story of what happens when Maasai leave their old homesteads in the Mara ecosystem of southwestern Kenya. A few years after the family moves to find better grazing in another part of the savanna, the thorn fence surrounding the homestead and the old mud houses fall down. About this time, a special grass, called settlement or* manyatta *grass, sprouts in the dung of the old cattle, sheep, and goat corrals and covers the abandoned homestead. This grass is more nutritious and stays green longer into the dry season than grass in the surrounding savanna. The green homestead grass then becomes the favorite grazing spot not only for cattle, sheep, and goats but also for gazelles, wildebeest, and zebras. But the real action happens after dark—Maasai must take care walking on moonless nights to avoid the elephants and hippos grazing on the lawns atop their old settlements.*

THE ENRICHING AND SIMPLIFYING FORCES OF NATURE

This chapter and the next address the key issue of this book: When are pastoral people, their livestock, and wildlife compatible in savannas, and when are they not? And why? Both enriching and simplifying forces may be at work as savannas respond in different ways to grazing animals. This chapter explores if and when pastoral people tolerate or even enhance the wildlife and the savannas around them, compared with other rangelands around the world. The next chapter investigates when these forces break down in east Africa, leading pastoral people and their herds to compete with wildlife, thus simplifying the structure and diminishing the productivity of savannas.

What evidence is there of these enriching and simplifying forces? I use the term *enrich* here in its classic sense of "more," specifically meaning the concentration of nutrients in particular locations in ways that promote plant (and possibly animal) production, as illustrated in the dung-enriched homestead grass example above. Less precisely, I also sometimes interchange the term *enrich* with *diversify*, referring to an increase in the number and variety of different native animal or plant species or vegetation patches (or habitats) in a savanna. I use the term *simplify* (or *impoverish*) to mean the opposite: a decline in a savanna's native species, nutrients, productivity, and/or varied habitats used by wildlife, livestock, and people.

What makes "enriched" savannas, ones with more nutrients, species, and vegetation patches, "better"? Their value lies in the eye of the beholder—or user. Where concentrated nutrients allow cattle or topis to produce more milk, it is "better" from the perspective of both herders and wildlife. However, if this concentration of nutrients in one area robs other parts of the savanna of important nutrients, is it really better? Greater species diversity is useful to pastoral peoples and livestock for forage, food, and medicine, but there are many species in a biodiverse savanna—tsetse flies, for example, or lions—that make livestock-keeping difficult. More native species benefit overall biodiversity, but more nonnatives do not. And for settled herder-farmers, having no elephants or warthogs raiding the maize field is certainly a good thing.

Less obvious benefits of species-rich ecosystems may apply to savannas. As biologist Michel Loreau and colleagues explain, "Larger numbers of species are probably needed to reduce temporal variability in ecosystem processes in changing environments." This is analogous to diversifying the holdings in a financial portfolio to reduce variation in returns over time. Indeed, Serengeti grasslands rich in plant species produce grass more consistently over time (i.e., growth is less variable) and regrow faster in the wet season than grasslands with fewer plant species. This greater constancy and rapid regrowth, brought about by more plant species, is a good thing from many perspectives. At the landscape level, herders setting fires may open up bushland, creating a complex mosaic of patches of grass and bush and so increasing the variety of habitats and the variety of plants and animals across the landscape. But from the viewpoint of an impala looking for woody plants, the opening up of bushland by fire removes food and habitat. And what about more productivity? It may appear to us that savannas producing more plants and animals can only be a good thing, with benefits for all; however, there will be some organisms, somewhere (certain slow-growing plant species, for example), that will be hurt rather than helped by high productivity. Thus, even though the words *enrich* and *diversify* imply a universally positive outcome, there are trade-offs in every change that will be valued differently from different perspectives. Similar trade-offs apply in savannas that are "simplifying," as discussed in the next chapter.[1]

It is important, at this point, to clarify wording. I will use the term *irreversible loss* to describe situations where there is no reasonable prospect of recovery over any time frame. I will use the term *degradation* for changes that are difficult to reverse, that is, that would take decades to centuries to turn around (or that are often irreversible in a human lifetime). I will use the word *decline* in relation to situations in which savannas lose productivity but can recover, perhaps in less than a decade. Though sometimes, for simplicity, I will use the term *degradation* to cover all three situations.

Can models of other ecosystems help explain how pastoral peoples and their livestock enrich or simplify savannas? Some forty years ago, biologist J. P. Grime discovered something unusual about how plants respond to disturbance (though the finding normally gets credited to biologist Joseph Connell, who also wrote about the idea five years later). In sum, Grime and Connell found that grasslands, coral reefs, and rain forests that have been disturbed by some event, such as a grazing animal, a severe storm, a landslide, or a falling tree, support more species than those that remain undisturbed. Too much disturbance, however, caused the number of species to drop off again. (See Figure 19, humped response.) Connell called this phenomenon the "intermediate disturbance hypothesis" because the number of species was highest when disturbance reached intermediate levels (the hump), and lowest on either side of the hump, when disturbance levels were either low or high. While some scholars argue that this intermediate disturbance response is not as widespread as initially thought, I propose below that it is quite widespread in the savannas of east Africa, where it may explain the effects of grazing and burning on plant and large mammal species and on plant productivity. This disturbance response may also help explain the distribution of wildlife around pastoral settlements.[2]

Another model, from ecologist Gufu Oba and his colleagues, suggests that savannas of Africa decline (or are simplified) if they are *not* grazed, and improve up to the point of moderate grazing (enriching response in Figure 19). Savanna plant species can also be indifferent to grazing (Figure 19, no response) or decline under grazing (Figure 19, simplifying response; see also next chapter). These responses are generalized, of course; there is much complexity not covered here. All but the enriching response are like those proposed for plant diversity and grazing at sites around the world by ecologists Daniel Milchunas and William Laurenroth.[3]

Although intermediate and high levels of savanna disturbance can increase species diversity, the increased numbers of species do not always benefit overall biodiversity. That is because the new species appearing in disturbed landscapes are often common weeds. Take, for example, an ungrazed savanna in Africa that contains ten plant species and a moderately grazed savanna that has fifteen. Which one is richer or more diverse? The grazed savanna may have more species, but what if most of those species are weeds or nonnatives? Weeds and nonnative species are less valuable to overall (regional or global) biodiversity because they are found elsewhere, not just in savannas; local native species, conversely, which may be found only in savannas, contribute something unique to the Earth's biological life.[4]

African savannas (and the Asian steppe) differ from rangelands in Australia and some parts of North America because they have been grazed for very long periods of time (see Chapter 4) by large herds of large wildlife. Rangelands with a long evolutionary history of grazing by large wildlife tend to be less sensitive to subsequent grazing by livestock than rangelands that have been grazed for a short time or by fewer and smaller wildlife. Rangelands with a short or no history of wildlife grazing, indeed, can be very sensitive to grazing by domesticated livestock.[5]

A clear example of this contrast can be found in the United States. There, the drylands between the Rocky Mountains and the Sierra Nevada evolved recently with light grazing by pronghorn antelope and other small grazers, while east of

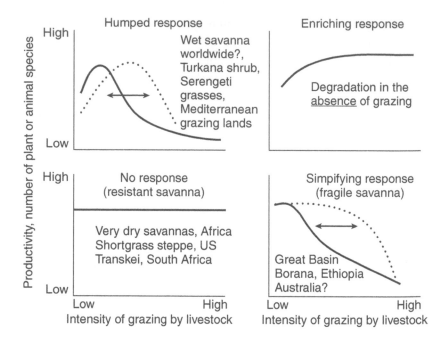

FIGURE 19.
The possible responses of savannas to livestock grazing are varied, as shown by the dotted curves. (Sources: Grime 1973, Connell 1978, Huston 1979, Milchunas et al. 1988, Milchunas and Laurenroth 1993, Oba et al. 2000b.)

the Rockies herds of millions of large-bodied bison once grazed the Great Plains. As farmers began settling these two areas in the late 1800s, building boreholes and livestock herds as they went, these rangelands responded differently to the introduction of livestock: whereas the Great Plains remained relatively unchanged, the intermontane deserts, with their former lighter grazing, shifted from tufted grasslands to woody shrublands.[6]

The savannas of Africa are probably as or more tolerant of grazing and fire than any ecosystem on Earth. Here, soils and plants evolved over millions of years along with wildlife and hominins. Most soils, plants, and animals in African savannas are adapted to the type and amount of disturbance that wild grazing animals create (with important exceptions). There is also good evidence that fires have been both frequent and widespread in Africa over the last 400,000 years, particularly during dry, glacial periods, and including fires that have been lit by people. Due to this long history of wildlife grazing and fire, it is plausible that savannas in east Africa were, in a sense, "pre-adapted" to the herders and their stock when they arrived from the north several millennia ago and started grazing and burning. There is no reason to suppose, however, that these savannas are adapted to more recent forms of disturbance, such as heavy and continuous grazing by livestock, replacement of many species of wild herbivores by a few species of domesticated animals, frequent burning, plowing for crop cultivation, permanent settlements, wells, and removal of pastoral people from protected areas for wildlife.[7]

OVERGRAZING OVERSTATED?

One of the most enduring public images of Africa (held by many Africans and non-Africans alike) is of a continent of expanding deserts, overgrazed by greedy herders with too many livestock, foretelling continuing cycles of famine and impoverishment. Two scholarly works gave legitimacy to these perceptions. In 1926, anthropologist Melville Herskovits described a "cattle complex" in east Africa, where herders, enamored with their cattle, accumulated more livestock than they needed. Then in 1968, zoologist Garrett Hardin identified what he called a "tragedy of freedom in a commons," where individual herders, benefiting from adding one more animal to a pasture managed by a group (the commons), have no incentive to conserve the common grazing land because "the effects of overgrazing are shared by all herdsmen." In the 1980s, ecologists Hugh Lamprey, Tony Sinclair, and John Fryxell argued that herders and their livestock were overgrazing Sahelian savannas at the southern edge of the Sahara, fusing heavily grazed areas around settlements and water points and causing the desert to spread south. They suggest that savannas are fragile systems and that livestock, by overgrazing, damage their productivity over the long term (Figure 19, a simplifying response). Grazing was seen as the dominant force in these systems, with strong feedback or "coupling" between livestock and vegetation. The logical conclusion was that livestock numbers should be adjusted to a carrying capacity defined by each savanna's specific productivity, which is determined by local rainfall, soils, and other factors. This is known today as the "equilibrium" view of rangeland dynamics.[8]

New evidence in the 1980s indicated that, while equilibrium dynamics exist in wetter, climatically more predictable savannas (semiarid to humid), the drier and more variable savannas (arid to semiarid) are harder to degrade. Why? While doing research with our team in the dry savannas of Turkana, Kenya, James Ellis and David Swift saw that frequent droughts, once or twice a decade, killed large numbers of livestock and thus prevented their numbers from reaching levels high enough to damage the vegetation. After each major drought, there were few livestock for a year or more, allowing the Turkana savannas to "rest" as herders rebuilt their herds. Thus Ellis and Swift concluded that drought (or climate more generally), not overgrazing, controls the number of livestock in dry savannas like those of Turkana. The savannas with these dynamics, showing little feedback or coupling between livestock and vegetation, are called "nonequilibrium" systems. This distinction is important because it implies that herders can stock as many animals as they want in dry savannas with no fear of degradation or loss of savanna productivity. Many scholars today take neither a strict equilibrium nor nonequilibrium view, but rather think that African rangelands fall along a continuum extending from those driven mostly by grazing to those driven mostly by climate. In equilibrium rangelands, we may be able to distinguish thresholds, or "tipping points," beyond which a decline caused by grazing becomes long-term degradation or irreversible loss, an idea explored in the next chapter.[9]

In productive ecosystems, research from lakes to savannas is beginning to show that moderate levels of grazing by wild species as different as limpets and wildebeest *sometimes* increase the number of plant species above that found where there is no grazing (Figure 19). Once these ecosystems are heavily grazed, the number of species either declines (supporting a humped response) or remains high (supporting an enriching response). For example, in productive ocean tidal ecosystems, ecologist Jane Lubchenco found support for the humped response, where more algae species thrived under moderate grazing by periwinkle snails, with fewer algae species when snail grazing was either high or low. In contrast, it appears that productive grasslands in the Great Plains of North America are more resilient to grazing then these ocean tidal systems, because both moderate and heavy grazing by bison increases the number of plant species present (enriching response). Similarly, in Tanzania's Serengeti, ecologist Joy Belsky found more plant species in short, medium, and tall grasslands that had been grazed heavily by wildlife (enriching response) than in grasslands protected from grazing by fencing. Less productive systems, however, appear to respond very differently to grazing than productive systems. Here, grazing initiates either no change or a loss in species (no response or a simplifying response).[10]

How does grazing boost plant diversity, and why does it do this mostly in productive ecosystems? Wet, nutrient-rich ecosystems are often dominated by a few strongly competitive plant species. Grazing prevents these dominant species from garnering most of the system's resources by removing more tissue from them than from the rarer species. In a sense, grazing "levels the playing field." As grazing becomes heavy, it can either lead to the removal of species (humped response) or encourage the establishment of more species (enriching response). In less productive ecosystems with fewer competitive dominants, grazing diminishes plant diversity more often than it encourages the establishment of new species (simplifying response).

Plant productivity can also be boosted by wildlife grazing. Ecologist Samuel McNaughton, working in the Serengeti for more than thirty years, found that moderate (intermediate) amounts of grazing by wildlife can double the productivity of some plant species but that productivity declines at higher grazing levels. For example, the grasslike plant *Kyllinga nervosa,* when experimentally clipped short every day, produced five times more than when it was clipped once a week. This promotion of productivity by moderate grazing can increase a savanna's carrying capacity, or the number of animals it can support.[11]

One of McNaughton's colleagues, Douglas Frank, found the same grazing effect in America's Yellowstone National Park where elk and bison graze grasslands. Frank and his team discovered that this tight interaction between grazers and their food goes even further: it can stimulate grass roots underground to grow more quickly. This means that grazing can increase the amount of carbon stored in the soil, which could help offset CO_2 emissions from fossil fuels.

But not all plants produce more when grazed, and the availability of water and nutrients changes their responses. Ecologist Michael Coughenour and his colleagues found this out when they experimentally clipped one sedge (*Kyllinga* above), two grasses, and one shrub at different intensities, adding nitrogen and water in different combinations. They found that the sedge grew more when grazed (humped response), while the two grasses and the shrub showed no response to light grazing but declined at heavy grazing (delayed simplifying). Coughenour and his team then created a computer model to test plant responses over a wide range of nutrient and water levels. They found that all species were more resilient to grazing when they had more water and nutrients, but when either nutrients or water was scarce, grazing tended to affect the plants faster and more negatively. Further, the grazed shrubs, even though they did not produce more when grazed, were more nutritious than ungrazed plants because grazing stimulated roots to take up more nitrogen from the soil for deposit in the leaves and stems. Thus, sometimes grazing promotes faster growth and production of plants, and sometimes it does not, depending on the species. And abundant water and nutrients tend to make plants more resilient to grazing.[12]

McNaughton also found that wild grazers change the architecture of grasslands, creating short "grazing lawns" where the concentration of nutritious food available is greater than where grass is tall and ungrazed. Consistent grazing by wildlife, like fire, keeps grass in a young state, which is easy to digest. This may seem trivial, but it is one of the biggest benefits that repeated grazing provides for the next grazer, and it is sustained as long as sufficient moisture and nutrients are present in the soil to allow the grass to keep regrowing. This grazing lawn effect is more important for smaller grazers (like Thomson's gazelles), which require more nutritious grass to meet their metabolic needs, than for larger grazers (like Cape buffalo), which can survive on taller, less nutritious grasses (see Chapter 2). In general, a moderate amount of grass (or intermediate biomass) interspersed with patches of short and tall grass may provide the most energy and nutrients to a range of grazers from small to large.[13]

Livestock grazing differs from wildlife grazing in several ways. Because they are herded, livestock tend to graze in tighter groups than wildlife (though some wildlife also cluster in groups). McNaughton says livestock do not create grazing lawns like wildlife do; however, I have often observed extensive grazing lawns created by livestock near pastoral settlements, for example. Livestock herds do not include the very largest of herbivores (like giraffes, elephants, rhinos) that generally remove large quantities of grass or trees and shrubs, strongly modifying savanna vegetation. Also, because livestock trek back and forth daily between pastoral settlements and grazing lands, the land around the settlement is particularly heavily grazed; such grazing intensity is rarely reached by wild herbivores except in unusual situations, such as the annual million-strong wildebeest migration in the Serengeti-Mara (see Chapter 8). Wildlife do, however, create moderately grazed lawns of short nutritious grass where predators are more visible when they gather together in multispecies herds (or associations; see Chapter 2).[14]

Livestock graze more like wildlife in soft-boundary savannas (Map 12), where herders and livestock move often to find forage and water, heavily grazing one place but then moving on to allow savanna recovery. It is different in hard-boundary savannas, where livestock graze heavily and continuous near permanently settled herders. Control of diseases and provision of water for savanna livestock lead to their higher and less variable stocking rates than those of savanna wildlife. Biologists Johan du Toit and David Cumming point out that livestock are more abundant now (particularly in nutrient-poor savannas) than wildlife have been historically;

this has a homogenizing effect on the structure of the vegetation, with most places heavily grazed. Thus livestock are most likely to enrich or diversify savannas that have soft boundaries, where they (and their minders) move about regularly.[15]

Despite their obvious differences, both livestock and wildlife can provoke similar vegetation responses in terms of plant diversity. In the semiarid Kaputiei Plains (see Chapter 9), ecologist Helen Gichohi, having fenced out livestock from a plot of savanna land for three years, found more plant species in the unfenced savanna heavily grazed by livestock and a few wild animals, than in the fenced, ungrazed savanna. The difference in plant diversity only widened over time, supporting the enriching-response hypothesis (Figure 19). In arid Kenya and semiarid Niger, ecologists Gufu Oba and Pierre Hiernaux found that plants responded to livestock grazing with a humped response. Savannas with sandy soils in semiarid Somalia responded the same way to grazing by cattle and goats. Livestock grazing, either light or heavy, had no effect on the number of plant species in arid Namibia and South Africa. These studies suggest that livestock grazing provokes various changes in savanna plant diversity, sometimes a decline in species, but no evidence of a universal loss of species.[16]

Where the number of plant species is unaffected by grazing, the kinds of species may change a lot. As we will see in Chapter 7, savannas differ not only in the number of plant species they contain but also in the traits and qualities those plants possess, such as whether they are leafy herbs (shrubs) or woody plants (trees); short- or long-lived; palatable or unpalatable to different kinds of grazers; native or nonnative. These distinctions matter—to herders, conservationists, and wildlife. Livestock grazing may encourage the invasion of weeds or plants that livestock find unpalatable, and heavy grazing can also encourage bush growth in grasslands, a change observed in parts of east Africa (all described in Chapter 7).[17]

Interestingly, African savannas don't always follow the prediction that livestock grazing has no effect on plant species in unproductive savannas. Oba, for example, observed an enriching response to livestock grazing in an arid, unproductive savanna in northern Kenya, whereas humped-response findings in Niger and Somalia were in wetter, semiarid, nutrient-poor savannas. It is possible that livestock grazing in African savannas, unlike in savannas elsewhere, tends to increase the diversity of plant species in both wet and dry savannas.[18]

Can livestock boost plant productivity like wildlife do? In general, livestock grazing sometimes boosts (humped response), sometimes has no effect (no response), and sometimes diminishes the standing mass of plants (simplifying response), especially when grazing is heavy. In arid Kenya and Namibia and semiarid Zimbabwe, livestock grazing had no impact on plant mass. In semiarid Somalia and Kenya, plant mass was greatest with moderate grazing (humped response). Working in Mali, ecologist Pierre Hiernaux speculates that grazing can either increase or decrease productivity depending on when grazing occurs: repeated clipping in the wet season likely diminishes plant productivity (simplifying response), as found by Oba, too, with *Indigofera* shrubs in Kenya.[19]

From the evidence above, it is hard to distinguish the effects of livestock and wildlife grazing on plant biodiversity and productivity in east African savannas. There may be two reasons for this. First, people and wildlife have lived side by side in east Africa longer than anywhere else on Earth, for thousands of years (Chapter 4). East African savannas were thus in a sense "preadapted" for livestock grazing by their long history of wildlife grazing, as the Great Plains of North America were preadapted for livestock grazing by their long history of bison grazing. Second, many pastoral people in the soft-boundary savannas of east Africa still move their livestock frequently to track changes in rainfall, a practice that, to some extent, mimics wildlife grazing.[20]

Savannas in east Africa appear to be quite resilient to both wildlife and livestock grazing. And the effects of livestock grazing on plants are probably more similar to those of wildlife than conventionally thought. Maasai herders certainly believe so: they observe that wildebeest can remove as much or more plant mass as their cattle in Ol Tukai, Tanzania, causing the grass "to become finished." Productive savannas are where much of the action is: here, where grazing strongly affects vegetation (and also in a few unproductive savannas), grazing at moderate levels appears to boost plant productivity and diversity, but decreases productivity at high levels. In the next chapter we will see other, more subtle, often negative effects of grazing on savanna vegetation.[21]

FIRE, HUMAN'S OLDEST TOOL: A DIVERSIFYING AND ENRICHING FORCE?

Another way that pastoral people affect savannas is by setting bushfires. For thousands of years, as we saw in Chapter 2, people created more grassy savannas by burning patches of bushland and woodland. Burning can bring diversity to a landscape by creating a tapestry of burned and unburned vegetation of different sizes and shapes, with a mix of patches burned at different times in the past. Part of what determines the level of diversity a fire adds to a landscape depends on how extensive the fire is. Very large fires can simplify a landscape by removing vegetation across large areas all at once. Smaller fires set at different times can increase savanna complexity.[22]

East African pastoralists say that they burn the savanna mainly to promote new grass growth for their livestock (and to remove tall, old grass so they can see and avoid snakes underfoot). They also set the savanna alight to reduce the number of ticks, which feed on the blood of their animals and can transmit livestock diseases, and to open up woodland vegetation that is home to tsetse flies, which can transmit trypanosomosis to their cattle. These are the answers I have gotten to the question "Why burn?" from Oromo pastoralists in Ethiopia, Turkana in Kenya, Maasai in Tanzania, and Bahima (grouped under Nkole, Map 5) in Uganda.

How does fire, whether lit by people or by lightning, affect savannas at a local scale? Once fire passes through an area, it clears away tough old plant material, promoting new fresh growth. The fire leaves behind ash on the ground's surface but usually leaves the roots of the grasses, trees, and shrubs largely untouched. With sufficient moisture in the soil after the burn, grass will resprout almost immediately, forming a brilliant green fuzz of short grass a few days or weeks after the fire passes.

Why does the grass regrow, green and luxuriant, after fire? The answer may be partly about nutrients, partly about water. Much of the nitrogen and sulfur in the leaves and stems of plants (and sometimes soils) literally goes up in smoke when burned. Most of the nitrogen discharged into the atmosphere eventually returns when it rains. One study of a savanna woodland in Zambia (south-central Africa), for example, found no significant loss of nitrogen or organics from the soil after fifty years of annual burning. The plant tissue of grass growing in a recently burnt area often contains more of several nutrients important for plant growth and grazers than unburnt grass. Burning also stimulates the movement of nutrients from the roots to newly growing stems and leaves. Even so, these brilliant green patches and their elevated nutrients are short-lived, the nutrients lasting only one to three months, on average, after a fire passes.[23]

Fire often increases the numbers of species of native plants, but in dry or infertile savannas many of the new species are common weeds. In the Serengeti, burning has been found to increase the number of grass species but change the types of species growing. Savannas burned occasionally contain more species than those burned often or not at all, supporting the intermediate disturbance or humped model. Soft leafy species become scarce in savannas that are burned frequently, whereas species like grasses that are designed to survive fires become more abundant.[24]

Wildlife, livestock, and insects find the green lawn that grows after a burn very attractive. Grasses that resprout after a fire are not only richer in nutrients but also easier for animals to digest than those in nearby unburned patches, so these are particularly attractive to smaller grazers who need higher-quality food. For this reason, wildlife gain weight faster grazing on recently burned patches of savannas than on unburned areas. They also can graze more calmly in burned areas, which have fewer ticks and flies. Kenyan Maasai say their cattle prefer recently burned grazing areas for the same reason. We know little about the impact of fire on most soil organisms (bacteria, fungi, worms) in east African savannas, although it is likely to be far less than that above ground because the fire's heat penetrates into the soil only a short distance.[25]

Even the most optimistic tawny-colored lioness knows that green grass, short enough to expose small stones on the ground, will provide her with little camouflage for stalking prey. But the same lioness knows that grazers will cluster on a burnt patch, so it is a good place to look for a meal—but only if there is good cover like shrubs or tall grass next to the burn for this sit-and-wait predator. In general, burnt patches are safer for grazing animals than the tall grass in unburnt patches, especially for small grazers that might not see stalking predators in tall grass. Grazers appear keenly aware of where the edge of the short-grass burn is, ready to sprint away from the tall grass at its edge if a predator dashes toward them.[26]

Do pastoral peoples diversify or simplify savannas by setting fires? The answer to this question depends on when and how often people burn a particular piece of savanna, how hot and widespread the fires are, and how the plants in the savanna react to fire. One example of traditional people diversifying landscapes using fire comes from northern Australia, where very large and uncontrolled fires are becoming the norm—but not in land managed by Aboriginal people. Indigenous Australians consider those who allow a large fire to burn out of control to be poor fire managers. Instead, they burn carefully, in certain seasons and in small patches, not only to improve habitat for kangaroo but also to protect special fire-sensitive plants. Aboriginal lands support as many plant and animal species as nearby national parks do, and very few weeds or invasive plants. Similarly, hundreds of years ago, Native Americans diversified landscapes in coastal California by burning dense shrublands, creating a mosaic of scrub and grass that attracted various types of wildlife. In general, then, where people burn a savanna too often or with very hot or very large fires, some or many plant species are likely to disappear. But where people burn carefully and irregularly, one place one time and another place another time, with fires sometimes cool and sometimes hot, the result may be a diversity of landscapes, as in the humped response in Figure 19.[27]

Do east African pastoralists diversify savannas with fire like Aboriginal people do in Australia? Perhaps, although little research exists to document this. In a typical dry season in the wetter savannas of Tanzania and Kenya, people burn many areas, mostly in small patches, each at a different time. This is particularly true in a landscape managed by pastoralists, where burnt patches are interspersed with heavily grazed grass near settlements that do not burn because there is not enough fuel to sustain a fire. Fires in savannas without pastoralists, as in a national park, sometimes burn larger areas and are stopped only by roads, rivers, or other firebreaks. In sum, although it is likely that east African pastoralists actively use fire to create more diversity in savanna landscapes, we simply don't know for sure.[28]

Pastoral people certainly burn areas to improve grass for their animals, a custom that also benefits some wildlife species. Ecologist Victor Runyoro and his colleagues at Tanzania's Ngorongoro Crater saw how a 1974 ban on pastoral grazing and burning of the savanna on the crater floor affected wildlife. Once these activities ceased, large grazers (buffalo) became more common and smaller grazers more rare. It is logical to conclude that previous burning by herders, and probably also their livestock grazing, kept grass young and nutritious, attracting small and medium-sized grazers such as Thomson's and Grant's gazelles and wildebeest. Without regular burning, grass became tall and tough, which is difficult for smaller grazers to digest.[29]

PASTORAL SETTLEMENTS: A LEGACY OF HOTSPOTS

Some of the most intriguing interactions between people and animals in east Africa happen around pastoral settlements, called *bomas* in Kiswahili. As early as 3,000 years ago, some people subsisted on livestock in savannas, as shown by the dominance of cattle, sheep, and goat bones found by archeologists such as Fiona Marshall and her colleagues in old settlement sites in the Mara ecosystem. Even in these very old settlements, pastoral people were clearly corralling their livestock for part of the day, presumably at night to protect them from predators such as lions, hyenas, and leopards. Today, the *bomas* that the Maasai people live in (which they call *enkangitie* in their Ki-Maa language) traditionally consist of a ring of dwellings, constructed with sticks and plastered with mud and cow

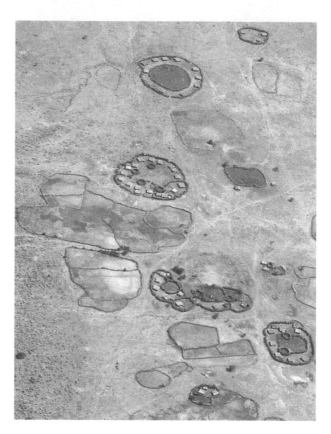

dung, that encircles several corrals for confining livestock at night. Each *boma* is enclosed by a thorn fence to protect the livestock—and their keepers—from predators (see Figure 20 for an aerial perspective). Pastoral people like those near Lake Turkana in remote northern Kenya live in even simpler settlements so they can pack up and move more easily to follow the region's very scarce and patchy rainfall.[30]

Where pastoralists construct their homesteads determines where their greatest impact on savanna vegetation and soils will be. In the driest savannas, pastoral people place a premium on having shade and being a close walk to water for themselves and their animals. In Turkana, for example, herders usually build their settlements near or under tall acacia trees that grow along dry streambeds. In a slightly wetter savanna such as Amboseli, the Maasai often build their settlements at the base of small hills, just where the land begins to flatten out into the surrounding plain, so that water drains away from their homesteads in the wet season. In the wettest savannas, such as the Mara, the Maasai build their settlements also at the base of hills and along river courses but often 100–300 m (about 330–1000 feet) uphill from the river on sandier soils to improve drainage.[31]

Pastoral people move their settlements when they need to find greener pastures, to escape ticks, or because there is simply too much dung underfoot for comfort. As described in Chapter 2, rain usually does not fall everywhere at once in savannas. Particularly where it is dry, this patchy rainfall requires that people be flexible and, like wildlife, follow the rains to catch the grass while it is young, green, and nutritious. In very dry savannas such as Turkana (<250 mm [< 10 in] annual rainfall), herders move once a month or so in search of the best pasture. Because they must be able to pitch or unpitch a homesite within a single day, the structures they build are simple. In wetter savannas, pastoralists can afford to move once a season or only once every few years, and so they build structures meant to last longer.

Once a pastoral family moves, they leave behind their homes, their livestock corrals, some refuse, and the protective thorn fencing surrounding their *bomas*. Years later, you can often still see the three stones of the cooking fires that warmed the families at night inside their huts. Even more noticeable will be the ring of shrubs or trees growing where the thorn fence once stood, or a bright green patch of grass where the livestock were corralled. Old settlement sites are most noticeable just after it begins to rain in the wet season or just as the dry season starts; at both times, the grass in the old corrals is greener than that in the surrounding landscape. Remarkably, these abandoned settlement sites are often visible from space using satellite imagery.[32]

Pastoral settlements leave long legacies in less noticeable ways as well. Immediately after a family abandons a homestead, levels of nitrogen, carbon, and phosphorus in the soils of the livestock corrals are extremely high. As time passes, these nutrients are either emitted in gaseous form (much of the nitrogen), move slowly into the soil (much of the phosphorus), or are respired away by bacteria (most of the carbon). Nitrogen disappears the fastest, phosphorus the slowest. In Amboseli, we found that phosphorus levels in the soils of settlements abandoned more than a century ago remained three times higher than those in soils of the surrounding savanna.[33]

Our observations in the Mara and Amboseli suggest that nutrients from livestock dung last longest in the soils of dry savannas and in settlements where pastoral people lived for long periods. Where rainfall is plentiful, it can wash away or leach many nutrients. In all savannas, the longer a pastoral family lives in a *boma* and the more animals they hold, the more dung is deposited in the livestock corrals. I have seen old, recently abandoned, settlement sites with dung piles in the central cattle corral reaching 2 m (over 6 ft) high, taller than I am.[34]

These nutrients are absorbed by the grasses and herbs growing on the abandoned sites, making them much richer in nutrients than plants growing outside the *boma* site. Initially, no plants grow on top of the dung piles because they are too deep and dry out quickly in the heat, which kills young seedlings. Both livestock and wildlife tend to avoid the first plants that grow on top of the corrals, presumably because they are too nutrient rich. Maasai in the Mara say that their cattle, especially in the wet season, will become bloated if they graze on old *boma*s that have been recently abandoned because "the grass is too fat" (with nutrients). They prefer instead to graze their stock just beyond where the thorn fence used to be, on grass that does not cause bloat, where the livestock once milled mornings and evenings waiting to be milked, to start the day's walk, or to enter the settlement.[35]

After the pastoral family has left a settlement, a remarkable succession of plants and animals uses the pile of old dung. The only plants that can grow on the raw dung for the first few years are large leafy weeds. Then the *manyatta,* or settlement, grass comes (*Cynodon plectostachyus*), covering the *boma* in a thick green mat of highly nutritious grass for five to thirty years. *Manyatta* grass grows wherever animals rest and drop feces—under trees, in old settlements, and sometimes around water points. Ecologist Nick Georgiadis, while completing his doctoral work in the Amboseli ecosystem, found that *manyatta* grass was the most nutritious grass in his sample plots, but under very heavy grazing it became poisonous, with high levels of cyanide.[36]

Despite the potential danger (cyanide can be poisonous, especially for ruminants), wildlife and livestock prefer to graze on the brilliant green *manyatta* grass on top of the old settlements rather than on grass nearby. In so doing, they appear to maintain the grass over time, lengthening its life on the landscape. Eventually, the *manyatta* grass atop the old settlements starts to disappear, taken over by small shrubs and trees. Wildlife are still attracted to the old settlement, but this new vegetation attracts different species than before, mainly those such as elands and impalas, which eat shrubs and trees. Birds, too, congregate around old *boma*s, finding the evolving landscape mosaic attractive. Up to a century after a family has moved away, it is still possible to make out the old settlement from the circular patch of acacia trees in Amboseli.[37]

In Chapter 2 I described how Turkana herders in northern Kenya leave behind trees in their old settlement sites, the circular patterns of which can be seen from the air and on the ground. Recently abandoned settlements are impenetrable thickets of young trees, whereas a settlement abandoned thirty or so years ago will have a ring of five to ten trees 3–4 m (about 10–13 ft) tall at the edge of the old livestock corral. Ecologist Jim Ellis and I calculated that most, if not all, of the acacia trees currently in south Turkana originated in livestock corrals. Trees also grow in old settlements across Maasailand, but not nearly to the extent they do in drier Turkana.[38]

Any phenomenon that promotes the establishment of large trees in savannas is likely to be important to overall biodiversity. Large trees are like small ecosystems unto themselves, supporting a dazzling array of species—many more than small trees support, and proportionately more than can be explained by the host tree's size alone.[39]

Do pastoral settlements have the same effects on vegetation all over Africa? My colleagues and I have observed altered vegetation on abandoned settlements in Ethiopia, Kenya, Uganda, Tanzania, and Niger. Scientists in South Africa believe that Iron Age tools found in circular patches of acacia trees may be from old settlements, suggesting that this phenomenon leaves long-term legacies in savannas. In central Kenya, "glades" of grass where wildlife prefer to graze are also old pastoral settlements. In places where they are common, old settlements visibly cover only 1–20 percent of the savanna, but there probably is not a patch of savanna in east Africa within walking distance of water that did not have a settlement on top of it at one time in the past.[40]

But are old *boma*s really enriching landscapes? Might the *boma* phenomenon be a waste of nutrients, robbing other parts of the same landscape of limiting nutrients? Ecologist David Augustine, working in a woodland in central Kenya, showed that herders, through their livestock, are "mining" particular nutrients from pastures where livestock graze during the day and piling them up at night in livestock corrals. For some nutrients, such as nitrogen, this may not matter in wet savannas, because most of the nitrogen turns into a gaseous form and becomes part of the atmosphere, returning to the savanna in the next rains. But in dry savannas, livestock probably do concentrate nitrogen in corrals more permanently, since less of the nitrogen gets redistributed more widely through rainfall. Other nutrients, such as phosphorus, however, are "sticky," so that once moved by live-stock, they stay put.[41]

The piling up of nutrients in old pastoral *boma*s may have an unexpected and potentially large benefit for grazers. Grasses in east African savannas are often nutrient poor, which is a problem for cattle and wildlife alike. Anything that lengthens the period that grass remains green or otherwise improves grass nutrients substantially benefits grazers. Particular grazers, such as pregnant or lactating females, need more nutrients than others. Fulfilling this need is critically impor-tant for pastoralists because the amount of milk their cows produce determines the health of family members and calves. Augustine showed that the plants growing on *boma*s have enough phosphorus and other micronutrients to meet the require-ments of lactating cattle and wildebeest but that the grasses in the surrounding savanna do not. He also found that impalas were ten times more likely to graze on top of old *boma*s than in the surrounding bushland in the wet season, when grass there is especially nutritious. They even preferred old *boma*s in the dry season, but perhaps for a different reason: to see approaching predators at night in the short *boma* grass. By implication, then, herders are creating key resources in savannas that improve the productivity of their own cattle and wildlife. By the same token, when people stop abandoning *boma*s and settle down, it could be that wildlife and cattle suffer.[42]

These pastoral settlements may also be the origin of some nutrient hotspots found inside today's protected areas, hotspots where large, diverse wild herds now cluster. McNaughton found that resident wildlife may graze in the same patches of vegetation in the Mara-Serengeti ecosystem year after year, enriching these fa-vorite spots with their urine and dung for decades. This creates the same effect as old *boma*s, nutrient-rich hotspots that attract grazing wildlife. Herders lived in the parks before they were created, and their old settlements may be one nucleus for these hotspots. Animals creating nutrient hotspots that then attract other wildlife is a phenomenon that has been studied in the Serengeti, South Dakota (for bison and prairie dogs), and Canada (geese).[43]

In the 1990s, a group of us set out to find out how currently occupied settle-ments affect the distribution and abundance of wildlife in the Mara ecosystem of Kenya. Our initial hypothesis was that wildlife would avoid human settlements because the pastoral livestock would compete with wildlife for forage and Maasai dogs would scare wildlife away. We wanted to know specifically how different species react to people and how strongly people affect wildlife.[44]

What we found surprised us. Most of the wildlife clustered around settlements, some distance away but still nearby. A few animals, such as elephants, seemed to avoid people altogether, while a few others, such as baboons, preferred to live close to settlements. But most species, including wildebeest, Grant's and Thomson's gazelles, zebras, and topis, seemed to be at once attracted and repelled by people and their livestock, clustering neither near nor far from settlements (analogous to the humped distribution of Figure 19).[45]

Why would wildlife cluster at moderate distances from occupied Maasai *boma*s? To answer this question, it is useful to know something about the social organization of grazers in east African savannas. Wildlife biologists in the Serengeti have proposed at least three reasons why some wildlife species cluster in tight knots across the savanna instead of spreading out more evenly: (1) to create and take advantage of short grazing lawns created by animals grazing together in the same spot; (2) to lessen the chance of being attacked by a predator; and (3) to graze in hotspots that are particularly nutrient-rich due to decades of repeated grazing. Because these groups often contain many species, ecologists Michael and Judy Rainy dubbed them "multiple species associations," or MSAs.[46]

We think wildlife cluster neither near to nor far from occupied settlements for many of the same reasons that wildlife group together in multiple species associations. In a sense, herders and their livestock become members of these larger MSAs, even their nucleus. Like wildlife grazing, livestock grazing creates moderate to short grass that is more productive and nutritious than ungrazed grass. This grass grows just where the wildlife gather, neither near to nor far from settlements. The relatively short grass lawns are safe places for grazing animals, for they can easily see predators. Wildlife may also cluster somewhat close to human settlements simply because people tend to occupy favored (resource-rich) places.[47]

Maasai all over Maasailand in Kenya and Tanzania describe how wildlife move in to graze and rest next to their *boma*s at night. As people and dogs retire to the compound at sundown, Thomson's gazelles, zebras, and other animals come close to the *boma* fence, followed quickly by hyenas on the hunt for these wildlife or late-arriving small stock. Later in the night, people hear elephants and hippos grazing nearby. The wildlife, herders, and dogs all listen for any sign of a predator, with herders bounding out of bed when they hear warnings from the grazers, birds, or their dogs. People and dogs thus make settlements relatively safe for grazing wildlife. Maasai near Lake Manyara in Tanzania, however, point to the downside of having wildlife close to the *boma*s at night: they attract lions.[48]

It is possible that wildlife and livestock can "facilitate" or provide mutual bene-fits to each other, perhaps more so in the wet season than dry. In a rare experiment exploring competition in Laikipia, Kenya, ecologists Truman Young, Wilfred Odadi, and colleagues used electrified "dingle dangle" fences to allow certain types of wildlife to graze in enclosures that they later grazed with cattle to understand how grazing by different wildlife species affects cattle. They found that the wild grazers improved the quality of the forage and thus weight gain in the cattle, espe-cially in the wet season. The reason may be that wildlife eat down the dead grass stems, allowing cattle to access better food more often. I think it is possible that cattle help wildlife gain weight in the same way, though this is unknown. In the dry season, in contrast, Odadi and team found that cattle and wildlife compete for the same food and cattle lost more weight if they grazed with wildlife than if they grazed alone. They also found that the facilitation that wildlife provided cattle in the wet season was greater than the competition provided in the dry season.[49]

On a broader scale, grazing by large wildlife may make savannas safer and more nutritious for smaller wildlife. Cattle may be involved here as well. About fifty years ago, biologist Desmond Vesey-Fitzgerald, working in the Serengeti, observed that some wild animals move into grasslands to graze in a particular order, which he called a "grazing succession." In an otherwise ungrazed area, large grazers (elephants, buffaloes, rhinos) tend to consume the tall and coarse grass first because their large rumens can digest enough poor grass to sustain them. These large grazers are also powerful enough that they are in little danger of being captured by lions and other predators hiding in the tall grass. As the large grazers shorten the grass, it becomes difficult for them to get enough to eat, so they move on to other tall-grass areas. Medium to small grazers (gazelles, wilde-beest, topis) that need less but more nutritious grass then take over where the big herbivores left off, grazing on the shorter grass. Cattle grazing in Amboseli serve the needs of gazelles in a similar way, by cutting grass down (see Chapter 9).[50]

Predators also adapt to herders and their livestock. Spotted hyenas responded to expanded livestock grazing within the Mara Reserve by extending their home ranges and becoming more nocturnal. Their populations held steady over time, suggesting successful adaptation to herders and their livestock. Such behavioral modifications may be more common than we think; Ethiopian wolves, for exam-ple, are known to hide behind cattle herds when hunting rodents in the Bale Mountains.[51]

We found that while some species of wildlife tolerate herders, their livestock, and their dogs well and are even attracted to human settlements, others, such as rhinos and hartebeest, do not. Conservation of a wide range of species may thus best be done by interspersing protected areas within larger, actively used pastoral landscapes. This "parks and people" model is new, only recently embraced by proponents of both conservation and pastoral development. It is also the type of hybrid strategy that may be most successful in savannas of east Africa.[52]

WHEN PARKS EXCLUDE PEOPLE

By accident, a widespread test of this question is going on in east African savannas today. Between the 1940s and 1960s, east African governments set aside about 10 percent of their savannas in protected areas: national parks and reserves, and conservation areas. People and their livestock were excluded from most of these protected areas and strongly discouraged from using the remaining ones. The parks, meanwhile, were largely left to manage themselves. This laissez-faire approach made sense at the time because most of these systems were open, with wildlife moving freely in and out of the parks. However, with no people to light fires, fires became rare; with no people to build *boma*s, the nutrient hotspots began to fade (although old *boma* scars remain in many parks).

By excluding people, didn't the parks return to a "natural" state that allowed them to recover from decades or centuries of damaging use? The preceding chapters have given us a glimpse of what is truly "natural" in these landscapes. People have been part of these landscapes for epochs, ever since hominins started walking beside their stunning array of wildlife more than 3.5 million years ago. People have likely been using fire to manipulate African savannas for hundreds of thousands of years, the only place on Earth where we are reasonably sure this is so. And livestock grazing is ancient here also, perhaps 5,000 years old.

Two ecologists, David Western (see Chapter 9 on Amboseli) and Helen Gichohi (see Chapter 10 on the Kaputiei Plains), were among the first to measure and write about what happened when people and their livestock were excluded from savanna parks in east Africa. Western collected information suggesting that mixed livestock-wildlife systems are more diverse and productive than systems supporting only livestock or only wildlife. He and Gichohi reasoned that they could test this idea by counting the number of species of wild animals and plants inside parks, on the edge of parks, and outside parks in pastoral areas dominated by livestock. In 1993, they published results strongly suggesting that their hypothesis was correct: there are more species of plants and more elephants on the edges of parks than either inside or outside the parks. Work by our team in the Mara ecosystem

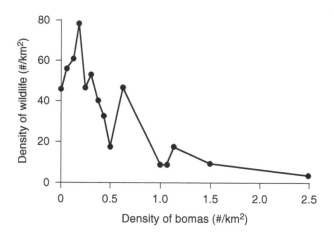

FIGURE 21.

Wildlife remain abundant with moderate numbers of herders but decline in number as human populations grow in the Mara, Kenya. (Source: Reid et al. 2003.)

supports their findings: the greatest numbers and diversity of wildlife are not in the Mara Reserve, where there are no people, nor are they well outside the reserve where there are many people; rather, they are right in between, where there are moderate numbers of settlements, people, and livestock (Figure 21). These patterns are as tantalizing as they are surprising, but more experiments are needed to test alternative explanations for their occurrence.[53]

If savannas where livestock and wildlife are mixed support more species than wildlife-only savannas, it follows that the Western practice of separating people and wildlife by way of wildlife parks may be not only socially unjust but also ecologically unsound. This is not to say that people and wildlife can coexist in harmony under any circumstances. Rather, it means that our approach to encouraging and conserving biodiversity in these magnificent landscapes is missing a critical ingredient: people. East Africans may need to rethink how to better manage savannas—not only by keeping people in them, but by giving the people who maintained these landscapes' diversity over millennia a major voice in how best to conserve them.

In summary, pastoral herders may enrich savannas by burning, engaging in moderate livestock grazing, and concentrating nutrients in their traditional settlements. It is not clear if livestock grazing can increase the number of plant species or the speed of nutrient cycling, as wildlife grazing does in the Serengeti, but I think it is likely. It is also unclear if the species encouraged by grazing tend to be natives and valuable to overall biodiversity. Livestock appear to attract medium-sized and small grazers to short-grass patches or grazing lawns. Human settlements may inadvertently be protecting these same grazers from predators, particularly at night. It appears that moderate livestock grazing may be able to increase the productivity of savannas in east Africa, but to what extent remains unclear. We haven't begun to understand more subtle effects of livestock grazing, such as those on small animals, or belowground, or on water resources.

What we don't know about people enriching savannas is considerable, partly because we are just beginning to ask the right questions. Under what conditions do people diversify savannas, and when do these diversifying forces break down? How does diversification really work, and what can we learn about these processes for conservation and human development? Are reports of decline, degradation, and loss exaggerated or, conversely, underestimated? Should people be separated from nature (are they part of the problem?) or should they be recontextualized within nature (are they part of the solution?)—or both? The following chapters attempt to answer these questions.

CHAPTER SEVEN

When Coexistence Turns into Conflict

After thousands of mornings herding sheep and goats, this one was different for Ole Shani. A "good" buffalo that he and his family had seen around their boma *(home-stead) for months had suddenly gone "bad" and attacked him. That evening, as we gave Ole Shani first aid, I asked what a "good" buffalo was. Ole Shani explained that some buffalo will live around Maasai settlements for months and only occasionally chase someone, rarely hurting them. Although he and his family are still wary of "good" buffalo, they tolerate them as part of the herding life in the Mara, as herders have here for millennia.*

This day started no differently than others, with Ole Shani grazing his flock as usual among the small, shrub-covered hillocks that dot the open grassland of the Mara land-scape near his home. The shrubs grow only a bit taller than Ole Shani, often closing together in thickets.

This day, Ole Shani rounded the edge of a smallish patch of bush with his flock of sheep and goats and walked straight into the buffalo. With the beast about to charge, Ole Shani ran as fast as he could, with the wind, down and away from the buffalo, hoping to conceal his scent from the charging animal—which weighed up to 2,000 pounds, about the same as a very small car, and carried deadly horns. Cape buffalo are short-sighted, so running downwind from them is an escape strategy that can work. But it didn't today. The buffalo kept coming, and Ole Shani could see it was gaining. He knew that lying on the ground with his head down might work, but only if he lay completely flat so the buffalo could not lift any part of his body with its horns. He dove to the ground, remembering how many people he knew who had been killed in similar attacks by buffalo.

Attempting to get Ole Shani up and running again, the buffalo stomped first on his hip, then on his shoulder, and finally on his head. The buffalo then abruptly stopped and wandered off, leaving Ole Shani bleeding but still breathing.

That evening, as we cleaned and bandaged Ole Shani's bloody head wound, he told us that other Maasai had not been so lucky: a buffalo will usually persist until it succeeds in crushing its victim to death. It does this first by stomping on its victim, as this one had. If that doesn't work, the buffalo licks its victim with its rough tongue until the victim's skin becomes raw. The buffalo may then urinate on the abrasion, which is blindingly painful. Most people cannot resist running to try to escape this onslaught—and then they are done.

PASTORALISM AND THE SIMPLIFICATION OF EAST AFRICAN SAVANNAS

Remarkably, no consensus exists as to whether, when, where, or how much livestock grazing causes long-term degradation or irreversible loss of savanna productivity in east Africa (see definition of degradation in Chapter 6). In some fragile savannas, to be sure, livestock grazing can diminish productivity, but even here few cases suggest that this diminishment is permanent. In more resilient savannas, evidence shows that livestock grazing can even improve productivity—up to a point, at least. And some savannas seem impervious to grazing, even when heavy. Certainly, some evidence now counters the once-prevalent idea that livestock in African savannas are responsible for vast environmental degradation and desertification. We must, therefore, look more carefully at this issue, and on a case-by-case basis.

The last chapter examined when and why pastoralism is compatible with, and sometimes even beneficial for, wildlife; it also presented some of the positive effects livestock have on plant species and their productivity. This chapter examines the flipside: when pastoralism simplifies savannas, leading to human-wildlife conflicts, declining savanna productivity, or the disappearance of savannas altogether. We will focus first on direct conflicts between herders, their livestock, and wildlife through predation, disease, and competition for forage and water. Then we will look more broadly at pastoralism and its role in the decline, degradation, and loss of savanna function, examining if and how herders and their livestock simplify vegetation in savannas through burning and grazing. Third, we will examine what happens to wildlife and savannas when pastoralists settle down or take up farming, or when migrants expand farmland into savannas, creating hard boundaries and loss of savanna lands. We conclude by assessing whether pastoral peoples, livestock, and wildlife in east African savannas remain in a state of coexistence today or are degenerating into conflict.

DIRECT CONFLICTS BETWEEN HERDERS, THEIR LIVESTOCK, AND WILDLIFE

We [Maasai] know the character of lions. If
a lion eats cattle, it will keep eating cattle
until you kill it. It will run to eat cattle until
you kill it or you move.

OL TUKAI, *Maasai elder, Tanzania*[1]

There is a long tradition of hunting among pastoralists, but often more as an occasional, opportunistic way to obtain food when times are hard. As described in Chapter 5, many pastoralists survived catastrophic livestock losses from rinderpest at the turn of the twentieth century by hunting wildlife, to the extent that elderly Maasai still refer to wildlife as their "second cattle." After their herds recovered, however, they switched back to using livestock for food. Ecologist David Western says that "few pastoral groups regard wildlife as purely competitive; Turkana take wildlife as an auxiliary resource, Maasai use wildlife in Amboseli as an emergency food supply during major droughts." In places where Maasai have abandoned the tradition of becoming warriors, they no longer kill lions or birds for their mane and feathers, items their elders used to construct warrior headdresses. And even in places near herders where wildlife poaching is common, like Tarangire National Park in Tanzania, the lion's share of the poaching is done by farmers, not herders.[2]

Where pastoralists have access to and hunt with guns, however, wildlife suffer. In northern Kenya, wildlife populations declined by more than half in just twenty years, between the 1970s and 1990s, and hunting by herders carrying guns is one likely cause. As I described in Chapter 1, hunting prowess was widely admired when I lived in Turkana in the 1980s. In Karamojong, in northeast Uganda, ecologist Richard Lamprey describes the effect of armed pastoral militias starkly:

"In conservation terms, the wildlife reserves of Karamoja are completely devastated. These areas are used by heavily-armed Karamojong herders for the pasturing of over 150,000 cattle and 100,000 sheep and goats." During the 1994 genocide in Rwanda, Akagera National Park lost some 90 percent of its wildlife due to unregulated hunting by Rwandans and Tanzanians, many of whom were pastoralists. Young Maasai boys hunt wildlife with their dogs during the day for sport and for meat to feed to their dogs. Herders' dogs in Ethiopia often kill wildlife for food. Even though eastern Cushitic herders (Galla, Afar, Somali) traditionally do not eat wild meat, involvement of Somalis in ivory poaching suggests that they do kill wildlife for sale.[3]

Similarly, predators create conflicts with herders by killing and injuring people and their livestock, as seen in Chapter 3. Lions, hyenas, and leopards sometimes injure people while breaking through the fences surrounding pastoral settlements at night to kill and consume the livestock kept there; less often, carnivores attack people as they herd their animals during the day. Herding families lose 1–3 percent of their livestock and usually some dogs to carnivores each year. Even though lions actually kill few livestock, Maasai disproportionately resent lions because they attack their prized cattle, whereas other carnivores focus more on less prized sheep and goats.[4]

Nonpredators, like elephants and buffaloes, also kill people when surprised or threatened. Lone (and often drunk) men are likely to be caught unawares by elephants in the night as they weave their way home from local drinking establishments. The presence of elephants limits where herders graze cattle, keeps women and children from venturing far from their homesteads, creates competition with other herders for water, and means families need to guard livestock that are kept in their homesteads. Many of my male age-mates from pastoral communities tell stories of the danger of being charged and gored by a rhino when they walked to school as young boys thirty or forty years ago, a threat much diminished today because of dramatic losses of rhinos, even in the most wildlife-rich savannas.[5]

Some herders retaliate and kill wildlife that endanger their families or their livestock. It is not uncommon for Maasai to spear an elephant that kills or injures a person near Kenya's Amboseli (see Chapter 9). Herders may also poison predators to protect their livestock. Unfortunately, a widely available and inexpensive pesticide, carbofuran (or Furadan), rapidly kills lions as well as vultures that feed on dead lions. In 2007, for example, several lions were poisoned in Kenya's Mara, and it is possible that poison was responsible for the death of twenty-five wild dogs in Loliondo, Tanzania, more recently.[6]

Livestock, dogs, and wildlife can pass diseases to each other, some of which are devastating. Wildlife serve as hosts for many diseases of livestock, including tsetse-transmitted trypanosomosis (called sleeping sickness in people), tick-transmitted East Coast fever, viral foot-and-mouth disease, and malignant catarrhal fever (MCF), which is transmitted by nasal secretions from newly born wildebeest. Livestock initially spread the deadly rinderpest in Africa (see Chapter 5), but fortunately it was eradicated in 2011. Domestic dogs pass rabies and canine distemper to African wild dogs, Ethiopian wolves, and lions, often reducing their populations severely.[7]

As described in Chapter 6, wildlife and livestock both facilitate and compete with each other. Grass-eating cattle, sheep, and goats may compete with wild grazers, while woody plant–eating goats and camels may compete with wild browsers. Herders around Tarangire, Tanzania, cite competition between their domesticated animals and wild animals for water and forage as one of the main problems of living with wildlife. Our pastoral informants in Kenya's Mara (see Chapter 8) and Kaputiei Plains (Chapter 10) consider livestock-wildlife grazing competition a big problem, particularly in the dry season and during droughts, when forage is limited. Experiments by ecologist Wilfred Odadi and his colleagues confirm these observations, showing that competition is greatest in the dry season. In North America, ecologist Tom Hobbs and colleagues found that elk both facilitate and compete with cattle by at once improving the quality of grass regrowth and removing significant quantities of grass. They conclude that competition is strongest when forage is scarce and cattle numerous, but that coexistence is possible at other times.[8]

Herders and their livestock also compete with wildlife for access to water, creating heavily impacted "piospheres" (from the Greek *pios,* to drink) around water points that can be seen on satellite images taken from space. The busy activity of people and livestock around boreholes, dams, and springs scares wildlife away from such locations during daylight hours, affecting in particular small to medium-sized grazers that prefer to drink during daytime when they can more easily see approaching predators. As water becomes more scarce in the dry season, Maasai herders in the Kaputiei Plains actively chase wildlife away from water points. Indeed, aerial surveys made by scientists at Kenya's Department of Resource Surveys and Remote Sensing show that wildlife are rare near water points during the daytime across wide areas of dry northern Kenya. This "pushing out" of wildlife by herders and their livestock is well recognized in the Kalahari of Botswana as well, and becomes worse as people build more water points. In rangelands of Africa and Australia, a concentration of people and livestock around water points also removes plant cover, including rare plants, though at the same time it encourages the growth of trees and shrubs, and weeds; it compacts soil as well, and introduces excess nutrients into the ground via livestock urine and dung. These changes shift the types of insects, mammals, birds, reptiles, and amphibians near water points.[9]

Herders may have one final negative impact on wildlife, by "overburning" the savannas. We saw in the last chapter, however, that small, low-intensity fires, set at different seasons, can in fact promote a diversity of habitats, plants, and animals, while large and intense fires tend to do the opposite in Africa and elsewhere. When, then, might fires be so large or intense in east Africa that they reduce diversity? In my experience, the only savannas with sufficient unburned biomass to maintain large or intense fires are inside parks, where it has sometimes been the custom of park managers not to burn savannas (especially in Kenya). Outside of parks, pastoral grazing often removes much of the grass fuel for fires, limiting their extent and intensity. These savannas are burnt rarely, which favors bush encroachment and a more simplified savanna.[10]

OVERGRAZING UNDERSTATED?

General assemblies of the Boran are called Gumii Gaayo. . . . At the 37th Gumii Gaayo held in 1996, the leadership acknowledged the declining welfare of the Borana society in general. . . . The "cattle problem," as viewed by Gumii Gaayo leaders, is seen as a reduced productivity per head due to high stocking rates and environmental degradation.

G. HUQQAA, *describing a meeting of Borana pastoral people that has occurred every eight years since 1696 in southern Ethiopia*[11]

Adding to these direct conflicts among herders, livestock, and wildlife is the widespread idea that livestock cause decline, degradation, or loss of rangeland function where grazing is heavy. Among scholars there have been strong evidence and arguments for and against the idea. As seen in Figure 19, savannas seem to respond to grazing according to four patterns: enriching, humped, simplifying (or degrading), and no response. Which one of these patterns occurs in a particular place depends partly on how productive (wet, fertile) the local savanna is, with productive savannas sometimes showing more of a humped or enriching response and dry savannas showing the other responses, with important exceptions. We also saw that different species, in a single place under the same conditions, can respond differently to grazing.[12]

Although the evidence in Chapter 6 suggests that the negative impacts of grazing have been overstated, I also think that recently the pendulum has swung too far in the opposite direction, with some arguing that pastoralists rarely degrade savannas in Africa, if they do so at all. This could be called "overgrazing understated." I think there are three main reasons for these pendulum swings and for the confusion about decline, degradation, and loss: (1) any one change in a savanna can be viewed as insignificant by some people and negative by others; (2) heavy grazing sometimes does and sometimes does not have negative consequences (as shown in Figure 19); and (3) savanna change is rarely measured with long-term experiments that are able to separate out the effects of grazing from those of climate and other confounding factors. Since the evidence is skimpy, I will use examples from all across Africa here.

Much of the debate over what constitutes decline, degradation, or loss is equivalent to comparing apples with mangos. Those interested in livestock production typically judge rangeland change in terms of the effect it has on livestock proper. As Roy Behnke and Ian Scoones put it, "Vegetation change is of no intrinsic interest unless it provides reliable evidence of changes in livestock productivity." This means that livestock "degrade" their environment only when their herders feel the impact, through less milk or meat on the hoof—a human-centric point of view (Figure 22, top). Those interested in rangeland ecosystems, in contrast, take an ecocentric view and consider degradation a negative change in the ecosystem itself, such as a decrease in the number of native plants, invasion by weeds, soil compaction, the shifting of wildlife populations (Figure 22, bottom), regardless of whether these changes then affect livestock. Most herders can be considered human-centric: while they are concerned with environmental changes, their greatest interest is in those that directly affect their livestock and livelihoods. In other words, they use livestock condition as an indicator of degradation and manage the savannas to keep the milk flowing.[13]

Herders have in-depth ways of measuring change in their rangelands, and their observations are often day-to-day, month-to-month, and year-to-year. They assess trends in vegetation at relatively local scales, from small patches to larger

A human-centric point of view

Livestock production

This doesn't matter unless...

this happens too

Vegetation production

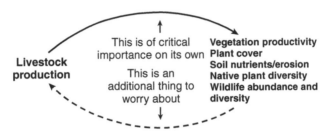

An ecocentric point of view

Livestock production

This is of critical importance on its own

This is an additional thing to worry about

Vegetation productivity
Plant cover
Soil nutrients/erosion
Native plant diversity
Wildlife abundance and diversity

FIGURE 22.

What constitutes "degradation" depends partly on your point of view. (Sources: Lamprey 1983, Sinclair and Fryxell 1985, Homewood and Rodgers 1987, Behnke and Scoones 1991, Brockington 2002, Sullivan and Rohde 2002, Vetter 2005, Oba and Kaitira 2006.)

landscapes of hundreds of hectares (or acres). Ariaal herders in northern Kenya, for example, assess the condition of their rangelands by identifying thirty-nine different landscape patches and six landscape types, and classifying plant species by whether grazing causes their abundance to decrease, increase, or remain stable. In northern Tanzania, Maasai classify landscapes in similar detail, and describe landscapes as degradable (*orpora*), or sensitive to overgrazing and soil erosion, and nondegradable (*orkojita*). To reduce these landscapes' vulnerability, they graze *orpora* terrain, for example, only in the wet season when such lands are less sensitive to grazing.[14]

Pastoralists have observed overall changes in the landscapes they graze over the last few decades. As the leaders of the Borana people, who live in a savanna with equilibrium dynamics in southern Ethiopia, recognized in the quote above, too many cattle can cause environmental change. As recorded by ecologists Waktole Tika, Gufu Oba, and Terje Tvedt, Borana herders recalled their landscapes in the 1960s as follows: "When we were younger, we played 'hide and seek' in the long grasses. Today . . . we see bare ground (*baarbadaa*) everywhere . . . [while the grazing lands are] covered with bush." Borana described how the grasses in the past gave them more milk and more and faster-growing calves. The reason for the loss of grass is complex, however, involving more than just too many cattle, as the Gumii Gayo suggested. Cattle stocking has increased as political conflicts compressed Borana herders into a smaller part of their former territory, and also because of land loss due to the expansion of crop cultivation, commercial ranching, and range enclosures. Traditional rules that used to ensure rest for certain pastures seasonally have broken down, and new rules now forbid burning of invading woody vegetation (which would restore grass). Families have also shifted from traditional to peri-urban settlements.[15]

Other herders see similar changes as the Borana do. In another part of Ethiopia, the Middle Awash Valley, most of the Oromo and Afar herders say their range is in poor condition for two reasons: overgrazing and more frequent droughts. The Maasai in northern Tanzania see more woody plants and less grass in their grassland too, and attribute this change to crop cultivation cutting off traditional grazing routes and herders settling down and grazing intensively around more permanent settlements. Pokot herders in Kenya describe how heavy cattle grazing encourages shrubs to spread, even far from the heavily grazed areas around waterholes. In the 1980s, Turkana herders in northern Kenya described to me how their landscapes had shifted from open savannas to woody savannas in just thirty years since the 1950s. They did not know why the vegetation had changed, but they were sure their livestock were not involved. I think that less fire and increasing CO_2, which favors C_3 plants, including trees and shrubs, could also be responsible for all these changes.[16]

While the reasons for these changes are complex, grazing plays a role—though often not because of the growth of herds, as one might expect, but because pastoralists are being constricted into smaller areas. Why might grazing cause bush to encroach? As described in Chapter 2, heavy grazing encourages bush encroachment by reducing fuel for fires that kill woody plants and by removing grasses that compete with woody plants for water and nutrients. Bush encroachment generally happens more often in productive, equilibrium savannas, where abundant rainfall encourages the growth of woody plants, and around grazing impact points like settlements, water points, and salt licks. Around boreholes in the Kalahari, for example, observers describe "sacrifice zones" where the grassland is gone, replaced by woody plants, but these are usually confined to areas within 200–500 m (650–1,600 ft) of the boreholes.[17]

Does it matter that grasslands are transformed into woodlands, whether by heavy grazing or for other reasons? The soils of savannas relatively rich in woody plants tend to be more compacted and eroded, and have less infiltration of water, than the soils of grassy savannas. However, woody soils often store more nutrients and carbon than grassy soils. Bush encroachment in savannas matters to herders because it may change their options: as grass production falls, for example, there is less for grazing cattle and sheep to eat but more for browsing goats and camels. On the other hand, woody plants can slow wind erosion during drought. Thus, more woody savannas may be good for some herders but not for others.[18]

Beyond the loss of grass and the expansion of trees and shrubs, grazing can also cause a more subtle change: a shift from a savanna dominated by perennial (long-lived) grasses to one dominated by annual (short-lived) grasses. This happens in the (relatively wet) semiarid savannas of Niger, Mali, and, Somalia as well as in the drier savannas of South Africa. Even heavy wildlife grazing can cause this kind of shift. But in an arid (resistant) savanna in Namibia, heavy livestock grazing did not cause a loss of perennial plants.[19]

In the end, it depends on one's point of view whether a change from perennial to annual plants counts as degradation. From a human-centric perspective, some annual plants, which produce from seed each year, contain more protein for livestock than perennials and thus can improve livestock forage. As anthropologist Michael Bollig observes regarding Kenya's Pokot and Namibia's Himba herders, any "loss in biodiversity and the change from perennial to annual grasses do not affect or interest them as long as the range remains productive." Indeed, herders are concerned when they see ephemeral annuals growing only when it rains and fading quickly when rain stops, providing little forage for extended dry seasons. From an ecocentric point of view, in contrast, a decline in perennial plant species because of sensitivity to grazing could represent a local or even regional loss in biodiversity, which would certainly be considered degradation or irreversible loss.[20]

Further, livestock grazing can cause the decline of plants that livestock eat and are able to digest and the invasion of weeds. This may be an indication of degradation from both points of view. Heavy grazing has caused palatable plants to disappear from savannas in Ethiopia, Kenya, Mali, Somalia, and South Africa. Some unpalatable plants are even toxic to grazers and spiny, as they are in South

Africa's arid Karoo. Weeds can also invade heavily grazed savannas, as they have in South Africa and coastal Tanzania. Generally, a decline in palatable plants and the invasion of weeds are some of the more common consequences of grazing.[21]

Can grazing affect soils directly, through trampling and deposition of dung and urine? Livestock seem to have more direct impact on soils in wetter than in drier savannas. In the wetter, semiarid Sahel, heavy grazing by livestock removed microbiotic crusts and compacted soils. It also lowered soil pH (that is, increased soil acidity), organic carbon, nitrogen, and phosphorus. Loss of biological soil crusts is significant because these are critical for soil fertility and stability. But in the drier savannas of South Africa's Namaqualand, heavy grazing had no effect on nutrients or moisture in soils. Similarly, in the Namib Desert, soils in heavily grazed communal lands had the same nitrogen, phosphorus, and plant growth as those in nearby lightly grazed commercial ranches.[22]

As for erosion of soil caused by grazing, the evidence, again, is equivocal. In Borana, Ethiopia, livestock, crowded onto smaller rangelands, are partly responsible for the increased erosion of surficial soils in some places, but not others. In a South African wetter savanna, there was no difference in erosion when comparing among heavily grazed communal land, lightly grazed commercial cattle ranches, and wildlife game reserves. A comforting projection is that it would take another 400 years of grazing to erode the soils away in the rangelands of Botswana.[23]

Because many African savannas appear to be resilient to grazing, the idea that livestock populations are declining due to overgrazing seems not to be supported, at least over broad regions. However, caution is required here because decline, degradation, and loss, and their impacts on livestock, are both difficult and expensive to measure. Scholars reason that if livestock populations have been holding steady or growing over the last few decades, they cannot have been "degrading" either soils or vegetation, because that would show up in decreased livestock numbers over time. Indeed, in Zimbabwe, Kenya, Namibia, and the Sahel, livestock populations have gone up and down over the last few decades, with little change in overall trends. Of course, if numbers of animals go up but their productivity goes down, that would be cause for concern, but very little data exists on either population numbers or productivity. The Borana of Ethiopia clearly see their cattle productivity declining, partly because of higher stocking in smaller rangeland areas. In South Africa, livestock populations declined markedly in forty-five districts of Cape Province from 1855 to 1981, a possible sign of degradation or loss (although stocking policy was another possible cause).[24]

In a book called *Do Humans Cause Deserts?*, ecologists Andrew Ash, Mark Stafford-Smith, and Nick Abel review the evidence worldwide and conclude that "there is remarkably little evidence quantifying a decline in secondary [livestock] production as semi-arid systems degrade and primary [plant] productivity decreases." In other words, savannas may decline and degrade, but this does not necessarily have an effect on livestock. Why is this so? The authors reason that rangelands, where rainfall and topography create highly heterogeneous landscapes that are constantly changing, provide opportunities for livestock to find forage if they can move. If we factor humans into the equation, even more opportunities for adaptation present themselves. People are constantly adapting to changing savannas by shifting from grass-eating cattle to shrub-eating camels, or by moving animals great distances when it is dry, or by importing forage from forested areas (or even city roadsides) during drought. Herders, masters at adaptation, have many more options than just the back pasture to keep their livestock alive. Of course, at some point, a limit to dodging degradation or loss may be reached, even in resilient savannas.[25]

It is entirely unclear how widespread degradation caused by livestock actually is. As Andrew Ash and colleagues caution, "The lack of evidence of declining secondary production may have more to do with the difficulty of detecting change than with its real absence." What is of most concern are those places where the effects of livestock grazing are difficult to reverse (degradation) or irreversible (loss), thus compromising the long-term viability of the savanna. Once these sensitive landscapes are degraded, reversing change is expensive, labor intensive, or requires uncommon political will.

The effects of grazing are also difficult to reverse where savannas are particularly sensitive to the very activity of grazing. Maasai of northern Tanzania call these sensitive parts of savannas *orpora* and have traditional rules for grazing them with caution to maintain their productivity. They use observations of soils and vegetation to judge how sensitive a savanna landscape is to degradation. Plants can be sensitive to grazing because of how long they live and the ways in which grazing impacts their biology. Annual plants, which grow anew from seed every year, can recover from grazing quickly, as long as grazers have not eaten a large proportion of the seed, which ecologist Pierre Hiernaux found unlikely in west African savannas. Perennial grasses, which regrow from existing rootstocks each year, usually take longer to recover, from four to twenty-five years. Long-lived shrubs can take even longer, especially if grazers remove most of their flowers or seeds. Biologists Thorsten Wiegand and Suzanne Milton estimate that even sixty years of rest from heavy grazing will not allow a shrubland in the arid Karoo of South Africa to recover to good condition because of the biology of the main shrub species. Bush-encroached savannas, moreover, can get "stuck" in their new bushland state—for sixty to one hundred years in the case of Borana, Ethiopia, which, in the context of a human lifetime, is virtually irreversible.[26]

Going back to the four models of how grazing affects plants (see Figure 19), the evidence so far best supports three of them: savannas either do not respond (are resistant), degrade (are simplifying), or respond first positively then negatively (have a humped response). The fourth response, enriching, may actually not be a different model at all; rather, it may be the first stage of the humped response. And wetter, more fertile savannas tend to respond more strongly, whether positively or negatively, than do dry, less fertile savannas. Unfortunately, responses in any savanna are complex, so different plant species in the same savanna may follow different models. Despite this complexity, the models are useful to help us weigh the preponderance of evidence for change in different aspects of savanna landscapes, before judging whether or not they are degraded.

The fact that grazing sometimes enriches and sometimes degrades savannas does not strongly support the quite common idea that most rangelands are overgrazed, degraded, and desertified. It does, however, suggest that African savannas are relatively resilient to grazing, especially given their long evolutionary history supporting grazing animals. I also think we are a long way from clearly answering the question of how often and where livestock do degrade savannas. In a recent review of evidence from around the globe, ecologist Gregory Asner and colleagues showed that for wetter savannas, woody plant expansion is common, but it is hard to pin down the cause, and its full extent is unknown. In west Africa, the common perception is that the Sahel is overgrazed and degraded, but here also the scientific evidence is not clear. Decades of work to assess Sahelian degradation, for example, shows little widespread vegetation degradation as measured by satellite imagery, but new evidence suggests that these analyses need to be revisited. Ecologist Jim Ellis and our team analyzed livestock numbers and savanna productivity across Africa and found that only 19 percent of the driest savannas, and 8 percent of the wettest, were overstocked—that is, with more livestock than the vegetation could probably support. In all of these global and regional assessments, not only do we not know if the areas are degraded, but when we do see change, as in the Asner example above, we do not know what role livestock play compared to other likely causes of degradation, such as CO_2 rise, rainfall decline, and expansion of other land uses like farming.[27]

These rather moderate conclusions might be seen to lend support to the human-centric view that herders rarely, if ever, cause degradation. But caution is needed. It is clear that grazing can cause bush to encroach, palatable plants to disappear, weeds to invade, and soils to change in African savannas—all signs of the simplifying response, that is, decline, degradation, or loss. Also clear is that irreversible loss occurs when herders settle and farm. For many places, however, the reality is closer to the middle ground between herders not causing any degradation at all and grazing necessarily causing irremediable loss—which is hard to accept, because many of us want clear, unequivocal answers. In the future, I hope to see more pastoral herders, from all over the continent, bringing much-needed nuance and sophisticated local understanding of savannas to this discussion.

THE TRANSITION FROM SOFT- TO HARD-BOUNDARY SAVANNAS

These moderate conclusions suggest that savannas with soft boundaries, wet or dry, may be resilient to grazing impacts. Of real concern, however, is where people are settling, many of them taking up cropping, a process that started in many areas more than a century ago. Here, savannas are being lost and hard boundaries are increasingly widespread (see Map 12).

Pastoralism itself, when it maintains soft boundaries, is generally compatible with wildlife, but this changes if settlements are large or widespread, or if they prevent wildlife from accessing key resources such as swamps or riverine areas (see right half of Figure 21: declining wildlife as human populations rise). Water points, which can be separate impact points from settlements, can repel wildlife, at least during the daytime. Such settlements and water points provoke the hardening of boundaries in savannas. My team found in Kenya's Mara, for example, that wildlife populations start to decline when pastoral populations rise above about 6 people per km^2 (15.5 per mi^2) or 1 *boma* per 5 km^2 (1 *boma* per 1.9 mi^2). At 39 people per km^2 or about 1 *boma* per km^2 (101 people per mi^2 or 2.5 *bomas* per mi^2), only 25 percent of the wildlife are left on average; at about 83 people per km^2 (215 people per mi^2 or 7 *bomas* per mi^2), this figure falls to 5 percent. In an earlier study in Kenya, biologists Parker and Graham estimated that elephants go extinct when human populations rise above 82.5 people/km^2 (214 people per mi^2).[28]

In the still relatively few pastoral places in east Africa where land is being privatized, herders are beginning to build fences in part to exclude wildlife, thus forming the first tangible boundaries built by herders in this region's savanna landscapes (see Chapters 8 and 10). The resulting fragmentation of the landscape is severing access for both wildlife and livestock to their traditional migration and herding corridors and preventing animals from reaching critical water points. It is also removing key habitats, such as Namelok Swamp, which used to support wildlife in Amboseli's drylands.[29]

What finishes savanna wildlife is people turning savannaland into farmland, which likely leads to irreversible loss for many wildlife species (see Figure 9c). As pastoral people settle and establish farms, they report more and more types of conflicts with wildlife than they did when they restricted themselves to livestock herding. Farmers, of course, grow plants with abundant seeds, fruits, and leaves for human and livestock consumption; these food and feed crops are highly attractive to many species of wildlife. In addition, as farms expand into the drier savannas, farmers first cultivate the best watered land, such as that near rivers and swamps, resources that are equally vital for wildlife and pastoral livestock. A shift from herding to crop farming in an area thus usually serves to break the long-standing coexistence between people and wildlife, creating in its place a situation intolerable to crop growers, who then erect fences and other hard boundaries and harass wildlife to keep them out of their fields. This tips the former tolerance extended by pastoralists toward their wild neighbors into conflict (see Figure 10).[30]

When farms cover a major part of a savanna, most wildlife disappear. Ecologist Hervé Fritz and colleagues measured how the size of farmers' fields affected wildlife access to rivers in the Zambezi Valley of Zimbabwe. Fewer wildlife used river water near large crop fields (especially those larger than 3 ha, or 7.4 ac) than near small fields or near land with no fields at all. This was true for carnivores and all herbivores, although elephants could tolerate farms ten times larger, up to 32 ha (79 ac) in size. In another, nearby area of Zimbabwe, elephant populations declined when farmers converted 40–50 percent of the land area into crop fields and villages, according to biologists Richard Hoare and Johan du Toit. In Kenya's Mara, scientists estimate that if current trends continue, wildlife will disappear when farmers' fields cover 25–30 percent of the savanna.[31]

Some wildlife are more affected by farming than others. Animals with large home ranges, such as African wild dogs, slow reproductive rates, such as elephants, and low population sizes, such as black rhinos, are particularly sensitive to increased hunting by people or loss of their habitat when farms expand. The opposite may be true for much smaller animals, such as birds and butterflies. In fact, there can be as many or more species of plants, birds, and butterflies in landscapes with scattered farms than in landscapes left open for wildlife, with no farms. An exception to this pattern occurs in bird populations of the Serengeti, where farm-land outside the park supports many fewer birds than savanna inside the park.[32]

HERDERS AND WILDLIFE TODAY:
COEXISTENCE? CONFLICT?
CONFLICTING COEXISTENCE?
A SUMMARY

So what is the bottom line? What characterizes the relationship between herders and wildlife in east African savannas—conflict or coexistence? Both sets of forces have been part of savanna life for millennia here, at first as herders, livestock, and wildlife mixed with hunter-gatherers, and more recently as they have mixed with (and become) farmers. As we saw in Chapter 6, herders and their livestock some-times change savannas in ways that promote coexistence with wildlife by improv-ing production and nutrient flows through burning and grazing, by providing safe havens for herbivores around settlements at night, and by improving food quality in old settlement sites. At the same time, however, overexploitation through hu-man hunting or poaching can severely damage wildlife populations, even extermi-nate them, with the hunters sometimes being local herders but more often being nonherders. In turn, elephant and predator attacks on livestock provoke human animosity toward wildlife. Competition for food and water can hurt pastoralists and wildlife alike. And livestock can transmit to wildlife, and wildlife to livestock, diseases that cause chronic, and sometimes fatal, illnesses.

In my view, this conflicting evidence makes it clear that extreme positions, whether human-centric or ecocentric, having little explanatory power, are funda-mentally unhelpful. They are also inaccurate: the facts simply do not support the belief that livestock cause widespread degradation of savannas or, conversely, that there is no need to worry about livestock damaging savannas. Rather, it is impor-tant to recognize that some of the diversity we ecocentrists so admire in these savannas has been maintained, and indeed partly created, by human herders. And we human-centrists (for I count myself in both camps) must recognize that as people settle, herders take up farming, populations grow, protected areas expand, and land becomes privatized, the savannas are being simplified. In the debate about pastoral overgrazing, I believe that the pendulum has swung both ways, initially with biologists and conservationists overstating the case for human deg-radation of the landscape, and now with social scientists and development profes-sionals understating it.

The bottom line in many places is one of conflicting coexistence. Even if tolerance between pastoral people and wildlife populations appears to be on the decline, such tolerance does exist and can be strengthened. And where pastoralists share fairly in the profits stemming from wildlife conservation, conflicts can be turned back into coexistence. We return to this idea in the last chapter. Meanwhile, the next four chapters show how these forces are playing out in four ecosystems in the wildlife-rich parts of Maasailand in southern Kenya and northern Tanzania.

CHAPTER EIGHT · The Serengeti-Mara

"Wild Africa" or Ancient Land of People?

Ole Nasipa walks toward his boma *with his cattle, proud that his reading lessons are paying off and he can read the headlines in that day's* Standard *newspaper. He always wanted to go to school as a boy, but it was his place in the family to herd livestock while his younger brother attended school. He can see that his little brother does not always know why a cow is sick or which plants are best for medicines. He thinks about his young daughter, who is the first girl to ever attend school in his family, and wonders what the Mara will be like for her and her children.*

As he pulls the acacia thorn gate closed to keep the cattle safe in the corral for the night, one of the thorn branches pokes an ugly scar that is healing on his head. Ole Na-sipa winces, remembering the two long days he spent waiting at the local clinic, part of the bone of his skull exposed to the air, a large flap of skin loosely attached to his skull with dried blood. He really thought that was it, the leopard had won. He and his cousins had taken a stupid chance, thinking they could flush the leopard from some Euclea *bush and kill it, in retaliation for all the sheep the leopard had killed the night before. When the leopard leapt on him from behind, he hardly felt it—the wound was so deep and the shock so fast. His friends saw him fall and lie motionless, saw his exposed skull and all that blood. Assuming he was dead, they ran back to the* boma *to tell the news. Ole Na-sipa thought he was as good as dead too, so he remained still and then slowly dragged himself into a position ready to die, face down with his head pointing toward the rising sun. After some time, he realized that maybe he wasn't going to die, so he crawled and then walked back to the* boma, *to the amazement of his family.*

When I first met Ole Nasipa, his scar was the length of his skull, and his herding skills and generosity were renowned throughout the Mara.

A JEWEL IN THE CROWN AND A CHALLENGE TO HUMAN RIGHTS?

The Serengeti-Mara: no other place on Earth has so many kinds of large grazing animals moving together in such great numbers. Depending on where you are in the Serengeti-Mara ecosystem, which straddles the Tanzania-Kenya border, you can see quintessential scenes of "wild Africa," with seemingly endless views of savanna and millions of animals. Many conservationists from Tanzania, Kenya, and elsewhere would argue that this place is an irreplaceable benchmark of wild-life grandeur, a flagship of global conservation, the noblest example of human re-straint as society hurtles toward development, ever accelerating its consumption of natural resources. Looking at the bounty of gate fees flowing into private and public hands from the Serengeti in Tanzania and the Mara in Kenya (estimates range from $6 million to $15 million a year), the economics of this place makes sense as well, from significant local employment to enriched national treasuries. Conservation efforts are remarkably timely too, as Kenya's Mau Forest disappears, river flows diminish, wheat farms engulf wildebeest calving grounds in Kenya's Loita Plains, the Tanzanian government considers a highway that will bisect the Serengeti, the western Serengeti groans with human populations that are past the million mark, illegal hunters snare some 160,000 animals of all species each year in the Serengeti, and Kenyan pastoral land gets cut up into private parcels. From this perspective, the Serengeti-Mara seems to be a classic win-win situation: jewel-in-the-crown conservation with high financial returns, preserved just in time, and in perpetuity, for the children and grandchildren of Tanzanians, Kenyans, and the global community. And luckily, it is one of the most studied ecosystems in the world, so there is scientific evidence for managers to use to conserve this pre-cious place.[1]

The Asi, Nata, Ishenya, Ikizu, Ngoreme, Ikoma, Sukuma, Kuria, Tatog, Sikazi, and Maasai peoples lost a great deal of their land when the colonial and postcolo-nial governments created the Tanzanian and Kenyan parks and reserves, which was done sometimes with (Sukuma, Maasai) and sometimes without (western Serengeti peoples) consultation. The Serengeti-Mara has been home to humans and their ancestors for millions of years and to hunters, herders and farmers for thousands of years. If you look closely, the signs of people are everywhere in these parks—in rock paintings, old pastoral settlements, stands of trees, pieces of worked obsidian, and Maasai and Asi place-names. Only fifty to sixty years ago, colonial governments evicted local people from the land that is now called Maasai Mara National Reserve and Serengeti National Park, but they still live around the edges of these parks as farmers, fishers, hunters, and herders. And as elsewhere, the vast majority of the costs of these protected areas fall on the shoulders of local, nearby people, while the vast majority of the benefits flow to people far away in places like Dar es Salaam, Nairobi, London, or New York. In the rare cases where significant benefits do flow to local communities, as in the Mara, the wealthy pastoral families keep most of the profits, leaving the poorer herders and farmers with less gain. Benefit sharing has improved with the movement toward community conservation and conservancies since the 1990s, but with mixed success.[2]

Perhaps the biggest myth about the Serengeti-Mara, one started by incoming European colonists about a century ago and continued by some tourism entrepreneurs today, is that it is one of the last examples of a "wild Africa" untouched by modern development. To be sure, the Serengeti-Mara *is* a rare example of a large, unpopulated savanna landscape and so, like other parks around the world, provides a benchmark against which to measure the impacts of the human hand. But, as I hope previous chapters have convinced you, the Serengeti-Mara is *not* an example of "wild Africa." That Africa is a mythical landscape, in that it lacks one critical group of ecosystem engineers: humankind. First hominins and then humans have been shaping this ecosystem for thousands, if not millions, of years. Over the last century, during the colonial period, the Serengeti-Mara has been strongly shaped by a succession of events, most caused by the influx of people, that determined how many and what kinds of wildlife live there, the mix of trees and grass that populate the land, and how the savanna functions. This chapter tells the story of how this landscape came about.[3]

Another true fact about this ecosystem is the grandeur of the wildlife migration it supports. Around the world, migrations such as this one are some of the hardest to protect, particularly when people build obstacles to deflect them or stop them altogether. Globally, we have lost 25 percent of all of our large-scale land migrants, like the buffaloes in the North American plains and the springboks in South Africa. But here, in the Serengeti-Mara, the largest mass flow of large land mammals that still exists on Earth continues to this day: the migration of over a million wildebeest, zebras, gazelles, and other species. This chapter tells their story as well, and the story of how people affect their movement across the savanna.[4]

PEOPLE AND WILDLIFE: HARD EDGES, ENDANGERED FLOWS

Baselines that exclude (or at least reduce)
the impact of humans, as in protected areas,
play a vital role in highlighting the negative
impacts of humans on their own ecosystem.

    ECOLOGIST TONY SINCLAIR *and colleagues,*
    2007[5]

The Serengeti-Mara ecosystem, straddling the Tanzania-Kenya border, is extraordinary for its productivity and size: covering 25,000 km^2 (almost 10,000 mi^2), it seems huge, about a fifth of the size of England and about half the size of the state of New Jersey (Map 14). While it is large among the world's 100,000 protected areas, it is still forty times smaller than the world's largest, in Greenland.[6]

The greater ecosystem is defined as comprising the savannas grazed by the wildlife migration. This includes two central protected areas, Tanzania's Serengeti National Park and Kenya's Maasai Mara National Reserve; their surrounding game reserves, Maswa, Ikorongo, and Grumeti; the Ngorongoro Conservation Area and Loliondo Game Controlled Area (GCA) in Tanzania; and the ranches neighboring the Mara Reserve in Kenya. This complicated landscape has an even more complicated array of rules for how people can use natural resources. No hunting, settlements, farming, logging, fishing, or grazing is allowed in the Serengeti or Mara; limited hunting, logging, and fishing are allowed in the Tanzanian game reserves; Tanzania's Loliondo GCA and the Kenyan pastoral ranches both allow settlement, grazing, and cultivation, and hunting is also allowed in Loliondo. Ngorongoro is a special case for a conservation area, allowing grazing, limited cultivation, and settlement inside most of its borders (see Chapter 11 for a detailed view).[7]

While three-quarters of the Serengeti-Mara ecosystem lies in Tanzania, the Kenyan part has outsized importance because it provides a safety net of food and water for migrating herds during the dry season and drought. The northern part of the ecosystem is part of the Mara River watershed, which extends about 80 km (50 mi) upstream from Kenya's Mara Reserve to the top of the Mau Forest Complex, flows down through the pastoral ranches, the Mara Reserve, across the border into the Serengeti, and then out to the west through Tanzania's Mara wetlands and into Lake Victoria, as one of the lake's ten tributaries. This

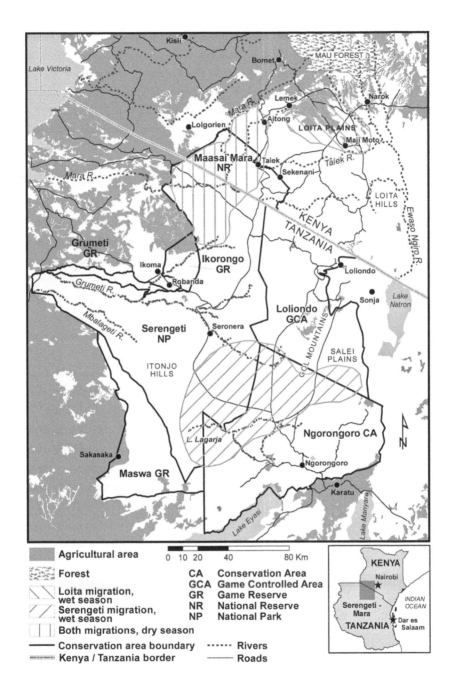

MAP 14.
Serengeti-Mara region in Kenya and Tanzania. (Map by Shem
Kifugo and Russell L. Kruska.)

watershed is immensely important to the people all along its banks: loggers, hunt-ers, and farmers in the Mau forest; farmers below the forest; the Maasai and their livestock; millions of migrating wildlife in the Serengeti; and farmers down-stream, northwest of the Serengeti. The ecosystem's other two main watersheds, the Grumeti and Mbalageti, mostly drain areas inside Serengeti National Park.[8]

Within this large ecosystem, we find a stunning array of animals, large and small. The Serengeti alone is home to 34 large mammal species (> 18.1 kg or 40 lb), more than 600 species of birds (two-thirds as many as in all of North America), at least 20 frog species, 80 grasshopper species, and 100 species of dung beetles. Many, many types of animals (most very small and in the soil) have yet to be found or counted despite the huge amount of research done in the Serengeti. While other savannas in Africa have a similar diversity of large mammals, only here are there so many of them: together, the Serengeti and Mara have about 1.1 million wildebeest, 360,000 Thomson's gazelles, 200,000 zebras, 7,500 hyenas, 2,800 lions, 850 leopards, 500 cheetahs, and about 350,000 other large animals. The ecosystem also has two completely different migrations of wildebeest: the million-strong Serengeti migration that spends most of its time in Tanzania, and the smaller (30,000) Loita migration, which stays mostly in Kenya. (To see the most recent movements of the Loita population, along with wildebeest in Am-boseli and the Athi-Kaputiei Plains, look for our daily updates, from 2010–2012, of their movement tracks at www.nrel.colostate.edu/projects/gnu/index.php.)[9]

This eye-popping abundance and diversity is overshadowed by the spectacle of the great migration. This migration exists because of two opposing gradients, one of rainfall, the other of nutrients. A strong rainfall gradient runs from 500 mm (19.7 in) in the southern short-grass plains of Tanzania's Serengeti to 1340 mm (52.8 in—more than Seattle or London receive) in the rich tall-grass plains and woodlands across the border to the north in Kenya's Mara. As the short grass in the south dries out and surface water evaporates and becomes salty in May and June, migrants move northwest to the Serengeti's western corridor. As the dry season persists from July to November, they continue north to the northern Serengeti and the Mara to find thick grass and the running Mara River. When it rains again, they move back south to the rich soils and nutritious grasses in the short-grass plains from December to April, completing the cycle. Smaller popula-tions of wildebeest and other species (called residents) do not migrate, but the vast majority do move with the seasons.[10]

Migrating wildebeest have some unusual biology. In the Serengeti, 80 percent of wildebeest mate all at once, in about a three-week period, so that their calves are all born in January and February just when food (and water) are most abundant. This mass calving floods the short-grass plains in Tanzania's Serengeti and Kenya's Loita Plains with newborns and overwhelms the ability of predators to capture all of them. If a wildebeest calf is born within a large herd such as this, it has an 80 percent chance of survival; if born outside a large herd, it has only a 50 percent chance. Once born, a calf can walk and run in an average of seven minutes, and in as little as three. The wildebeest body shape (large shoulders, smaller rump) allows the animals to run with a rocking motion, similar to the American bison, making for an efficient way to move and forage over large areas. Maasai herders know that wildebeest move as soon as it rains and often use their movements as an indicator of where the grass is green for their cattle.[11]

More than a million people live to the northwest of the ecosystem, between the Serengeti-Mara and Lake Victoria. In 1995, the people surrounding the Serengeti lived in the poorest country on Earth (Tanzania), and people in the western Serengeti were even poorer than the national average. Since 2000, Tanzania's economy has been heralded as a major success story because it has grown at about 6–7 percent per year, but this growth has primarily benefited those in urban and not rural areas of the country.[12]

This concentration of people to the west of the Serengeti are mostly farmers who create hard boundaries to the west and northwest of the ecosystem, while to the north, east, and southeast in lightly populated Maasailand the boundaries are softer. Hard boundaries occur where people modify the open nature of savanna landscapes by erecting fences, clearing farms, constructing towns, and the like. You can see this quite easily from space: the shapely outline of the ecosystem is obvious on the west where the farmed land meets the parks and very indistinct on the other side, where Maasai graze their livestock. Southwest of the park, farmers are clearing the forest, making the boundary of Maswa Game Reserve also clearly visible from space.

But even Maasai are creating some harder boundaries in small trading centers on the northern edge of the Mara Reserve in Kenya. These Mara villages are an example of a "honey pot effect," where people settle at the edge of parks to reap the benefits of tourism jobs or wildlife payments. This effect is seen clearly in the Mara region, where the density of pastoral settlements grew, on average, an astounding 9 percent per year within 10 km (6.2 mi) of the park between 1967 and 2003—more than double the national average. Growth of populations far from the park, meanwhile, was slower than the national average, at 2.8 percent. As people concentrate along the park border, more and more herders herd their livestock illegally at night in the Mara Reserve (this occurs during drought as well).[13]

Farming usually damages wildlife populations and their habitat much more than does the creation of settlements; indeed, sparse settlements can even attract wildlife (see Chapter 6). In contrast to the areas to the west and northwest of the Serengeti-Mara, where more people means more farms, more hard boundaries blocking wildlife movement, and less wildlife habitat, in the areas to the north, east, and southeast more people often just means more settlements (though it sometimes means more farms). Once human populations hit a certain moderate level, however, they cross a "tipping point" where more and more people and settlements means fewer and fewer wildlife (see Figure 21, Chapter 6). Between 1977 and 2009, for example, a period of significant population growth and settlement, the pastoral area north of the Mara Reserve lost about 82 percent of its resident wildlife (like topis, impalas, warthogs). It is not clear if the settlements themselves are the cause; I think larger villages bring more opportunities for poachers to take wildlife. In addition, the effects of more people and more pastoral settlements (and more poaching by non-Maasai) extended into the park, with almost as much wildlife lost (74 percent) inside the Mara Reserve as outside on the pastoral ranches.[14]

Expansion of wheat farming particularly hurt Kenya's Loita wildebeest migration. This migration favors the nutritious grass of the Loita Plains, about 40 km (24.9 mi) north of the Mara Reserve, in the wet season for calving. In the 1980s, Maasai leaders started to lease this land to commercial wheat farmers, a move that ecologist Wilber Ottichilo and colleagues suggest was intended to control transmission of the deadly malignant catarrhal fever (MCF) from wildebeest to Maasai cattle. Over the next fifteen years, the 130,000-strong Loita wildebeest migration lost 100,000 animals, partly because of fences blocking their path to their favorite calving places. Although the farmers did not plow all of the Loita, they did plow the productive northern area, where the best wildebeest calving grounds and the best farmland coincide. Because the vast wheat fields, which Maasai vacated when they leased the land, were "empty," it was easy for poachers to kill and remove wildebeest carcasses undetected. In the past, Maasai in the Loita, like Maasai living with rhinos in Ngorongoro Crater (see Chapter 11), may have inadvertently protected this migration by their presence.[15]

West of the Serengeti-Mara, some people hunt for subsistence, others for the market, and still others for trophies; many hunt legally, many illegally. In Kenya, hunting was prohibited in 1977, so hunters there have to kill animals surreptitiously. Over the course of thirty-five months from 2001 to 2004, the staff of the Mara Conservancy, who manage the western third of Kenya's Maasai Mara National Reserve, arrested 278 poachers and recovered 1200 kg (2,646 lb) of meat from nine wildlife species. By late 2011, the number of poachers arrested in the previous ten years had climbed to 1,700, and more than 14,000 snares had been collected.[16]

Across the border in Tanzania, peoples of the western Serengeti had hunted for millennia in what is now Serengeti National Park before the colonial government restricted hunting by natives in 1921. Today, some hunters kill animals as the wildebeest migration spills out of the park into villages and farmland each year, and as many as 50,000 people hunt illegally inside the Serengeti and surrounding reserves, snaring and transporting carcasses of 40,000 wildebeest and up to 160,000 animals of all species each year. Western Serengeti peoples hunt for food, especially when crops fail during drought, or cash, receiving about a third of their income from bushmeat. When the migration passes near village land, families that normally consume one or two meals of meat a week now consume meat almost every day as prices for bushmeat fall by half. This practice takes its toll on wildlife: the Serengeti's neighboring Grumeti and Ikorongo game reserves and Ikoma Open Area, with many bordering villages, for example, support 70 percent fewer impalas per unit area than the Serengeti itself. Ecologist Simon Mduma and his colleagues estimate that an offtake of 40,000 animals is sustainable, but if doubled to 80,000, the wildebeest population would decline to unsustainable levels over the course of twenty-five years.[17]

Between the 1970s and 2005, the dry-season flow of the Mara River dropped by 68 percent. Part of the reason is that in about 2001 commercial farmers north of Kenya's Mara Reserve began pumping some three-quarters of the river's dry-season water flow for irrigation of their high-value horticultural crops. At the top of the watershed, meanwhile, about 90 km (56 mi) upstream from the Mara Reserve, people have been clearing the Mau Forest for charcoal, farms, tea estates, schools, churches, and roads. And trees are falling fast: between 1973 and 2000, people cut about 34 percent of the forest, and another 1 percent has been felled every year since then. Like other rivers in deforested watersheds worldwide, the Mara now alternately floods and dries out faster than it did in the past. The deforestation of the Mau is a long-term political issue in Kenya because of contested land rights and political corruption, and there is strong pressure to restore the forest. Farming is also blocking access of wildlife to drinking water in Lake Victoria and the Mara wetlands along the lower Mara River.[18]

Infectious diseases have far-reaching effects on wildlife as well as on how people use the land in this ecosystem. Domestic dogs from the western Serengeti were the source of rabies that killed all the wild dogs in the region by 1992 and of canine distemper, which killed 30 percent of the lions in 1994. Malignant catarrhal fever, transmitted by wildebeest calves, is deadly to cattle. This means that herders must give up prime grazing land for three months after the wildebeest calve and graze their animals on poorer pastures in the highlands and woodlands where other diseases, like East Coast fever and trypanosomosis, can sicken and kill cattle (see Chapter 11). And historically, rinderpest and the measures taken to control it entirely reshaped this ecosystem (see below).[19]

Wildlife predators and large dangerous animals like elephants and buffaloes create strong conflicts when they kill or injure people or livestock. Dickson Kaelo, a pastoral ecologist, mapped when and where people encounter elephants. He found that lone drunk men walking at night often suffer fatal attacks by elephants in the pastoral ranches north of Kenya's Mara Reserve. In 2003 in villages to the northwest of Tanzania's Serengeti, farmers lost an average of 4.5 percent of their livestock to hyenas, leopards, baboons, lions, and jackals, equivalent to a fifth of their cash income. Hyenas killed 98 percent of the livestock lost. Whereas the cats only killed livestock in homesteads close to the park, hyenas killed livestock more than 30 km (18.6 mi) from the park's edge. In the Mara, Maasai have a saying, "You are as hungry as a hyena in the wet season," and biologists Joseph Kolowski and Kay Holekamp found that hyenas do indeed kill more Mara livestock in the wet season, when the Serengeti migration is still to the south in Tanzania. When the migrants return to Kenya in the dry season, fewer Maasai livestock die in the jaws of hyenas, because the hyenas have plenty of wild animals to hunt.[20]

Proximity to Maasai and their livestock sometimes repels predators, though not necessarily. In the wet season of 2003, ecologist Joseph Ogutu counted the lions, hyenas, and jackals that were attracted to a broadcast recording of the sounds of hyenas and lions squabbling on an animal kill and the bleats of a dying wildebeest calf. To his surprise, no lions responded to the recording on the pastoral ranchlands, but many did in the nearby Mara Reserve—even though there were more hyenas and jackals (and nearly three times more of their prey) on the ranches than in the reserve. In the previous dry season of 2002, our "Mara Count" team conducted daytime counts of lions and found a number of lions on the pastoral ranches. This suggests that lions sometimes vacate the pastoral ranches for the reserve, or they may simply not respond to recordings on the ranches to avoid being detected by the Maasai. Holekamp and her team found that Mara hyena groups that regularly encounter herders and their livestock hide from herders in the daytime, spend twice as much time being vigilant for threats, travel longer distances from their dens, and occur in smaller groups in the open grasslands than hyenas isolated from people. By contrast, in Loliondo, east of the Serengeti, wild dogs and cheetahs appear to survive better on Maasai lands than in the nearby park because Maasai actively chase away competing lions, as they do in Kenya's Mara.[21]

Unfortunately, predators and scavengers are under a growing threat of death by poison. Maasai reported that they occasionally poisoned hyena clans in Kenya's Mara in the late 1980s. Several lions died in the Mara in the late 2000s from eating carcasses laced with a cheap and readily available insecticide, carbofuran (or Furadan). In 2007, poison was involved in the death of twenty-five wild dogs in Loliondo, east of the Serengeti. Poison is also one likely cause of the 60 percent decline in the Mara vulture populations over the last twenty-five years. Maasai targeted and successfully killed hyenas with poison in Tarangire, Tanzania.[22]

As described in Chapter 6, in the wet season many wildlife species prefer to graze on the pastoral lands rather than in the Mara Reserve. Anthropologist Tom Maddox found the same to be true when comparing the pastoral areas of Tanzania's Loliondo with the Serengeti and Ngorongoro, with the fewest wild animals occurring in the people-free Serengeti in both wet and dry seasons (here, access to water may be a key factor). In the pastoral areas, the grass grazed by livestock is likely more nutritious, and the short grazed grass probably makes predators easy to spot. (Grass in the Mara Reserve can grow up above your waist when not grazed—ideal lion-hiding habitat.) Near settlements, herders chase predators at night, which may explain the nightly movement of some wild grazers toward settlements in the evening.[23]

When looking across the whole of the Mara, it seems that there is an ideal, moderate number of people of about 7 people per km^2 or 1 *boma* per 4 km^2 (18 people per mi^2 or 1 *boma* per 1.8 mi^2) that maximizes the number and diversity of small to medium-sized grazers, such as gazelles and wildebeest, which need highly nutritious forage and are also favored prey of large predators (see Figure 21). If people are fewer than this ideal, there are fewer wildlife, repelled by the tall grass. If there are too many people and settlements, dogs and people harass the wildlife; in addition, livestock (often sheep) graze grass too short to support many wildlife. Some species, like lions, elephants, and rhinos, however, prefer the long grass of the reserve, away from pastoralists. This pattern of attraction and repulsion persists today, despite the declines in the total number of wildlife in the Mara over the last thirty-five years.[24]

Today, the Serengeti-Mara ecosystem is governed by a variety of models of conservation, principally differing by who has the power to decide who uses the land and how, who bears the cost, and who garners the benefits (profits) from wildlife. In these models, most of the power is in the hands of local (Mara Reserve) or national governments (Serengeti National Park and surrounding reserves), and the bulk of profits like gate fees or tourism income flow either to these governments or to private-sector tourism businesses (see Figure 23, p. 259). But governments are sharing some of these benefits with adjacent pastoral communities and landowners. In the 1980s, Tanzania's Serengeti and Kenya's Mara started to distribute some 10–20 percent of their annual budgets to surrounding communities. The Transmara County Council in Kenya currently contracts management of the western third of the Mara Reserve to the Mara Conservancy, and large portions of the profits flow back to the county council. In the case of the Narok County Council, which manages the eastern part of the reserve, most of the profits in the past flowed into the pockets of wealthy Maasai, with the majority of poorer Maasai seeing very little. For the last two decades, tourism operators in both countries and hunting operators in Tanzania have created conservation incentive programs and shared some of their profits with community members in Maasailand outside the protected areas in Kenya and Tanzania.[25]

The flow of profits to local people is now more equitable as Maasai landowners join in partnerships with tourism businesses all over the Mara. As Maasai in the Mara have shifted land ownership from group ranches to private holdings, the prospect that the land would be fragmented into fenced parcels, creating more hard boundaries, was of great concern to all who profited from wildlife and support conservation. Instead, privatization has created an unexpected window of opportunity, providing new ways to keep the land open and improve the flow of profits to herders in the form of wildlife conservancies—agreements with private tourism companies to grant access to Maasai-owned land in return for a share in the profits. In most cases, Maasai landowners agree to move off their land to create large, fence-free landscapes mainly for wildlife use and tourism (and some livestock grazing in the dry season). In return, each month private tour operators deposit part of their profits into the landowners' bank accounts.[26]

Whether these new conservancies will persist into the future is unknown; they can only be maintained if the benefits outweigh the costs, there is strong and honest leadership, and government regulations support them. As Maasai move off their land to make way for conservancies, unexpected things may happen. In moving,

Maasai have to crowd into other parts of the landscape, with the prospect of creating harder boundaries in the destination landscapes. And conservancies may encourage herders to take their animals into the Mara Reserve at night to maintain their agreements not to graze in the conservancies, as suggested by Bilal Butt, a human geographer working in the Mara.[27]

Are the Mara conservancies protecting wildlife? We aren't sure yet, but local observations give us an indication of their viability. My Maasai colleagues describe the changes they have seen in the Mara landscape since establishment of the conservancies as follows: As soon as the Maasai families abandoned their homes and left the new conservancies with their livestock, the lions recolonized the area, along with a flood of wildlife. The wildlife clustered around the wetland areas and in places where the Maasai used to live, where grass growth is rich thanks to the accumulation of dung in old livestock corrals. Some time later, as the grass grew taller, much of the wildlife moved out of the area, following the people and their livestock to the north. In the dry season, however, when the grass is grazed low near pastoral settlements, the wildlife return to the wildlife conservancies to feed on the more abundant grass there. The tourism companies are now mimicking what the Maasai used to create around their settlements: patches of grass of different heights, which attract wildlife of all sizes and encourage them to stay in the conservancies. This is a classic demonstration of the value that Maasai land management strategies can bring to wildlife.[28]

Similar changes are taking place in Tanzania, not on private land but on village-owned land. Wildlife Management Areas (WMAs) were designed by the Tanzanian government in 1998 to allow rural communities and private landholders to benefit from the local wildlife. Maasai and western Serengeti peoples, however, tend to be suspicious of the government in such ventures, partly because of their long history of losing land to the central government, even though this is not the governmental intention with WMAs. Nonetheless, some villages have circumvented this problem by making independent agreements with private tourism companies, not unlike their brethren over the border in Kenya. In Loliondo Game Controlled Area, directly east of the Serengeti, for example, Maasai and private tourism companies developed a community-based wildlife management plan that has proved highly profitable for the villages involved. However, this partnership is illegal under the WMA statutes, so its continued existence is uncertain.[29]

HOW DID WE GET HERE? IT ALL
DEPENDS ON HOW FAR BACK YOU GO

While the Greater Serengeti Region
includes some of the world's most important
wildlife conservation areas, it is also home
to tens of thousands of indigenous people,
whose livelihoods have been largely ignored
by conservation efforts. As a result, local
communities have developed deep-rooted
feelings of antagonism towards
conservation.

> MORINGE PARKIPUNY, *former Tanzanian*
> *minister of Parliament, commenting on the*
> *establishment of Serengeti National Park, 1997*[30]

The influence of humans and their ancestors on this landscape is ancient. Just southeast of the Serengeti is Laetoli in the Ngorongoro Conservation Area, the place with the oldest evidence that hominins walked on two legs almost four million years ago. As early as three thousand years ago, people subsisted on livestock here, as shown by the dominance of livestock bones found by archeologist Fiona Marshall in old settlement sites in Kenya's Mara ecosystem. For much of the last two thousand years, the Asi hunted and gathered food from the Serengeti plains, the Tatog grazed their livestock from the western corridor of the Serengeti to Ngorongoro, and some Bantu people cultivated crops near Seronera in the central Serengeti. But grazing and farming were not extensive because there is no permanent water in the central and south Serengeti (only north along the Mara River) and the heavy clay ("black cotton") and alkaline soils made cultivation difficult with digging sticks, the farming technology at the time. In the mid-1800s, the Maasai took over the plains and made the Asi and Ndorobo hunter-gatherers their dependents. The central Serengeti plains now served as a relatively empty buffer between the western Serengeti peoples and the Maasai to the east. During colonial times, the Maasai used the Serengeti for dry-season grazing as far west as Seronera and the Moru Kopjes.[31]

It is possible that humans—hunter-gatherers and herders—actually created this open Serengeti-Mara landscape. This is an important notion because if people created the Serengeti-Mara ecosystem in the first place, it could suggest that the best way to sustain the ecosystem is for people to continue to use and maintain it in a similar, traditional way. Our best evidence suggests that herders, hunters, and farmers did modify the ecosystem by building water points (which elephants use); burning woodlands, thus creating and maintaining the Mara grasslands; creating nutrient hotspots in their old settlement sites; and protecting grazing wildlife from predators. While no evidence yet exists to suggest that this ecosystem was vastly different before people had enough technology (fire, livestock) to manipulate it, we cannot be absolutely sure. What is clear is that, by using it only moderately, they sustained it, intentionally or unintentionally, over millennia, and that is why the ecosystem is as rich as it is today.[32]

By the mid-1800s the Maasai were fully entrenched in the ecosystem, the ivory trade was at its height (with one trade route passing just south of today's Serengeti), and the region was probably full of people, wildebeest, and trees, but not elephants. Then the *Emutai* (meaning "to wipe out" in the Maasai language, Maa) came in 1890 as the Italians unleashed a new disease, rinderpest, on the African continent (see Chapter 5). Geographer Oscar Baumann, walking across the Serengeti in 1892, estimated that 75 percent of the people who once lived in the region had died from rinderpest and smallpox in the preceding two years. Nearly 95 percent of the cattle, buffaloes, and wildebeest died also. From 1890 to 1950, rinderpest returned every decade or so, until a campaign vaccinating cattle in settlements around the Serengeti eradicated the disease in 1964, controlling it in wildlife as well. The result was immediate: first-year survival of wildebeest and buffalo doubled, and their populations grew rapidly. In the next fourteen years, the wildebeest numbers grew from about 200,000 to over a million animals and then leveled off, limited by the amount of green grass in the dry season. This means that between 1890 and the 1960s the great migration that many consider emblematic of the wild Serengeti simply did not exist. Before 1890, it is anyone's guess what the wildebeest population was, but I would guess that the great migration had been around for centuries, if not millennia.[33]

The decline and then resurgence of the wildebeest population (an addition of about 135,000 tons of animals) completely changed the Serengeti-Mara ecosystem. Biologist Tony Sinclair spent much of his career finding out just how. As improbable as it seems, the spread of rinderpest and its subsequent control are responsible, in part, for the number of trees that exist in the Serengeti today. Wildebeest eat only grass, and by the time their population reached a million strong, they had consumed most of the standing grass. Grass is an essential fuel to support fire, so fires became rare. Add to that the fact that in the 1980s ivory-seeking poachers killed 2,000 of the Serengeti's 2,700 elephants, while 300 more of the pachyderms fled into Kenya, leaving the northern Serengeti with few elephants. With little fire and few tree-damaging elephants, trees grew rapidly and spread across the northern Serengeti. Thus Sinclair suggests that the million-strong wildebeest population, which rebounded after the imported livestock disease was controlled, together with ivory poaching, turned the northern Serengeti into the woodland we see today.[34]

Oddly, across the border in Kenya, Maasai and elephants were busy creating the opposite: a landscape full of grass, a story told by ecologists Holly Dublin and Richard Lamprey. As rinderpest came under control, the Maasai, seeing an opportunity to access more grazing land, started moving down from the north. When they got just short of the Mara Reserve, they burned the tsetse fly–infested woodlands to destroy the pest's habitat (and that of trypanosomosis, the cattle disease it transmits) and open up more grassland. Between 1960 and 1975, the cover of woodlands dropped by about 85 percent in the Mara, from extensive patches of forest to a few scattered trees. The Ki-Maa (Maasai language) name Mara, meaning "spotted," refers to the *Croton* woodlands that covered the hillocks, with grass in the valleys, describing the landscape as it appeared before the 1960s. In just six years, between 1961 and 1967, the Maasai burned out a third of these woodlands. Burning became rare again in the 1980s, but now elephants, escaping ivory hunters in the Serengeti, crowded the Mara and grazed young regenerating trees to the ground. Thus Maasai-lit fires, followed by poacher-fleeing elephants, rendered the Mara into the grassland that exists today. In certain places along the international border that separates the Mara and Serengeti, you can clearly see the woodlands to the south in Tanzania and the grasslands to the north in Kenya.[35]

The initial idea behind a new Serengeti national park in the 1940s was to ensure the coexistence of people and wildlife, but it ended a decade later with colonists segregating the landscape: half for wildlife, half for people. British preservationists allied with the Society for Preservation of Fauna of the Empire pushed for a Serengeti without people, while African farmers, hunters, and herders, along with some local colonial officers, resisted this idea, supporting African land rights. Historian Jan Shetler describes how the western Serengeti peoples were left out of the negotiation for the park boundaries because they did not have a unified political voice and were considered poachers. Only the Maasai (herders) and Sukuma (farmers) were able to negotiate boundary changes to their benefit. In 1956, with the approval of the park board of trustees and the Maasai Council, the Legislative Council of Tanganyika issued a controversial white paper outlining a compromise to meet the needs of wildlife and people, allowing Maasai to use all land except the western Serengeti and the Ngorongoro and Empakaai Craters. Many in the global conservation community were strongly against this proposal, arguing that wildlife would not be conserved unless the park was much larger to enclose the entire wildlife migration.[36]

In compensation for the loss of the Serengeti, the nearby Ngorongoro Conservation Area (see Chapter 11) was set aside for joint use by Maasai and wildlife. Most of the Maasai leadership agreed to this trade, once they were promised water development and veterinary care in Ngorongoro, but one of the twelve Maasai traditional leaders who signed the agreement says they were forced to move out of the Serengeti. Moringe Ole Parkipuny, former Tanzanian minister of Parliament from Ngorongoro District, summed up the Maasai and other peoples' long period of resistance to the park: "The eviction of indigenous people from their land is a barbaric act of alienation of those people who have been the vanguards of conservation." Focusing on the Maasai, he further writes that "with the creation of the national park in 1959, the Maasai of Western Serengeti and Loliondo lost vast grazing areas, salt lick grounds and permanent sources of water which were critical to the viability of their pastoral economy. These rangelands, although partially infested with tsetse flies, provided an important livestock refuge in times of drought."[37]

Even though the Mara Reserve was established at about the same time as the Serengeti (1940s to 1960s), it, along with several other Kenyan reserves, was a new model of conservation for east Africa. This model arrangement devolved the profits from wildlife to local Maasai governments, even though the Kenyan national government retained control over the wildlife themselves. In 1956, as Maasai lost access to the Serengeti, the colonial government in Kenya proposed a policy to allow district councils to adopt game reserves as "African District Council Game Reserves," including the district next to the Serengeti in the Mara. From this time until hunting was banned in Kenya in 1977, trophy hunters paid their fees to this local district and not the national government, as was the case in Tanzania. This African district reserve became the Maasai Mara Game Reserve in 1961. After independence in 1963, the Maasai-run council in Narok District took over the management of the Mara Reserve, turning a profit by collecting gate and camping fees and lodge concessions. Thus, today the Mara is not run by the central government hundreds of kilometers away, as Serengeti is, but by the local county (district) council, like more than half of Kenya's protected areas. Local Maasai have much more say in the management of this reserve than Maasai have in management of any protected areas in Tanzania.[38]

This local control makes the Mara Reserve management more sympathetic to Maasai concerns like access to grazing during the dry season and drought. Even though on paper the reserve rules do not allow herders or their livestock into the park, in reality livestock enter the reserve almost every night and often during the day, particularly when grass is exhausted in the pastoral lands outside the reserve. This access to the protected area reduces the stress faced by herders and their cattle during drought. Although it is possible that this practice creates competition for forage with the migrating wildebeest, this has never been measured, but it should be.[39]

THE FUTURE: WHITHER THE SERENGETI-MARA?

Even if, for example, the whole of the Serengeti were to be handed over to the local Masai, this enormous, relatively undisturbed ecocomplex could absorb the growth of the Masai population for only some forty years.

HERBERT PRINS, *biologist, 1992*[40]

It is clear that the current grand diversity of wildlife in the Serengeti-Mara is the legacy of two sets of peoples: (1) those in the past who chose not to extinguish the great migration and (2) others who thought the best way to continue that legacy was to lock much of the ecosystem up in parks and reserves. It is also clear that the Maasai have sacrificed huge swaths of land in the name of wildlife conservation, even though it is not clear that this was necessary. Rather, it was a way for colonial authorities, tourism operators, and then postcolonial governments to take power over both land and the profits from wildlife. The result has been largely to alienate herders and other peoples from the idea of conservation, though they are the very people who determine its success in these democratic societies. That said, now that ever more people and settlements are crowding around this ecosystem, we must recognize that the wildebeest migration might not still exist if the protected areas had not been created.[41]

What does the future look like for this shared pastoralist-wildlife ecosystem? One of the biggest current threats to people, livestock, and wildlife is that the only perennial source of water, the Mara River, might dry up in a severe drought. Even if the river stopped flowing, stagnant pools would remain available for wildlife, though it is conceivable that even these pools could eventually dry out completely in a protracted drought. If that were to happen, ecologist Emmanuel Gereta and his colleagues predict that 30 percent of the migrating wildebeest would die immediately, within two weeks. That is almost 400,000 animals! And if the drought continued, so would the die-off, at 30 percent more each week thereafter. This is an estimate, but even if it is only partly correct, this would be a massive loss. Of course, many other water-dependent wildlife would die also, like hippopotamuses and crocodiles, as would the Maasai cattle, sheep, and goats. The Mara River is also home to many fish species that have disappeared from Lake Victoria, like *Oreochromis* (a cichlid fish related to tilapia) and *Esculenta*.[42]

The competing needs for the Mara River water are complex and difficult to manage. Ultimately, all downstream water users—herders, farmers, wildlife— are entirely dependent on what happens upstream when the Mau Forest is cut, since the forest cover regulates the constancy of the river flow, or water is diverted for hydroelectric generation or irrigation. Deforestation is a national issue, and its resolution depends on a highly contentious and long-delayed process with a wide range of stakeholders. Some of this process may lead to a trade-off between the land rights of indigenous forest dwellers in the Mau and downstream users' rights to water. Significant progress has been made in monitoring the forest and making recommendations on how to remedy the situation.[43]

There is also a need for a Mara River Transboundary Management Plan, in which the governments of Kenya and Tanzania and local communities plan for the water needs of a wide range of powerful and less powerful stakeholders. In addition to regulation and coordinated planning, stakeholders in this watershed could establish a water services payment program where downstream users (tourism, irrigated farming) pay upstream Mau foresters to reforest catchment areas in the Mau Forest in order to restore more reliable flows. Another approach is to open up better access to water downstream. Allowing wildlife access to the small piece of land that separates the western tip of the Serengeti from the Speke Gulf in Lake Victoria, for example, would help Serengeti's wildlife, but local communities must agree and be compensated fairly.[44]

Climate change will only make any problems with the river more difficult. Joseph Ogutu's data from the Mara suggest that droughts are becoming more frequent. As described in Chapter 12, more droughts mean less nutritious forage and more evaporation of surface water. If these trends continue, the ecosystem may not be able to support as many livestock, people, or wildebeest in the future. But none of this is certain; ecologist Mark Ritchie, for example, predicts cooling in the Serengeti woodlands for the future.[45]

Another lingering threat to this system is the construction of a highway bisecting the Serengeti, under consideration by the Tanzanian government for the last twenty years. In mid-2011, the government announced it would not build an asphalt road in the park, but it may build one right up to the park boundary, with a gravel road crossing the park. Although a new road to serve the million people who live west of the Serengeti would bring much-needed economic development, it could also cause as much as a 40 percent loss in the migrating wildebeest population. This is not an idle fear: our GPS-collared wildebeest did not cross a major road in the Kaputiei Plains for eighteen months in 2010–2012 (see Chapter 10). In any case, an alternative road accessing the lands west of the Serengeti already exists to the south.[46]

In Kenya, if human population growth and expansion of farming continue at the same pace for another decade, Kenyan Maasai will find themselves well past the "wildlife tipping point," living in hard-boundary landscapes with little wildlife, right up against the border of the park. As described above, Maasai are working with the private sector to keep this from happening. Only time will tell if their conservancies will prove successful, from both a wildlife and a human perspective. If they do, new opportunities may arise for additional cash payments to those Maasai families who manage the land to maintain ecosystem services, such as carbon sequestration or clean water or increased biodiversity. Adding these new payments together with the profits from both wildlife-oriented tourism and pastoralism, landowners would find that pastoralism with wildlife is more lucrative than farming and other wildlife-incompatible uses of the land.[47]

More monitoring of the wildlife and of Maasai welfare, particularly of those groups who were excluded from the new Mara conservancy schemes, is a critical need. Now that the Mara is largely split up into hundreds of smaller, individually owned land parcels, landowners must coordinate with one another if land is to stay open for livestock and wildlife grazing. Land in view of the conservancies and the Mara Reserve is of very high value to locals and foreigners alike, and the incentive to sell for high prices is strong. Some good news is that the new Kenyan constitution (approved in August 2010) devolves decision making to the local level, and requires local authorities to carefully manage changes in land use. Questions remain for the future of the Kenyan part of this ecosystem, of course. How can the conservancy model be improved so that it meets the goals of equitable sharing of wildlife profits while also conserving wildlife? Does this model mean that Maasai will have to reduce the size of their herds to make space for wildlife? Will any Maasai still own land around the Mara Reserve twenty years from now? Or will the land be slowly sold off to rich non-Maasai, as is happening today?

For the Tanzanian part of the ecosystem, the questions are more of a policy nature. Will the Maasai and the western Serengeti peoples regain some voice in how the parks, reserves, and conservation areas are managed, or will their remaining village lands become mere extensions of these protected areas? Would the Serengeti be damaged if a co-management model were adopted, where both local and national representatives have authority and responsibility over this large landscape? If poaching by western Serengeti hunters is in fact driven by poverty, would economic development, more livestock production, and alternative employment slow the killing of wildlife? And finally, is there a real prospect of new profit streams from carbon sequestration and other ecosystem services for families who continue to manage the land in wildlife-compatible ways?[48]

CHAPTER NINE

Amboseli

"Cattle Create Trees, Elephants Create
Grassland" in the Shadow
of Kilimanjaro

Mt. Kilimanjaro is the tallest free-standing mountain on Earth, a massive water tower
60×90 km (37×56 mi) at the base. Few people walk all the way from its frozen icecap
in northern Tanzania to the hot and dusty plain of Amboseli in southern Kenya, a verti-
cal distance of almost 5 km (3 mi), and 35 km (22 mi) as the crow flies. But let us imag-
ine such a walk by three people with contrasting views of this region: a Chagga farmer,
an Ilkisonko Maasai herder, and a park ranger. The three start in Tanzania at the
mountain's peak, Kibo, wrapped in their parkas against the freezing temperatures and
gasping for breath at 5895 m (19,341 ft) in altitude. They are about to walk for two very
long days down the northern slopes of the highest mountain in Africa, across the border
into Kenya, through foothill farms, and out onto the savanna, dotted with swamps, that
drapes the mountain's northern side like an apron.

Standing on the 50 meter–deep (164 ft) icefield, the Kilimanjaro National Park
ranger describes the visible loss, in her own lifetime, of the icefield, which has recently
been melting down by as much as a full meter (3 ft) each year. Yet even though the ice-
cap is melting, the savanna below is seeing less water flowing into it because the forest on
the mountain's flanks is being cut down. As the group picks its way around boulders and
across a heath, she laments that the upper edge of the mountain's treeline is also receding
downslope, by as much as 10–15 m (33–49 ft) each year. She thinks trees are receding
downslope because the drying climate desiccates the fuel lit by honey gatherers, poach-
ers, and timber cutters (pitsawyers), causing their fires to burn out of control and destroy
the upper edge of the forest within the park.[1]

A day later, now below Kilimanjaro National Park, the group reaches forest planta-tions of fast-growing cypress and pine, which occupy lands where rich, indigenous mon-tane forest once stood. They are almost off the mountain when they cross from Tanzania into Kenya, stopping for chai (milky tea) in the small town of Oloitokitok. As they reach the mountain's footslopes and walk onto the Chagga's farm, he tells how his mother's people have lived on the other side of the mountain in Tanzania for as long as anyone can remember. She, however, came to this northern side only thirty years ago, to marry his father, a wealthy Maasai. She cleared this land to farm, and at first they had to chase buffaloes and giraffes out of their maize fields, but now they see more elephants and baboons. The Ilkisonko herder, who still drives cattle on the dry plain below, remembers seeing the farms spread across the edge of the forest. At first he thought noth-ing of it. But after the government built many boreholes and he and his neighbors filled the plain with cattle, there was less and less room for livestock. He now herds while one of his wives, a Kikuyu, grows onions and tomatoes in Namalok Swamp, and another, a Maasai, tends the family's goats in Emeshanani, the dry rocky ridge north of the swamps.

The trio catches a short ride with some park rangers, driving northwest along the edge of the dusty plain and stopping at the headquarters of Amboseli National Park to meet the warden. The group leans forward and listens as the warden describes how the Am-boseli yellow fever acacia trees died, salt-loving plants spread, and elephant numbers grew. And he also tells of a mysterious flooding of the swamps. He says that the farmers create most of the conflicts with elephants because they grow maize in the elephants' favorite habitats, which the elephants used long before the farmers arrived. The herders only occasionally fight back at elephants, but will spear them in retaliation if they kill people or livestock. Over some nyama choma *(roasted meat), the trio agrees on the need to discuss collaboration, because they see that they are part of the same larger mountain system, connected from top to bottom by flowing water.*[2]

WATER, THE HEART
OF AMBOSELI TODAY

Without Kilimanjaro, this region (Map 15) would look like thousands of dry sa-vannas all across Africa. Because of the mountain, the montane forest on its southern slope receives nine times more rainfall (about 3000 mm [118 in] per year) than the savanna at the foot of the northern slope, in the mountain's rainshadow (350 mm [14 in] per year). At the top of the mountain, where the average tempera-ture is −7°C (19°F), mild frostbite is possible any day of the year; in Amboseli, or the "dry, parched land" on the mountain's north side, heatstroke is possible in the middle of most days, which average 37°C (99°F) year round.[3]

Because of the mountain, the dusty savanna on its northern side is punctuated by six major springs and swamps (and many smaller ones), which are filled year round with open water and lush green grass. Two of these swamps, Enkong'u Narok and Lonkinye, are inside Amboseli National Park, while the other four—Namelok, Kimana, Lenkati, and Esoitpus—are to the east of the park in the pastoral group ranches and on private land. Nolturesh spring feeds the upper reaches of the Nolturesh River. About 10 percent of the greater Amboseli ecosystem (which overall covers 5726 km², or 2,210 mi²) forms a small basin 600 km² (232 mi²) in size in and around tiny Amboseli National Park (388 km², or 150 mi²), established in 1974. Part of this basin, which has no water outlet, briefly becomes shallow Lake Amboseli after heavy rains. Hills dot the basin and the rest of the ecosystem, with the largest range of hills, the Chyulu Hills, to the east in Chyulu National Park. Much of the larger savanna ecosystem is within Kajiado District, and stretches across at least eight Maasai group ranches—Mailua, Olgulului/Ololarashi, Kimana, Eselenkei, Imbirikani, Kuku, Rombo, and Kaputiei South—which are currently on a trajectory from communally managed to privately owned land. (See Chapter 10 for a brief history of group ranches in Kenya.) Amboseli elephants and baboons, at least, do not respect the international border, but use the slopes of Kilimanjaro and Longido, just west of the mountain in Tanzania.[4]

Understanding how and where water flows is crucial to understanding the greater Amboseli ecosystem. The swamps, created by the mountain, are the lifeblood of this ecosystem, because there are no perennial rivers here, only a few springs like the Nolturesh. On the south side of Kilimanjaro, almost all the streams originate within or below the forest, with very few starting above the forest. This means that almost all the water flowing from the mountain to the surrounding lowlands, on all sides, originates in or below the montane forests. These wet forests certainly exist here because of higher rainfall, but the trees also probably help sustain the mountain's clouds through transpiration (evaporation of water from roots to leaves to the atmosphere). Trees also add to the mountain's rainfall by "harvesting" 10 percent more water in the form of fog from the clouds, which condenses on tree branches and leaves. This water then flows through underground lava tunnels or fracture conduits within the mountain and appears again around the mountain's base in springs to create the swamps and the Nolturesh River. Once here, water flows many places, some close, some far. For example, 60

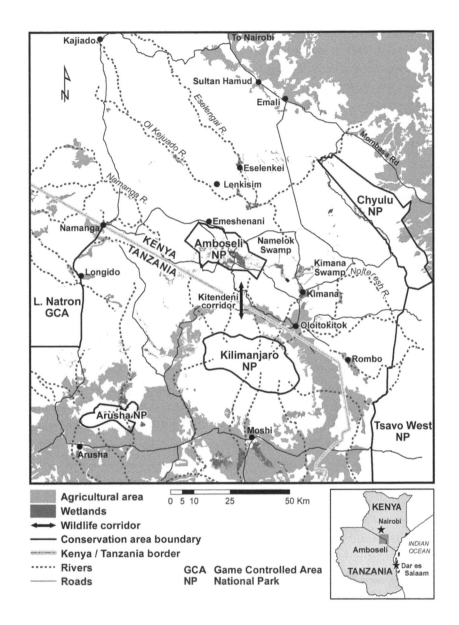

MAP 15.

Amboseli-Longido-Kilimanjaro region in Kenya and Tanzania.
(Map adapted from DMA 1992 and WPDA 2009 by Shem
Kifugo and Russell L. Kruska.)

percent of the Nolturesh spring, connected by a long-distance pipeline, is used by farmers and other people just outside Nairobi, more than 100 km (62 mi) away. And some of this water helps to grow flowers that are flown daily from Nairobi to markets in Europe.[5]

The mountain, swamps, plains, and their soils support an unusual diversity of habitats, over short distances, for people, plants, and wildlife in this dry Amboseli savanna. Ecologist David Western, whose life's work has been to understand this ecosystem and its peoples, places the origins of this diversity in a mixture of nutrient-rich volcanic soils, derived from lava that flowed north across the plains as Kilimanjaro formed about a million years ago, plus impermeable lake soils and nutrient-poor soils derived from the ancient African shield. Four habitats figure prominently in this story: the swamplands, with their luxuriant *Phoenix* palms and papyrus; the salt-tolerant *Suaeda monoica* shrublands around the dried lake beds; the *Acacia tortilis* woodlands on the mountain side of Amboseli National Park; and the patchy woodlands of yellow-barked fever trees (*Acacia xanthophloea*) that border some of the swamps. Beyond this core area, the wider savanna and *Commiphora* woodlands spread out to north, east, and west in Kenya and to the southwest in Longido, Tanzania.[6]

This diversity, with swamp, woodland, and savanna in close proximity, attracts an abundance and diversity of wildlife: wildebeest, zebras, gazelles, elands, elephants, buffaloes, giraffes, ostriches, oryxes, impalas, and hartebeest. More rare are waterbucks, lesser kudus, and bushbucks. The variety of vegetation and landscapes also attracts herders from the Ilkisonko section of the Maasai tribe (and the Matapato and Kaputiei Maasai at the edges of the system), who graze their cattle, sheep, and goats on various parts of the savanna at different times of year.[7]

Amboseli is most known for its elephant population, and rightfully so. This is the longest continuously studied group of elephants in the world, with over a thousand identified individuals living in sixty-four family groups and known family trees stretching back sixty to eighty years. Understanding and conserving these elephants has been the life work of author and biologist Cynthia Moss, a project now run on the ground by Maasai biologists Soila and Katito Sayialel and Nora Njiraini, with help from their colleagues Joyce Poole, Keith Lindsay, Phyllis Lee, Harvey Croze, and others. Thanks to the Maasai and Moss, the Amboseli elephants did not suffer the heavy poaching of the 1980s that decimated populations elsewhere.[8]

The swamps create unusual opportunities in this savanna, serving as key resources that sustain plant and animal life through the dry seasons and frequent droughts, allowing a rich assembly of people and wildlife to survive, and even thrive, here. Despite the swamp's life-giving role, Amboseli elephants, wildebeest, zebras, and herders all prefer to leave the swamps in the wet season, when water is abundant elsewhere, and disperse to pastures in the northern bushlands. They leave the swamps because the bushland grasses are almost three times more nutritious than swamp grass in the wet season. And once in this bushland, wildlife and livestock can also feed on the nutritious grasses growing in abandoned Maasai settlements.[9]

As surface water dries up in the bushlands at the beginning of the dry season, nearby herder and wildlife migrants return to take advantage of the permanent water in the swamps and springs. Elephants and buffaloes often return first and graze down the tough tall grass in the swamps, followed by smaller species that graze on the short and more tender grasses in the lawn that these mega-herbivores create (an example of a grazing succession; see Chapter 2). Even in this season, the swamp grass is only half as nutritious as forage in the nearby acacia woodlands south of the park, so wildlife and herders also graze away from the swamps. Elephants and some other wildlife nightly migrate out of the park to graze, returning to the swamps again each morning.[10]

When the dry season is in full swing, and particularly during severe dry spells, herders and wildlife find themselves on a collision course at the swamps. In 1977 Maasai agreed to leave the core area of Amboseli National Park free of livestock in exchange for construction by wildlife authorities of a water pipeline and water storage tanks north of the park. The park authorities, being often short of funds, do not maintain the pipeline well, forcing Maasai to use swamps in the park, particularly during drought. Serious droughts, where rainfall is half of the average, occurred three times between 1976 and 2000. During these times, herders, livestock, and elephants come into closer contact when trying to access scarce water and forage, which often results in more conflicts and deaths than normal. During the drought of 1973–1976, according to geographer David Campbell, about a third of Maasai herders reported hunting wildlife for food after their livestock died. And in the droughts ending in 1976 and 1984, more elephants died from Maasai spearing than in wetter years.[11]

In Maasai tradition, elephants are thought to have the same origin as humans and are treated with respect. According to biologist Kadzo Kangwana, "On passing an elephant skull, the Maasai put grass into its orifices, as they would do to a human skull, in reverence of the elephant's human origin." More recently, however, Maasai attitudes toward elephants have become more negative because they perceive that elephant populations have grown and that they kill more people and cows. Settled and farming Maasai generally see all wildlife in a more negative light than herding Maasai. Although elephants tend to avoid Maasai, coming close to their settlements only at night, giving them a wide berth at water points, and running away when they hear Maasai cow bells, if they do come in close contact, elephants can kill. As Kangwana observed, "Elephants, when they get caught in herds of livestock, toss animals in the air with their tusks." In retaliation, Maasai spear elephants that kill people or their livestock, but this has gone way down (though Maasai still kill elephants for other reasons, like shows of bravery) thanks to the consolation program started by Moss and her team. They also spear other species of wildlife to prove their bravery or as a form of political protest over their loss of land. When Maasai warriors (*moran*) come of age during a drought year, the elephant spearings can be particularly acute, as Cynthia Moss and Joyce Poole observed during the drought of 1984. This may in turn initiate a vicious killing cycle, since elephants, like people, become more aggressive in adulthood if they experience a trauma, such as the death of a family member, at a young age.[12]

For several hundred years, Maasai herders were the main people using the land in the greater Amboseli ecosystem and the lower slopes of Kilimanjaro; before that, they presumably pushed out hunter-gatherers and earlier herders. Now the Maasai are outnumbered by new arrivals who, along with the Maasai, farm the best land. These farmers, most of them from the Kenyan highlands to the north where land is scarce, now grow crops in the foothills of Kilimanjaro near Oloito-kitok and are contemplating spreading to the slopes of the Chyulu Hills. It is easy to see why many of the swamps outside Amboseli National Park are now farmed (often by Maasai and non-Maasai in partnership). Profits from irrigated farming (about $550 on average per ha per year, or $1,358 per acre) are ten times those received from livestock production ($50 per ha per year or $124 per acre) and twice those of rainfed lowland farming (maximum of $260 per ha per year or $642 per acre). However, swamp cultivation is risky. Irrigation water is not always available, input costs are high, and crop prices vary. Where rainfall is low outside the swamps, crop cultivation can fail three years out of every four, creating losses or, at best, no returns for farmers. Under these conditions, livestock production, if practiced over many hectares, can be a more viable way to turn sunlight into food.[13]

MELTING GLACIERS, FLOODING
SWAMPS, AND DISAPPEARING TREES

[From about 1851 to 1871,] during the period
when the Nyanjusi I age-set were warriors,
the woodlands were almost non-existent
and the swamps were far more extensive
than at present.

MAASAI ORAL HISTORY, *recorded by Western*
and van Praet, 1973[14]

Our picture of Amboseli is incomplete without an understanding of major changes
this savanna and its peoples have undergone since the late 1800s, including melting ice, flooding swamps, and disappearing trees. Perhaps more important is the
expansion of protected areas and farms in the region, along with the shift from
communal to private ownership of land. These changes, starting at different times
in the past but acting in concert today, profoundly affect the way this system
works.

We start with the melting of Kilimanjaro's icecap, which may be an unprecedented event in the last 11,000 years. The melting began about 130 years ago
(about 1880), as it did on glaciers on Mt. Kenya and the Ruwenzori Mountains.
Since then, most of Kilimanjaro's icecap has disappeared through sublimation, a
process in which ice turns directly into water vapor. Sublimation means that very
little of the icecap ends up recharging the groundwater. In 1880, the icecap was
probably 20 km^2 (7.8 mi^2) in area; by the year 2000 it had shrunk to 2.6 km^2 (1
mi^2). As our hypothetical ranger pointed out, the icecap is now melting and thin-
ning, receding up the mountain as much as a full meter each year. Earth scientist
Lonnie Thompson and his colleagues think that if current trends continue, the
"permanent" icecap will disappear entirely sometime between 2015 and 2020.[15]

Why is the icecap melting? This is under debate. It is unclear whether this
melting is part of glacier melting worldwide, which is partly caused by global
warming, or whether it is a unique local event, related to drying at the mountain's
peak. At the base of the mountain in Amboseli, during her life's work on Amboseli baboons, biologist Jeanne Altmann and her team measured rainfall and
temperature from 1976 to 2000 and found a remarkable 6° C (11° F) rise in the
maximum daytime temperature over this period, but no change in rainfall. At
the top of the mountain, however, the story appears to be different: here there has

been no warming since the 1970s, but it has become drier. While the cause of the melting is not yet clear, the fact that the icecap is disappearing is indisputable. It is not clear if this melting is permanent or part of a regular cycle.[16]

While Kilimanjaro's icecap has been melting, the band of forests at its midsection has been contracting from both below and above. On the lower slopes of the mountain on the Tanzanian side, as our fictional trio of walkers saw, forest plantations of pine and cypress now grow where native juniper forest once stood. Downslope from here, and across the border in Kenya, farmers started clearing trees at the lower boundary of the forest in the 1930s for agriculture. This farming did not expand rapidly until after Kenya's independence in 1963, when farmers from elsewhere in Kenya became free to move and establish farms here.[17]

The upper treeline, now situated at about 3000 m (about 9,900 ft), is shrinking downslope today. Just thirty years ago, Kilimanjaro's treeline was some 400 m (1,312 ft) higher and consisted of an extensive subalpine *Erica* forest. More than 80 percent (150 km², or 58 mi²) of this particular forest—or 15 percent of all the mountain's forests—was lost to fire between 1976 and 2000, as ecologist Andreas Hemp found when he mapped changes in the forest over this period. Hemp thinks fires now burn aggressively and frequently on the mountain because its climate has become drier and warmer over time, making the forest more flammable. Most of the forest fires are started accidentally by honey gatherers and poachers inside Kilimanjaro National Park. As the forest shrinks at its top and bottom, people are cutting trees in the middle of the forest too. In 2001, Christian Lambrechts and his team counted 8,000 tree stumps within the forest belt during aerial surveys.[18]

The predicted disappearance of Kilimanjaro's icecap, while symbolic to many, may not affect the herders, farmers, and wildlife in the Amboseli ecosystem at all. Indeed, if all the remaining ice were to melt in one day, it would be the equivalent of about 16 mm (0.6 in) of rain falling over the entire mountain, or one good downpour. Rather, it is the wet montane forests further down the mountain that most affect the region's water flows and levels, and thus it is forest changes that will most impact people and wildlife in the lowland savannas.[19]

When fires, farmers, or sawyers remove trees on the mountain, there are fewer trees to harvest fogwater and consequently less water in the mountain system. Typically, loss of trees means more water runs off into the watershed, because water that formerly evaporated continuously from the trees into the atmosphere now flows into the groundwater instead. Indeed, when hydrologists modeled what happens when farmers replace forest with cropland in Ethiopia and Malawi, they found that lake levels rose in the lowlands. Without trees, moreover, water runoff tends to be more variable, with higher highs and lower lows, as we saw with the Mara River in Chapter 8. If this is the case on Kilimanjaro, cutting its trees should lead to more, and more variable, flows of water into the swamps today, but less and less water over time due to declining rainfall, caused in part by climate change and in part by tree felling. If farmers and foresters plant more trees in Kilimanjaro's forests, the opposite should occur.[20]

This leads us to an event in Amboseli that is puzzling many scientists: the flooding of the swamps. David Lovatt Smith, once an Amboseli park warden, describes the surprise and delight of the Maasai and park authorities when, in October 1957, the swamps began to grow in size and overflow into parts of the parched landscape. Over the next thirty-five years, the water levels in a well at Ol Tukai in the park, at first a deep 11 m (36 ft) below the ground, rose to just 1 m (3 ft) shy of the ground surface. About a third of this rise happened in just three years, between 1961 and 1964. The abundance of water was particularly welcome during and after the 1955 and 1961 droughts, when conflicts between park authorities, herders, and wildlife were particularly acute. Since then, the water levels have begun dropping and rising in cycles, contracting during drought and expanding since the early 1990s. David Western thinks the swamp-filling in the 1960s was caused by a short-term rising trend in rainfall. This is supported by high rainfall region-wide and rapid filling of Lakes Victoria, Naivasha, and Turkana during that timeframe.[21]

If the rainfall rise was only short term, why are the water levels in the swamps continuing to rise? Lovatt Smith thinks the cause is farmers clearing forest to grow maize on the lower slopes of Kilimanjaro. Forest tree loss by fire and timbering must be another cause. Hydrologists Allard Meijerink and Willem van Wijngaarden suggest two further reasons: subsidence of Kilimanjaro and heavy grazing by livestock. Simply put, as, over time, the mountain sinks slowly into the ground, the water levels around it should rise. And heavy grazing, by removing plant cover, should allow water from rainfall to run more freely downhill across the bare ground and fill the swamps. All these events have likely been playing a role. Leading up to the 1960s, livestock numbers rose rapidly, farmers and sawyers were busy clearing the forest, and there were frequent fires. Then rainfall increased in the early 1960s, causing the swamps to swell. After the 1960s, livestock numbers still rose, though more slowly, and tree felling and fires continued. As described below, the decline of the acacia woodlands at the edge of the swamps may have played a part as well. As Western suggests, with fewer trees around the swamps to evaporate moisture into the atmosphere, swamps will retain more of their water.[22]

At the same time that the swamps flooded, the Amboseli basin lost most of its woodlands of yellow-barked fever trees (*Acacia xanthophloea*). Dense woodlands of fever trees, which in 1950 covered 20 percent (120 km², or 46 mi²) of Amboseli's 600 km² (232 mi²) basin, had mostly disappeared by 1987. Grasslands, open bush, swamp-edge plants, and the new swamps took their place. *Suaeda monoica*, a halophytic (salt-tolerant) shrub absent in 1950, first appeared in 1967 and now covers about 85 km² (33 mi²) of the basin, although it is recently in decline.[23]

Like the swamp flooding, several explanations are possible for the yellow-barked fever tree loss. In 1973, Western and van Praet reasoned that the rising water levels in and near the swamps had caused salt levels to rise in the soil, which inundated the acacia roots, killing the trees. The fact that the salt-tolerant *Suaeda monoica* began spreading at the same time the acacia trees were dying lent support to this hypothesis. Over time, Western has reconsidered salt as a cause and now thinks that elephants are responsible. Western came to this conclusion after running an experiment with his longtime colleague, David Maitumo, by excluding elephants from part of the Amboseli savanna. When trees regrew inside their enclosures and not outside, they concluded that it was elephants, not salt, climate change, or overgrazing, that kept this kind of acacia tree from growing back, particularly in the park. In contrast, ecologists Harvey Croze and Keith Lindsay argue that salt and elephants are both to blame for the loss of two different types of fever-tree acacia woodlands, those in dry sites and those along the edge of the swamp. They think salt is responsible for the death of the more extensive drier woodlands, whereas elephants are responsible for the lack of regrowth of the smaller swamp-edge woodlands. They emphasize that cycles of loss and regeneration of acacia trees are a natural part of savanna ecosystems, pointing to evidence of just such cycles from woodland to grassland and back in Lake Manyara, Tanzania, and elsewhere. In the future, new replacement tree seedlings may regrow along the swamps if there is a series of wet years that allow elephants to remain outside the park for long periods.[24]

Ecologists Truman Young and Keith Lindsay, however, posit another, and interacting, reason for the acacia die-off in the Amboseli woodlands: old age. They argue that the acacia trees could have become established all at once during a window of opportunity afforded by conditions favorable for seedling survival. This is known to occur for example in years with high rainfall. In Amboseli, such a window may have opened in the late 1890s, when wildlife that feed on trees or tree seedlings had been virtually wiped out by the twin catastrophes of the ivory trade followed by the rinderpest panzootic. Amboseli's acacia woodlands may have gradually disappeared, then, because the even-aged trees, perhaps weakened by years of damage from elephants *and* salt, died about the same time.[25]

The flooding of the swamps and loss of woodlands have strongly affected Amboseli's wildlife. Even though the vanished woodlands covered only 20 percent of the Amboseli basin, they were important habitat for browsing animals. On the other hand, the greater amounts of water and green grass in the expanding swamps probably sustained more animals through the dry season. As the acacia woodlands disappeared, wildlife dependent on this habitat, including lesser kudus, impalas, giraffes, vervet monkeys, leopards, and baboons, became rarer in the park. Patas monkeys disappeared altogether. Biologists Jeanne and Stuart Altmann and their colleagues found that some baboons moved southwest and out of the park, while others stayed in the small woodland created by fencing out the elephants around Ol Tukai Lodge, inside the park. Vervet monkeys declined steadily starting in the 1960s, with their populations crashing between 1977 and 1989 as the last of the fever trees disappeared. But grass-eating grazers, such as zebras, wildebeest, and gazelles, were unaffected by the tree loss and may even have benefited from it. Overall, the disappearance of the woodlands led to fewer wildlife species left in the park, though in turn the more extensive swamps attracted many migrating birds to stop over in Amboseli or to stay permanently.[26]

The establishment of the park continued a history of land loss for the pastoral Maasai. The Maasai had earlier been pushed out of some of the best highland savannas of east Africa by colonial settlement and farming. In 1911, the British moved them from productive highland savannas in the Rift Valley near Mt. Kenya to the less productive (and disease-ridden) southern reserve in Kajiado and Narok, thereby depriving them of 50–70 percent of their territory. As historian Lotte Hughes puts it, the Maasai were "driven there at gunpoint and corralled along with their cattle in a virtual human zoo." Colonial authorities then eyed the Amboseli Plains, teeming with wildlife, for tourism and conservation. Worried about the loss of wildlife to colonial hunting, in 1948 they established the large (3260 km^2, or 1,259 mi^2) Amboseli Game Reserve, stretching from Namanga east to Namelok, which eventually, in 1962, became a tiny 77 km^2 (30 mi^2) wildlife-only area around the swamps, and finally, in 1974, a 388 km^2 (150 mi^2) national park. Despite the tiny area conserved, the park sits on top of two swamps that the Maasai traditionally shared with wildlife, particularly during the dry season and drought. However, the park area far from covers the large seasonal wildlife dispersal areas that encompass both their wet and dry season ranges.[27]

When the park was established, the Maasai leaders and the conservation authorities engaged in thoughtful negotiation, in an effort to compensate Maasai for lost access to land and water. Still, promises were broken, and some Maasai hit back by killing wildlife. In 1951, the warden of Amboseli Reserve, driving for two hours around Ol Tukai, saw thirty-one rhinos. By the 1970s, herders were spearing rhinos and poisoning lions to protest the loss of their land to conservation, then represented by the independent Kenyan government. By October 1991, Maasai had speared three of the last five rhinos in the park, and they were all gone by 1994. But as of 2005, over fifty lions had recolonized the park.[28]

This collision between Maasai and park authorities was inevitable as pastures shrank, livestock populations grew, and more people, Maasai and non-Maasai alike, settled the region. Early on, between the 1940s and 1960s, Maasai livestock populations grew rapidly thanks to new water development and veterinary care. The new waterholes opened up new grazing lands, but these lands soon became filled with livestock. Livestock growth then slowed in the 1980s and leveled off. At the same time, in just forty-one years, between 1948 and 1989, human populations (Maasai and many farming immigrants) in the surrounding Kajiado District grew tenfold. By 1999, human populations were still growing 50 percent faster in Kajiado District than in the nation as a whole, partly due to the many new arrivals.[29]

Farming has grown markedly, increasing human conflicts with wildlife, as immigrant farmers moved to the Amboseli area and Maasai turned to farming. After losing many of their livestock in the severe drought of 1973–1976, even herders began to eye the best land for cultivation. For the first time, the Maasai began farming on the mountain slope, in swamplands, and along rivers outside the park. Farm irrigation, however, uses four times the amount of water as conservation or pastoralism, and today upstream irrigation is drying up Esoitpus Swamp, on Kuku Group Ranch. Also in the 1970s, farmers cleared land for crops in Kitendeni, now the last elephant migration corridor that connects Amboseli's lowlands in Kenya with Kilimanjaro in Tanzania. With farming came more conflicts with wildlife, so fences started to go up, first around farmland at Namelok Swamp, then around the Kimana springs south of its swamp. More recently, Maasai and others are establishing farms around the base of the Chyulu Hills and also along the leaky pipeline that goes north from the spring at Nolturesh. Some Maasai and non-Maasai farmers are even trying to grow crops in the dry savanna, away from permanent water, in what geographer David Campbell calls "expeditionary cultivation." Even though the Maasai are diversifying into farming and other sources of income, livestock still form the foundation for herder livelihoods here and across Maasailand.[30]

While Maasai herders lost land to farmers and protected areas, they also began settling and fragmenting the remaining rangelands. Since the 1970s, more and more Maasai families have chosen to stop moving their entire homestead, leaving many family members behind in settlements when the herds move. (More permanent settlements have tin roofs, which have replaced the traditional dung roofs.) In the 1970s, Kajiado District was divided into fifty-two ranches owned by groups of Maasai, each ranch averaging 340 km² (131 mi²). Twenty years later, some of the group ranches around Amboseli decided to subdivide their group land into private parcels, with one parcel, typically less than 1 km² (.4 mi²) in size, allocated to each Maasai group ranch member—a process that is still under way. It seems paradoxical that many Maasai would want to give up their large group ranches for smaller parcels of land; however, they fear that their group ranch land will be taken away or "grabbed" by powerful Maasai and non-Maasai, so they prefer to have a smaller parcel with a secure land title. Ecologist María Fernández-Giménez calls this the "paradox of pastoral land tenure," referring to her work with Mongolian herders.[31]

Unfortunately, private holdings and sedentary life present herders with substantial problems, as we will see in Chapter 10 (Kaputiei Plains). Ecologist Randall Boone and his colleagues, for example, tried to look into the future and estimate what will happen to Maasai families and their livestock if they subdivide their group ranch land into private parcels, fragmenting the land. Boone's ecosystem model shows that if Maasai herding families in Amboseli settle on small parcels of land just 1 km² in size, rather than staying on the larger group ranches, they will be able to feed only 75 percent of their herd. This limitation of grazing areas happens even without privatization of land, when pastoralists settle down together in villages and move their herds less. Livestock confined to such small areas can no longer track changes in the savanna, losing access, for example, to distant patches of green pasture where rain fell recently. Boone's model showed that livestock able to track resources over larger areas typically grow faster and survive drought better than those that do not. When sedentary herders lose access to the wider savanna, they thus lose an important way to cope with drought; when this happens, families become poorer or drop out of pastoralism altogether. The drier the savanna, the greater this problem becomes because good patches of green grass are more ephemeral and spread out across the savanna.[32]

Where people farmed or settled, wildlife disappeared, whether the land remained in a group ranch or became privatized. Biologist Moses Okello, for example, found fewer wildlife around farms and settlements in Kuku Group Ranch, pushed away by the permanent concentration of people there in the mid-2000s. This occurred even though it was still a group ranch because people settled down and lived in one place more permanently. Wildlife also decline when people farm swamps. Ecologist Jeffrey Worden found more wildlife in swamps within Amboseli National Park than those outside the park when he counted wildlife from an ultralight aircraft, in the early 2000s. Outside the park, significant wildlife remained in Kimana Swamp, which contains a wildlife sanctuary as well as grazing land and cropland, while wildlife disappeared in Namelok Swamp, which is fenced and cultivated.[33]

Privatization of land encourages sedentary life for herding families, and this is especially harmful to wildlife. Western and his colleagues compared the changes in wildlife over three decades on communally managed Imbirikani Group Ranch in the Amboseli area and on land held privately by Maasai north and outside of Amboseli. Over time, the number of people in the two areas did not change, but the density and permanence of their settlements did. When these herders privatized land, spreading their settlements more uniformly over the landscape, each on their own parcel, and no longer moving their herds seasonally, wildlife populations and grass production dropped dramatically.[34]

In contrast to the private land north of Amboseli, within the greater Amboseli area, which includes the park and the group ranches around the park, Western's historical data suggest that most wildlife populations were remarkably stable up until the drought of 2009. There were no major changes in the overall numbers of wildebeest, gazelles, kongonis, elands, and giraffes over this wide area, despite more local changes within the park itself. After the 2009 drought, however, wildlife counts by the Kenya Wildlife Service team and ecologist Jeffrey Worden and his colleagues showed massive losses of wildlife. In Worden's study, the ecosystem lost 92 percent of the wildebeest, 86 percent of the zebras, and 66 percent of the Grant's gazelles (plus 66 percent of the cattle and half the sheep and goats).[35]

In this same 2009 drought, Amboseli lost more than three hundred elephants. Before 2009, the Amboseli elephant population had grown remarkably over time. When Moss and her team began following Amboseli's elephant families in the 1970s, the park and surroundings had about 600 elephants. Although many died in the mid-1970s from poaching and drought, by 2008 the population had nearly tripled to about 1,500. The spreading swamps also probably helped elephant populations to grow. Elephant population growth here is all the more remarkable given the expansion of farming that occurred across Amboseli at the same time. In fact, elephant populations have grown at about the same rate as Kenyan human populations, 3–4 percent per year, since the 1970s. Moss and Western both attribute Amboseli's swelling elephant numbers (before 2009), in the face of massive elephant losses across the rest of Africa, to the goodwill of the Maasai, despite occasional Maasai spearing of elephants. Recognizing the rising potential for conflicts between herders and elephants, since 1997 Soila Sayialel, Cynthia Moss, and their Maasai colleagues have run a "consolation" scheme to pay for elephant-caused deaths of livestock and have reduced retaliatory spearing.[36]

THE FUTURE OF AMBOSELI

David Lovatt Smith titled his book on Amboseli *Nothing Short of a Miracle,* and I am inclined to agree with his sentiment. This ecosystem is a tough place for herders, farmers, and wildlife to live, much less thrive. And the pace and variety of change over the last century, from the mountain to the savanna, are difficult to comprehend and then manage. The formerly fluid landscape, where herders, livestock, and wildlife typically mixed, is evolving into a segregated landscape, with wildlife more concentrated in parks and herders and their livestock settled outside parks. Old people-wildlife compatibilities are starting to disappear, particularly as new residents, most of whom did not grow up moving with savanna weather, use more and more of the best land. The Maasai themselves are no longer just herders, but farmers, shopkeepers, and statesmen and -women, with new aspirations for themselves and their children. Some herders now see more profit from plowing the land, planting crops, and engaging in other diverse activities than from either herding livestock or conserving wildlife, and their attitudes toward wildlife can shift quickly from tolerance to animosity as they take up farming. Those who remain herders find it unfair that they who traditionally shared this land with wildlife rarely share in the large profits that tourism operators obtain. If this continues, there will be a profound loss of the ancient tolerance between herders and wildlife.[37]

Despite rising conflicts and scarcity of resources, there are some bright spots in Amboseli's future, but still no easy solutions. Ongoing discussion will be needed if people and wildlife are to continue to coexist in this ever changing ecosystem. Starting in the 1950s, Maasai and their supporters began turning a less-than-advantageous situation around to their benefit, such that Amboseli is today considered one of the birthplaces of community conservation. This occurred in spite of continuing land loss to conservation, farming, and other uses. In the past, virtually all the profits from wildlife flowed out of the ecosystem, either to foreigners, the national government, or the county council in far-away Kajiado town. When returns did flow into the ecosystem, they often ended up, as elsewhere in Maasailand, in the hands of wealthy Maasai elites, leaving little for poor herders. Now, though, park profits are flowing increasingly back to community members. Some of the ecotourism businesses are working hard to benefit locals (in the Chyulus and elsewhere), and some group ranches and other local agencies have set up, and are receiving profits from, community wildlife conservancies on their lands (Kimana, Elerai, Kuku, Eselenkei, Kitirua). While these efforts are laudable, only a small percentage—8 percent in 2001—of households participate. If averaged across all Maasai households, this income is minuscule, only 5 percent of total income. To be effective, more households need to participate and more profits from wildlife need to be shared.[38]

Some efforts have been made to reduce human-wildlife conflict and improve pastoralism and there are more opportunities for progress. Namelok and part of Kimana farms are now fenced to protect crops. Local government could encourage building of settlements and schools away from the park boundary to ease contact between herders and wildlife. Earthen dams outside the park could be refurbished so that Maasai can avoid conflicts with wildlife in the park for some of the dry season. The pastoral enterprise itself could also be improved, along the lines of the work my colleague, sociologist and inventor Leonard Onetu, is doing to make herding more profitable, by introducing new cattle breeds as well as camels. Given the growing demand for livestock products in Kenya and abroad, better marketing will help improve profits as well. Human ecologist Shauna BurnSilver and geographer Esther Mwangi found that private landowners in several former group ranches in the district are banding together to form grazing associations that will allow reciprocal grazing during drought and dry seasons. There may also be opportunities for future payments, from local businesses or philanthropy, to local herders for the ecosystem services they provide by managing land well for wildlife, carbon-rich soils, and clean water.[39]

All this, however, will not be enough. Some of my Maasai colleagues think the Amboseli Maasai need to move to a new level in how they think about, manage, and envision the future of this ecosystem. This is a future in which Maasai control more of the returns from wildlife, exploit high-value horticulture, and get paid for water taken from their springs. In this future, the ancient love of Maasai people for their livestock will still play an important role. Yet it will only succeed if they join hands with all of their fine collaborators to manage the ecosystem at the large scale needed to sustain mobile livestock and wildlife enterprises, which are connected to other systems, including those across the border, in Tanzania. This future

also requires the Maasai to take responsibility for elephant spearings, lion poison-
ings, and any damage to the environment that their settlements and livestock
cause, and to accommodate the aspirations of the many non-Maasai people who
have moved into their area. We will return to these matters in the last chapter,
which explores alternative futures for people, livestock, and wildlife in this ecosystem, across Maasailand, and in other savannas of east Africa.

The Kaputiei Plains

The Last Days of an Urban Savanna?

The scene is poignant, full of possibility, yet impossible. David, only the second man in his pastoral community village to receive a master's degree (and who now has his Ph.D.), grins broadly in front of my camera, resplendent in academic cap and gown, with bright Maasai beads criss-crossed over his chest. Over his shoulder, an ostrich grazes calmly, and behind the ostrich a high-tech farm raising cut flowers for export to Europe stands in the middle of a rolling savanna that extends from one horizon to the other. Behind me stretches Nairobi National Park, with only a fence separating us from Nairobi and its two million people. The savanna beneath our feet, next to David's homestead, is one of the last remaining wildlife dispersal areas of what was once the second largest wildlife migration in east Africa. It connects the park to the north to a wet-season calving ground in pastoral lands to the south, used once a year by herds of migrating wildebeest and zebras.

David has spent the last eight years working on a "land leasing program" that will reward them for allowing wildlife free access to their land. With this program, he and his colleagues have persuaded many of their Maasai community members to desist from killing the occasional lion, leopard, or other carnivore that attacks their livestock, and to keep their plots of land open (unfenced) to benefit wildlife (and livestock) grazing and migration. In exchange for these concessions, participating Maasai receive "conservation payments" three times a year, disbursed just when school fees are due. For those who receive them, these payments are an unqualified success for both herders and wildlife, increasing the annual income of the poorest families here by a third, allowing families to send their children to school (including many girls for the first time), and keeping the savanna open for wildlife and livestock alike.

But David and his colleagues know that, unless expanded, this land leasing program is neither big enough nor powerful enough to stem the rush of fence building and sales of savanna land to city dwellers. They know, too, that many wealthy elites in neighboring Nairobi are eyeing this savanna, wishing to exploit its current and future prospects, most of which are incompatible with continued cattle keeping or wildlife conservation. He and I speak candidly about this impasse only infrequently; it is painful to admit the likelihood that the program will fail. In the meantime, David and his community continue working every day to give the herders, livestock, and wildlife here a chance to survive together, in a functioning ecosystem, as they have for the last several millennia.

I snap the picture, full of admiration and hope for this man and his people. We finish our feast of roasted steer and goat meat and join the women at David's homestead nearby, where people give speeches and clap and sing to celebrate this big day and show how proud we are of this fine and gentle man. David listens solemnly to the praise and advice that are heaped upon him, knowing the difficult days that are ahead for his people here, their livestock herds and livelihoods, and the wildlife populations that are rapidly disappearing from his beloved Kaputiei Plains.

THE KAPUTIEI PLAINS OF THE PAST

David Nkedianye's homestead sits at the northern end of a pastoral-wildlife savanna called the Kaputiei Plains, also known as the Kitengela or the Athi-Kaputiei Plains. This savanna is one-of-a-kind: supporting a diversity of wild mammals and an annual long-distance wildlife migration, it is nevertheless located right next door to the bustling capital of Nairobi (Map 16), separated from it only by a wood and wire fence. Nairobi National Park, 117 km^2 (45 mi^2) in extent, occupies the northernmost tip of this 2456 km^2 (948 mi^2) wildlife-pastoral ecosystem, about half the size of the U.S. state of Rhode Island. Remarkably, the park's northern boundary is located less than 5 km (3 mi) from the central business district. South of the park stretches a large, flat, fertile plain, more than twenty times larger than the park itself, which provides essential forage for wildlife and livestock alike. Migrating herds of wildebeest and zebras use the park during the dry season for its water and abundant grass, then spill out into the open pastoral lands farther south when it is time for the wildebeest to calve in the wet season. Here, the Kaputiei Maasai live alongside many other ethnic groups, who together use

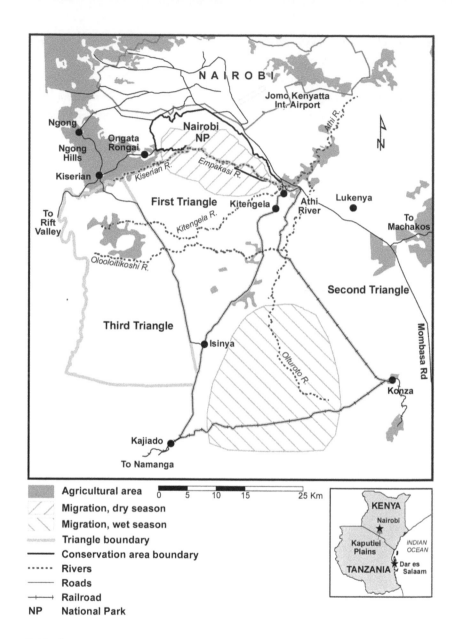

MAP 16.
Kaputiei Plains (or Kitengela, Athi-Kaputiei) and environs in
Kenya. (Map adapted from DMA 1992 and WPDA 2009 by
Shem Kifugo and Russell L. Kruska.)

the land for settlement, livestock grazing, cultivation, horticulture, quarrying, and export processing.[1]

Maasai herders arrived in this landscape in about the 1600s, pushing out hunter-gatherers and earlier herders. The richness of the area's wildlife was described by early European colonists who arrived in the late 1800s as the most spectacular concentration of wildlife in east Africa. In 1891, rinderpest and smallpox reached this part of Maasailand, killing animals and people in great numbers (see Chapter 5). The Kaputiei Maasai here were particularly hard hit by smallpox, so this area was probably thinly populated by herders when the British began to establish their settlements along the railway line, which was built between 1896 and 1901 to connect Mombasa, on the coast, to Port Florence (now Kisumu), on the eastern shore of Lake Victoria.[2]

Early descriptions of the area suggest it was teeming with wildlife. M. H. Cowrie, later director of the Royal National Parks of Kenya, described life in the railway camp of Nairobi at the end of the 1800s, with lions using the same watering holes as Ugandan railway crews, giraffes eating washing hung out to dry, and monkeys swinging from tent guy ropes. In 1902, the Uganda Railway was complete, and Nairobi had only one tin shop selling provisions, Jeevanjee's Soda Factory, one wood-and-tin hotel, and one street, Victoria. Even though this was barely a decade following the rinderpest panzootic and massive wildlife and livestock deaths, by 1902 wildlife herds were again grazing around the growing town of Nairobi and Maasai cattle herds had started to recover. Richard Meinertzhagen, a British soldier in his twenties, found the savannas where Nairobi's industrial area and Buru Buru Estate now stand good hunting. The difference between then and now is that he counted two to four times more wildlife than cattle in 1902, while nearly a century later, counts by scientists at Kenya's Department of Resource Surveys and Remote Sensing indicated the reverse, with livestock outnumbering wildlife by four to one.[3]

Just when the Maasai were at their weakest in the 1890s, the British and Germans cut Maasailand into three parts, creating German East Africa (to become Tanzania), the Protectorate of East Africa (the eastern two-thirds of what is now Kenya), and the Uganda Protectorate (the western third of Kenya and Uganda). High Commissioner Charles Eliot, one of the first leaders of the East African Protectorate from 1901 to 1904, very quickly began allocating the best land to European farmers and ranchers. Olonana, the British-supported leader of the Maasai, agreed in 1904 to two reserves for the Maasai, one in Laikipia, central Kenya, and one in Kajiado, to the south, even though these two reserves made up only 40 percent of the territory the Maasai had held just twenty years earlier. Five years later, Olonana agreed to move the northern Maasai in Laikipia to an expanded southern reserve. Thus the Maasai lost their best cattle grazing areas in the high grasslands of the central Rift Valley, which had long served them as essential grazing reserves in dry years and droughts.[4]

Even though observers in the early 1900s said these plains supported abundant wildlife, we know neither their numbers nor their migration paths. Animals probably moved between wet- and dry-season pastures in the open savannas of the Kaputiei Plains, situated between the two great peaks of Mt. Kenya and Mt. Kilimanjaro. It is probable, too, that the forests and swamps equidistant between the two mountains—where Nairobi is located today—were a favorite dry-season refuge for wildlife and pastoral cattle alike. Wildlife counts in 1902–1903 suggest that wildebeest and zebras did migrate because their numbers fluctuated widely in the Kaputiei Plains between wet and dry seasons. Seventy years later, during the 1973–1976 drought, biologists Jesse and Alison Hillman, studying populations of eland in this ecosystem, observed that individuals that they had recognized on sight had disappeared from the Kaputiei Plains in May 1974. The Hillmans spoke with ecologist David Western, who was studying wildlife south of the Kaputiei Plains in and around Amboseli at the base of Kilimanjaro about 100 km (62 mi) away. Western reported seeing eland populations increase suddenly, from 1,200 to 4,900 animals, during his wildlife counts at about the same time. By August of that year, eland populations in Amboseli had returned to normal, at which time the Hillmans saw their elands return to the Kaputiei Plains. Maasai say movements between these two systems are still common today.[5]

In the early 1900s, colonial authorities were already expressing concern about wildlife losses and environmental degradation, and many colonists had begun to call for the establishment of game reserves. East Africa's first national park was gazetted, just south of Nairobi city, in 1946. It was small ($117\,km^2$, or 45 mi^2, about twice the size of Manhattan Island), raising concerns that it would not be able to maintain viable populations of wildlife as human populations grew around its edges. Though relatively small for a game park, Nairobi National Park sits within an ecological transition zone between forest and savanna, a transition that largely determines what people do with the land and what kinds of wildlife exist on either side of it. A semi-evergreen forest in the western end of the park, near its main gate, at about 1800 m (5,905 ft) is a low part of the Kenyan highlands. This small forest faces a gentle downward-sloping plain that flows east and south into a vast savanna ecosystem that reaches Kilimanjaro and into Tanzania. These transition areas, called ecotones, are usually very rich in wild plants and animals and provide diverse options for people as well.[6]

After Nairobi National Park was established in 1946, the country's African population, mostly nonpastoral, began to get government titles to own land (although agricultural societies already had in place a sophisticated land tenure system that amounted to ownership), which further curtailed pastoral and wildlife movements. The savannas were relatively open, and my pastoral informants say their grandparents' families moved livestock to the drier savannas in the south in the wet season, retreating to the lusher pastures near the Ngong Hills and Old-oinyo Sapuk near Thika in the dry season. Migrating wildlife moved to similar parts of the savanna seasonally, calving in the wet season in the south, and moving to wetter pastures in the hills during the dry season as well as to the forage and permanent water in newly established Nairobi National Park.[7]

Beginning in the 1950s and 1960s, farmers and settlers, predominantly non-Maasai, took up land around the base of the Ngong Hills. By the 1970s, all the land north of the road leading south out of Nairobi to Lake Magadi had been settled and was thus no longer available for use by pastoral herders or wildlife. In 1963, the Royal Parks Board built a fence at the western and northern edges of Nairobi National Park, to separate the park and the city, effectively ending half of the migration of wildlife and also protecting park wildlife from the city residents. Four years later, the fence was extended to the eastern base of the Ngong Hills, further limiting movement to this part of the dry-season grazing reserve.[8]

After independence in December 1963, many Maasai feared that the new government, led by farming peoples, would give away their reserves to non-Maasai people. In the late 1960s, international donors and the Kenyan government proposed establishment of "group ranches" as a way to do several things at once: ensure Maasai ownership of their remaining traditional lands, encourage sustainable development of rangelands, and redress perceived environmental degradation. The formation of group ranches kept the land, which in most cases could not be sold, in the hands of pastoral peoples, although the new government did grant some pastoral chiefs and other Maasai elites title to individual ranches. The first group ranches were formed in the Kaputiei section of Maasailand in the 1970s, with 15 established for 1,300 families on 603,000 acres (2440 km^2 or 942 mi^2) of land. These group ranches probably slowed the spread of cultivation and the fencing in of wildlife-rich pastoral lands by solidifying pastoral control of these large areas.[9]

Beginning in 1986, the Kaputiei Maasai again led the way in Kenyan Maasailand by being the first to subdivide their large, communally owned group ranches into many small, individually owned ones. Interviews conducted by geographer Marcel Rutten revealed that in the case of the fifteen former Kaputiei group ranches, each member received title to private plots ranging in size from 51 to 298 ac (21 to 121 ha). This change in land ownership has had profound impacts on what the Kaputiei Plains look like today, how they are used, and how easily livestock and wildlife can move from one place to another in search of better pastures.[10]

THE KAPUTIEI PLAINS TODAY

Traveling by road out of Nairobi to the Kaputiei Plains today, it is hard to imagine that just one hundred years ago wildlife were so abundant here as to be described as spectacular. The road from Nairobi to the town of Athi River is crowded on either side with industrial enterprises and small commercial centers, and remains so as you turn right off the main highway and drive southwest toward the Tanzanian border and the bustling town of Kitengela. The Kaputiei Plains ecosystem is conveniently described as being made up of three triangles of land, bounded by roads, a railway, and the escarpment. Passing out of Kitengela town, the first triangle is on your right to the northwest; the second is on your left to the southeast; and the third is down the road ahead (Map 16). Only south of Kitengela town does the landscape finally open up somewhat, making it possible to picture the abundant wildlife of the recent past, but even here you will see so many fences, farms, and houses that it is hard to imagine herds of wildebeest still migrating to and from the calving grounds to the south. Away from the main road in the interior of these two triangles of land, there are many fewer fences and wildlife and pastoral livestock move more easily.

The clearest view of these patterns is from space. What you notice first when looking at a satellite image of the Kaputiei Plains is not the fences along the road but Nairobi National Park, standing out in the shape of a smooth-surfaced mango between the stippled area demarcating the city to the north and the spotted areas denoting the pastoral lands to the south. The boundary between the city of Nairobi and the national park is hard, with city housing estates and open savanna just a few hundred meters apart. But to the south, the park edge is soft, blurring into pastoral land and not always distinguishable from it. To the west of the park lies higher land at the top of the Rift escarpment, forming an abrupt topographic edge—the western side of the third triangle of the Kaputiei Plains. Rivers flow from this escarpment eastward, with permanent water only in the Empakasi River, at the southern edge of the park, and in the Kiserian River. The rest of the landscape's fingerlike riverbeds flow with water only when it rains. In fact, only 21 percent of this landscape is within livestock walking distance of permanent water. To the east, the pastoral land meets commercial ranches established by Europeans on former Maasai and Kamba lands in the early 1900s. Far to the south, open grasslands of the second triangle run into rocky, bushy woodland. Despite all the obvious boundaries observable on a satellite image, the Kaputiei Plains are by no means a closed ecosystem.[11]

What we cannot see from space are the details of how people, livestock, and wildlife survive in this landscape. From above, you notice the road only if you look carefully, and the fences, so important to migratory movement, are invisible, even on the finest aerial photographs. Much of the land here is owned by the Kaputiei Maasai, although people of many other cultural groups live in the towns and own land, particularly along the edge of the park (mostly Europeans) and along the tarmac road (many Kenyan ethnic groups). In terms of numbers, the Maasai are the minority ethnic group here, while still making up most of the large landowners. Whereas the Maasai are the focus of this chapter, the diversity of different peoples with different values living in the Kaputiei Plains greatly complicates people and wildlife issues. About 60 percent of the Maasai still live off the land—10 percent as pastoralists, 48 percent farmer-herders, and 2 percent farmers. The rest work in businesses, teach in schools, do research, or serve as soldiers, lawyers, doctors, engineers, or pastors. The average land holding is about 150 acres (61 ha). Families, many of them headed by a man with several wives, have an average of seven children. About 80 percent of the heads of households are married and 20 percent widowed. Many (42 percent) have no formal education, while 31 percent finished primary school, 16 percent secondary school, and 11 percent went on to a college or university. More than half the families make less than 1 dollar a day per person, a common definition of poverty. More than half the Maasai residents eat wildlife meat on occasion.[12]

The Kaputiei Maasai and their neighbors live in what ecologists call a "eutrophic" savanna, with rich soils and a correspondingly rich array of wildlife. The plain was created in the early Miocene (about 20 Ma) by volcanic activity associated with the formation of the Rift Valley to the west. This activity deposited phonolitic lava on the land, which has served as a cap protecting the plains from erosion ever since. Rainfall is moderate here, with 800 mm (31 in) falling each year in the northwest and 500 mm (20 in) in the southeast. Having a bimodal rainfall pattern (see Chapter 2), the Kaputiei Plains experience two rainy seasons a year, one or both of which can fail. Crop production is therefore risky, successful only one year in five. Twenty-four species of large mammals live on these rich plains (elephants, killed off before 1962, no longer being one of them), although their numbers are now fast declining. These plains are part of a wide swath of savannaland that rings Kenya's central highlands, forming a great ecological boundary between highland forest and grassy and woody savanna, a border clearly visible on a satellite image.[13]

People use these plains for many activities besides herding livestock, such as cultivating vegetables and growing flowers in greenhouses for export. A favorite wildebeest calving ground when David Nkedianye was a young man is now an export processing zone, where ten thousand workers make textiles for export abroad. Miners blast rock in the Kapio River valley to make gray building stones and gravel for construction. And the town of Kitengela itself, a growing metropolis at the edge of the plains, has expanded rapidly since the 1990s and is now home to thousands of people.

The critical obstacles in this landscape to movement of both livestock and wildlife are the fences that people, Maasai and non-Maasai alike, built around their plots after land was privatized. In 2004, our Maasai colleagues used GPS units to map all 6,741 fencelines in both the first and second triangles; five years later, the number of fences had grown to more than 10,000. All these fences arose in just twenty years; Nkedianye does not remember any fences when he herded here in the late 1980s. Even so, the majority of the plains is still rolling open grassy savanna, dotted with homesteads and a few trees along its streams and rivers, on which moving herds of livestock and a diminishing number of wildlife graze.[14]

Even though most of these plains have been subdivided into privately owned plots, pastoral herders still move their livestock in response to the seasons. When the rains fail, herders—despite fences, private lands, and farmed fields that bar much of their way—will walk their animals tens to hundreds of kilometers in search of better pasture. During Kenya's 2000 drought, herders drove their cattle 80 km (50 mi) north toward Mt. Kenya, 100 km (62 mi) east toward the town of Sultan Hamud, and more than 100 km south to northern Tanzania. Some herders took their animals to graze in the city of Nairobi on roadside grass. In 2005, herders from all over Maasailand crowded into the Kaputiei Plains because of the better rains there, which, owing to the high densities of animals, unexpectedly caused more deaths than occurred among livestock that stayed in areas with lower rainfall. Wildlife also move during droughts, of course, but because Nairobi

National Park is for wildlife only, wild herbivores do not have to walk such long distances to find food.[15]

Some wildlife prefer to spend much of their time outside Nairobi National Park, grazing in the first triangle, where livestock keep the grass short and nutritious. As early as the 1950s, managers of Nairobi National Park mowed and burned the park grass to encourage the wildlife to stay within. Today, the park grass is no longer mowed, and even though park managers burn small patches of land there each year, some wildlife still prefer to graze outside the park in spite of being chased away from water points in the dry season by Maasai herders trying to preserve the water for their stock. As we saw in Chapter 6, the smaller grazers, which require higher-quality food and prefer open vistas so they can watch for predators, are attracted to places where grass is naturally short or is grazed short. Short grasses in the Kaputiei Plains have 20–75 percent more protein per mouthful than ungrazed grass. Most of the grasses in Nairobi National Park, in contrast, grow in heavy black clay and become tall and tough unless the land is grazed, mowed, or burned. In the park's small patches of lighter soils with naturally shorter grasses, wildlife cluster at all times of the year. An area just south of the park border is another magnet for wildlife because livestock grazing here keeps the grass short throughout much of the year. It is not uncommon to find more wildlife on pastoral lands just outside the park than in the taller, less-grazed grasslands within the park (similar to the Mara, Chapter 8).[16]

Ecologist Helen Gichohi argues that Nairobi National Park has not always been a favorite grazing area for wildlife, but has become so only in the last thirty years when other grazing lands disappeared under settlements and fields. Once people fenced and cultivated the higher-elevation, better-watered areas around Nairobi, the park became one of the few remaining options left to wildlife for dry-season grazing. Given a choice, both pastoral peoples and wildlife prefer higher, wetter areas, such as the Ngong Hills or Oldoinyo Sapuk, a large hill north of Nairobi and to the west of Thika. With little evidence that either people or wildlife preferred to live or graze where the park now stands when these other choices were available, Gichohi and others refer to the Kaputiei Plains ecosystem as a "truncated ecosystem."[17]

Despite this truncation, according to Gichohi, even in the early 1990s thousands of wildebeest and zebras walked from Nairobi National Park to the southeastern part of the Kaputiei Plains to capture the ecosystem's richest but most ephemeral grasslands just after grasses sprouted, three days after the first rain. Here, in the driest part of the ecosystem, soils are less soggy in the wet season and the grasses are shorter and more nutritious for wildlife and livestock alike. As the shorter-grass areas became exhausted, these large herds then moved progressively north- and westward to taller-grass areas, sometimes just to the first triangle, but often all the way to Nairobi National Park.

As the open land shrinks in this savanna, conflicts between people over resources and between people and wildlife are becoming more common and, on occasion, violent. Years ago, cattle were a Maasai's most precious possession, but now land is also a pastoral passion. As pastoralists sell off their land in the Kaputiei Plains, people whose lifestyles are incompatible with free-ranging wildlife control more and more of the land. New arrivals and some Maasai choose to fence their land to conserve forage for their livestock and decrease the contact between livestock and wildlife that often transmits disease. With fewer places to graze their livestock, the pastoralists resent that wildlife are the only animals allowed to use the grass and water in Nairobi National Park, a place they used freely for centuries. Some Maasai solve this problem by herding their cattle into the park at night to poach grass, despite the danger of lion attack.[18]

As people build more and more fences, wildlife are squeezed into ever smaller pastures, where they must compete with domestic animals for what food and water remains available. Livestock grazing lands that are heavily frequented by wild grazers not only are depleted of forage relatively quickly but also attract predators from the park, which often then attack Maasai cattle. Some residents who see little benefit from wildlife feel compelled to act, as they did in June and July 2003, killing nine of the remaining twenty-two lions in the Kaputiei Plains ecosystem. And non-Maasai poach wildlife of any kind for dinner tables in Nairobi, often running wildebeest against fencelines to corner and kill them. More than half of Maasai residents have also eaten wild meat, preferring elands and wildebeest and avoiding unpalatable zebras, preferences that match the relative declines in these wildlife species.[19]

Such conflicts, along with floods and droughts, caused the wildebeest migration to collapse by 2001 and brought about major changes in the rest of the wildlife populations of the ecosystem, including in the park. In 1977, the wildebeest population stood at about 30,000; in 2007, it was below 2,000. Our team, led by ecologist Randall Boone, has been tracking twelve wildebeest remotely by means of GPS collars from 2010–2012 to find out if the migration is truly dead. In the 18 months after collaring, none of the wildebeest crossed the Athi-Namanga road that forms the boundary between the first and second triangles. If these wildebeest represent the rest of the population, then the longer-distance wildebeest migration from the park, which used to move out of the park, across the first triangle and the road and then on to the south end of the second triangle, is now dead.[20]

Unfortunately, between 1977 and 2007, populations of seven other major wildlife species fell by about 70 percent, in both Nairobi National Park and the Kaputiei Plains. Populations of migrants declined more quickly than those of nonmigrants, while park populations declined more slowly than those in the Kaputiei Plains. Inside the park, zebra and buffalo populations grew, much as they did when Maasai moved out of Ngorongoro Crater in 1974 (see Chapter 11). A population collapse is not without precedent: the hartebeest population crashed to low numbers during the 1973–1974 drought and only partly recovered, but just in the park.

Over the same ten years, the number of livestock (cattle, sheep, goats, and donkeys) in the Kaputiei Plains fell by almost half. In short, this urban savanna has lost most of its animals.

WHAT HAPPENS TO HERDERS AND WILDLIFE WHEN LAND BECOMES PRIVATIZED?

Privatization of land in the Kaputiei Plains two decades ago, along with other factors, initiated a succession of changes in the people and wildlife of the area. Because most of the group-owned land in Kenya (and some in Tanzania) is likely to be privately owned within the next twenty to thirty years, the Kaputiei Plains may be a harbinger of the change that the region may see. But, while important, land privatization is only one of the reasons for the changes we see in this ecosystem today. During the same period that the Kaputiei Plains became a favorite place for speculators, businesses, churches, and Nairobi residents to buy land, Kenya's human population grew as fast as any in the world and pastoral people increasingly invested in and received better health care and education.[21]

Why do pastoral people decide to subdivide their own communal land into their own private plots? Nonpastoralists tend to perceive land that people with low populations use for their livelihoods—as in the wetter pastoral lands of east Africa—as "unused" or "open." All over the world, indigenous land users have difficulty ensuring that their group rights to land are respected by other groups of people. Pastoral people of east Africa, however, see private land ownership as one way to ensure that nonpastoralists and the state do *not* take their land, for settlement, game parks, or other purposes. At the same time, pastoral families are fully aware that no one can survive as a livestock herder through the dry season or drought on a bounded piece of land—that movement is critical. Yet in the wetter savannas of east Africa, the need of pastoralists to have secure land ownership is beginning to outweigh the need to move. This is what I referred to as the "pastoral paradox" in the last chapter. When faced with privatization of land, pastoral people in the Kaputiei Plains hope that they will be able to make cooperative arrangements with their neighbors to move their livestock and graze over wider landscapes, as they have for centuries.[22]

How Maasai think and talk about land changes when they own it privately for the first time. When Marcel Rutten asked Kaputiei Maasai if they felt there was more or less cooperation after subdivision than before, they said that cooperation was about the same, but they made fewer communal decisions on where cattle should graze. Their language changes, too. My Maasai colleagues who have herded cattle for many years in Narok District started using the word *my* rather than *our* with reference to a *shamba* (farm) for the first time in my memory after their group ranch began to subdivide.

How did the subdivision of group ranchland in the Kaputiei Plains take place, who got the best parcels, and was the allocation of land fair? Some land was given by the colonial government to wealthy pastoral leaders in the mid-1950s, twenty years before the group ranches were formed in the 1970s. These were meant to be demonstration ranches where the wealthy would experiment with intensifying livestock production by building water points and improving veterinary care for their animals. In the 1970s, the now independent government ceded ownership to the group ranches. Group ranches then had a membership that was composed of one member per family. When group ranches in the Kaputiei Plains split up into private plots, one per group ranch member, in the 1980s, the wealthiest people again often received the biggest parcels of land and in the best places, such as near water. In Olkinos Group Ranch, the first to subdivide in the Kaputiei Maasai section, the group ranch committee members, composed of a select group of the entire group ranch membership, allocated themselves twice as much land, on average, as the rest of the group ranch members from each family received. According to Rutten, who interviewed people at the time, "Some people complained that some members of the group ranch committee and their relatives and friends had allocated themselves large parcels in the most favourable places." Nearby, in Kisaju Group Ranch, members decided to enforce equity and carefully allocated 300 acres to each group ranch member (David Nkedianye remembers the "300-

foot chain" that they used to measure each plot). Maasai author Naomi Kipury, meanwhile, describes how single, divorced, or separated women were disinherited after subdivision.[23]

What happens to pastoral people as they gain private ownership of land? Rutten figures that only 10 percent of all households in Kaputiei would have had enough grass to support their livestock on their own plots right after subdivision in 1986. The number of families that share a homestead also declines as families move to their own parcels of land. As a consequence, family members have less leisure time because there is less opportunity to share herding labor when traditional multifamily homesteads split up into single family plots.[24]

Rutten believes that private land ownership impoverishes families because for the first time people have a huge asset they can sell—and sell it they do. Maasai land owners, for example, had sold almost a third of their new plots after subdivision of Kisaju Group Ranch, mostly to non-Maasai, only six years after it was privatized. By the early 2000s, most of the Maasai from the Kaputiei Plains over fifteen years of age earned less than $1 per person per day. The average price of land in the Kaputiei Plains, however, was more than $2,500 an acre ($1,010/ha); with a 150-acre allocation, for example, a Maasai family could have access to considerable cash if they sold part of their land. It is very risky for pastoral families to sell all their land, because they lose the means to support their livestock, but some faced with a crisis did just that, moving to a smaller parcel or in with other family members. This is what Marcel Rutten describes as "selling wealth to buy poverty." Another risk of owning particular plots of land is if a family owns land in the wildebeest calving area because they have to remove their cattle from their land for three months (March to May) each year to prevent them from contracting malignant catarrhal fever (MCF) from infected wildebeest calves. In recent years, families have recognized the dangers of selling land, and sales today are much less common than in the past.[25]

Where do herders graze and water their livestock if they no longer have group access to most of the land? The answer depends on the kinds of agreements families come to after land becomes privately owned. In the Kaputiei Plains, herders cannot graze or water their animals on someone else's land unless they ask permission. On the face of it, it might seem to be to a landowner's disadvantage to allow a neighbor to graze or water stock on his land. But pastoral people develop uncommonly strong social ties, and these can serve as a safety net when things go wrong (for example, if one herder loses stock to disease, neighbors with whom he has cultivated strong relations will lend him stock to rebuild his herd).

These strong social ties, which are essential to the survival of pastoral peoples around the world, remain in the Kaputiei Plains despite land privatization. Nevertheless, the people here are no longer as free to move their livestock as they were before the group ranches became private. If a wealthy landowner chooses to build a fence and exclude his neighbors, large parts of a grazing area become inaccessible to herders and their livestock. Data collected by ecologist Mrigesh Kshatriya show that cows graze over smaller areas and move more and faster when they are in places heavily fenced than when they are in relatively unfenced areas.[26]

What happens to the landscape after families or individuals gain sole ownership of land? If people don't fence their land or grow crops, the ecosystem may change little. As soon as people start to build fences or to grow crops, however, the landscape becomes fragmented and wildlife move differently, with restricted access to water and other important rare resources. When wildlife managers fenced the western side of South Africa's Kruger National Park, wildebeest numbers plummeted 70 percent. In the 1980s, Botswana's veterinary fences, built to prevent transmission of diseases between livestock and wildlife, ended up killing some 52,000 wildebeest and 10,000 hartebeest by preventing them from reaching their seasonal pastures; within that decade, after a string of droughts, 90 percent of their numbers had disappeared. As we saw in the last chapter, ecologist Randy Boone created a computer model that predicts what will happen with subdivision in Amboseli. If a herder cuts a 300 km² (116 mi²) savanna into thirty pieces of 10 km² (less than 4 mi²) in the future, each new small parcel would support 25 percent fewer cattle. Even though the animals would have access to the same amount of grass, each plot would have less variable forage than the savanna as a whole did; when conditions change, therefore, the cattle would be restricted in their ability to reach food of good quality. Particularly in east Africa, where rainfall is often patchy, animals can find themselves trapped behind a fence, unable to access a nearby green patch of ground.[27]

Supporters of the privatization of grazing lands argue that ownership addresses the problem of the "tragedy of the commons." This idea of Garrett Hardin, published in the late 1960s, is that groups do not care well for land they hold in common because no individual has sufficient reason to conserve common resources. Hardin's publication provoked a long and heated debate, with opponents asserting that, conversely where land is held in common and governed by community rules of use, no one individual can take advantage of others in the group. Missing from the discussion is what privatization does to the land, especially for livestock herders. In east African savannas, a "tragedy of private ownership" may be more common than we think.[28]

WHAT LIES IN THE FUTURE FOR
THE KAPUTIEI PLAINS?

Are we seeing the last days of this "urban savanna" and its pastoral lifestyle, or do herders have a future here? The pressures on this savanna and its pastoral people are immense. Within a few years, this entire area could be fenced, with much of it cut up into quarter-acre plots, to relieve growing pressure for land in nearby Nairobi. Parts of the Kaputiei Plains could become as densely packed as Nairobi's crowded slums. That said, it is also possible that with strong support of local and national government in partnership with local communities, efforts to keep at least part of this savanna open for pastoral people and wildlife could succeed.

Certainly, if the pace of land sales does not slow, the Kaputiei Plains will become a suburb of bustling Nairobi. In 2004, about 14 percent of the land in the first two triangles was fenced; in the following five years, this figure increased by 52 percent, more than 10 percent per year. Nkedianye interviewed one hundred pastoral landowners to find out who fenced land and who intended to sell land in the future. In 2003, he learned, the average pastoral landowner owned about 150 acres of land, of which he or she had already fenced 11 acres, or 7 percent of the total. Many landowners intended to sell a small portion of their property by 2008 and to fence even more. Most of the people who buy land in the Kaputiei Plains are new arrivals, neither pastoralists nor Maasai, and Nkedianye found that most of these new landowners are fencing their land as soon as they can after buying it, so that it is securely demarcated. Currently, a large number of poor Nairobi residents have been settled in the middle of a major wildlife dispersal area near the town of Isinya (Map 16). In the future, private companies will likely want to expand their industrial facilities to this area. If these trends continue, people, wildlife, and livestock mixing freely will be but a memory very soon.[29]

For all these reasons, many argue that Nairobi National Park should be fenced as soon as possible. This action, they believe, will protect people and livestock in the Kaputiei Plains from predators that live in the park while also protecting the remaining wildlife in the park from illegal poaching and loss of habitat. The risk in this strategy is that, once fenced, Nairobi National Park would likely support much less wildlife partly because the wildlife cannot move to more nutritious grasslands outside the park in the wet season. And a near-empty wildlife park next door to land-scarce Nairobi would be easy prey for land developers. Proponents of fencing admit that the structures would be costly to erect and maintain.

What is less often articulated is how difficult and costly it would be to maintain diverse populations of large mammals in a relatively small, fenced park. Moreover, we simply do not know what would happen to the remaining wildlife populations in the park if they were cut off from the 90 percent of their ecosystem lying to the south. Even managers of Kruger National Park in South Africa and other large fenced parks have found it difficult to maintain wildlife populations that formerly migrated over large areas outside their parks. Once Nairobi National Park is fenced, it is likely that all the wildlife outside the park will be lost unless pastoral peoples establish game ranches there. Many argue that the key to protecting the Kaputiei Plains' wildlife is to provide local people with incentives for doing so in the form of more of the profits from tourism and other benefits that come from that same wildlife. As Nkedianye observes, "It is paradoxical that [pastoral herders] are expected to remain paupers as they live day-in day-out with gold—the wildlife."[30]

In April 2000, the Friends of Nairobi National Park and the Wildlife Trust (later run by the Wildlife Foundation) created the Wildlife Conservation Lease Program to ensure that wildlife in the Kaputiei Plains could move freely to their traditional habitats. Each participating household receives about $4 an acre if they refrain from fencing or subdividing their land, permit wildlife to move freely on their land, protect natural vegetation, report poachers, and do not poach themselves. The lease program expanded in 2007 through funds from the Kenya Wildlife Service, but it still covers too little land—less than 10 percent of the total plains region—and is not growing fast enough. To expand the current program to cover the whole of the first triangle would require about $650,000 per year, or an endowment of $16 million.[31]

Despite these limitations, this program is a rare example of what many have been calling for: a scheme that improves the welfare of pastoral peoples while also helping to conserve wildlife. The program pays participants three times a year, conveniently just when school fees are due. In years of low rainfall, when families need income the most, the payments are sufficient to double the incomes of the poorest landowning pastoral families who lease their land to the program.[32]

Another effort is in place to ensure that families do not have to bear the full cost of a predator killing one of their animals. From 2001 to 2008, lions, leopards, cheetahs, and crocodiles killed over 1,300 livestock in the Kaputiei Plains. The monetary cost to a family losing an animal to a predator is large, but the psychological cost of a lion or leopard leaping over a thorn fence into one's homestead is much greater. Maasai tell of evenings when a lion has jumped their thorn fence and killed ten to twenty sheep in an apparent killing frenzy. A "consolation" program to repay families part of the cost of losing an animal is now run by the Friends of Nairobi National Park. The program pays about $200 for a cow, $33 for a sheep or goat, and $67 for a donkey killed by wild animals.[33]

Are these land leasing and compensation schemes making a difference on the Kaputiei Plains? Participants in the leasing program have more positive attitudes toward wildlife, are more willing to share water and pastures with wildlife, and strongly support keeping the range open without fencing. These land leasers also strongly believe that Nairobi National Park should not be fenced. Nkedianye further recommends: (1) government zoning of the Kaputiei Plains for wildlife and livestock use; (2) the purchase of land to protect crossing points for migrating wildlife along the road south from Kitengela to the border town of Namanga; (3) greater sharing of the park's wildlife revenues with local pastoral communities; (4) a strengthening of livestock disease control, breeding schemes, and other improvement programs; and (5) recruitment of community scouts to monitor poaching. Helen Gichohi adds that there is a need for conservation land purchase, leasing, and easements to focus on tracts of land most important to wildlife.[34]

Recently, when it became clear that the leasing program was not going to slow land development in the Kaputiei Plains enough, several partners banded together to seek a more comprehensive solution. In early 2010 the Greater Kitengela Land Use Master Plan, created by the Kitengela Landowners Association and other landowner groups, the Kenya Wildlife Service, and the County Council of Olkejuado, working with the Ministry of Lands, was signed—the first land-use plan for a pastoral area in Kenya. From the beginning, Nkedianye was a leader in developing this plan, hoping that it could not only save the wildlife and livestock of these wonderful plains, but also, and more importantly, buy time as the younger generations of Maasai rise to become part of the fight for their land. The plan limits the sprawl of towns and settlements, restricts development on wildlife dispersal areas and in riverine vegetation, sets rules on land sales and land sizes in different zones, and designates grazing areas for livestock and wildlife. If this plan is implemented and the lease program is strengthened to keep fences down, there is some possibility that the speed of land fragmentation in the Kaputiei Plains can be slowed. However, the problem now is to implement the plan, in the face of the immense political and economic pressures to subdivide and develop land. Still, Maasai landowners from the Kaputiei Plains keep pushing back: in 2010, they stopped the expansion of Nairobi city into the Kaputiei Plains. Their efforts, while not entirely coordinated, are a model of collaborative action to make positive change in a pastoral savanna under immense development pressures.[35]

The fate of the Kaputiei Plains savanna matters in that it may foretell of development paths for other savannas. The pressures on the Kaputiei Plains for change are arguably stronger than they are on any other intact savanna in east Africa, but the choices that savanna people are facing are similar across the entire region. Even while they argue to keep their traditional grazing lands open, they are buying up land to ensure their future security. They well understand that cutting the savanna into pieces restricts the movement of their livestock and weakens their ability to survive the next drought. They know also that wildlife will not survive on small fenced plots and that the benefits they get from wildlife will disappear along with the wild populations. At the same time, rising material aspirations spur pastoral people on to making more money and engaging more with the government than they have in the past. Like most people, pastoralists want more for their children: they want them to be healthy, educated, and raised in solid houses, like the children of Nairobi, so close by. The question, then, is not whether east African savannas can stay as they are, but rather how people will change the savannas in and for the future. Will they include pastoralism and wildlife, or not?

CHAPTER ELEVEN

Ngorongoro

A Grand Experiment of People and Wildlife

> [The Ngorongoro Conservation Area (NCA)] is
> internationally renowned as a conservation area for its
> scenic beauty, its spectacular wildlife and its important
> archaeological and paleontological remains. It is also
> outstanding for its pioneering joint land use policy,
> which is dominated by conservation aims but at the
> same time maintains a large population of Maasai
> pastoralists living from traditional cattle and small stock
> husbandry . . . . The central management issue in the
> NCA is the conflict between conservation and
> pastoralist interests that has surfaced in many ways
> throughout Maasailand and elsewhere in Africa.
>
> ANTHROPOLOGIST KATHERINE HOMEWOOD
> AND ECOLOGIST ALAN RODGERS, *1991*[1]

Ngorongoro is a grand experiment, a ground-breaking effort to integrate the conservation of wildlife with the development of people. But is this experiment a success? The answer to this question depends on one's point of view. From a wildlife
and tourism perspective, Ngorongoro is a huge success for conservation and its
profits, generating half of all game viewing fees that the Tanzanian government
receives from its wildlife estate, usually more than the nearby Serengeti. Even so,
those with the interests of wildlife at heart worry that the expanding human populations living in Ngorongoro and their need to farm are damaging wildlife and
their habitat and that their livestock are overgrazing pastures. There are worries
that tourist lodges are depleting water sources and too many vehicles are visiting
Ngorongoro Crater. These concerns are large enough that UNESCO recently
questioned the status of the conservation area as a World Heritage Site.[2]

On the other hand, those concerned with the interests of pastoralists point out
that since pastoralists moved out of the Serengeti to Ngorongoro in 1959, they
have been impoverished by the conservation policy, which restricts their livelihoods without compensating them for the costs of these restrictions. (The interests of hunter-gatherers that live nearby are rarely mentioned.) And many Maasai
wonder why the policy has to be restrictive when they have lived side by side with
abundant wildlife since the 1700s.[3]

One solution to these opposing goals, as proposed in the 1996 Ngorongoro Conservation Area (NCA) Management Plan, was to provide different zones for people or wildlife, while acknowledging that some mixing of wildlife and livestock would be necessary for them to be successful over time, since they need to track forage and water with the seasons. But the situation is far more complicated than zoning people and wildlife. With cultural values, profits, and politics at stake, it seems that no one here gets exactly what they want; instead, it is a case of ongoing conflict, negotiation, and compromise.[4]

THE NGORONGORO TODAY: BURSTING AT THE SEAMS

Ngorongoro Crater, formed when an ancient volcano collapsed to form a bowl-shaped caldera about 2 million years ago, is the sixth largest of its kind in the world. Despite its outsized reputation, the crater itself actually makes up only 3 percent (250 km², or 97 mi²) of the Ngorongoro Conservation Area (8292 km², or 3,202 mi²). The rest of the NCA is equally spectacular, with the Serengeti Plains and Oldupai Gorge to the west, the Crater Highlands to the north and southwest, the Angata Salei Plains and Gol Mountains to the northwest, and the escarpment that falls down to Lake Eyasi in the southwest (Map 17). Two other craters, Olmoti and Empakaai, are north and northeast of Ngorongoro's main crater. Just north of Oldupai Gorge are moving sand dunes. The Serengeti National Park lies to the west of Ngorongoro's boundary, the Loliondo Game Controlled Area to the north, and agricultural communities to the south and east. The still-active Lengai volcano sits just off the NCA's northeast corner.[5]

Ecologically speaking, Ngorongoro is not unusual for having a huge migration or rare species (except for black rhinos and some plants), and it does not have snow-capped mountain peaks. Rather, Ngorongoro is special because, over short distances, the savanna changes from wet to dry, soils from rich to poor, and vegetation from forest to grassland. The Crater Highlands are wetter than the soggy parts of Kenya's Mara (1200 mm [47 in] annual rainfall). Nearby, the low rainfall in Oldupai Gorge is similar to that around Kenya's dusty Amboseli (300–400 mm [12–16 in]). The volcanic highland soils, deposited only a million years ago, are rich with nutrients, while just north in the Gol Mountains you find poor soil weathered from ancient basement rocks of the African shield that are hundreds of millions of years old. This variety creates a rich mosaic of vegetation, from the dry short grass of the Serengeti plains (50 percent of the NCA) to highland tussock

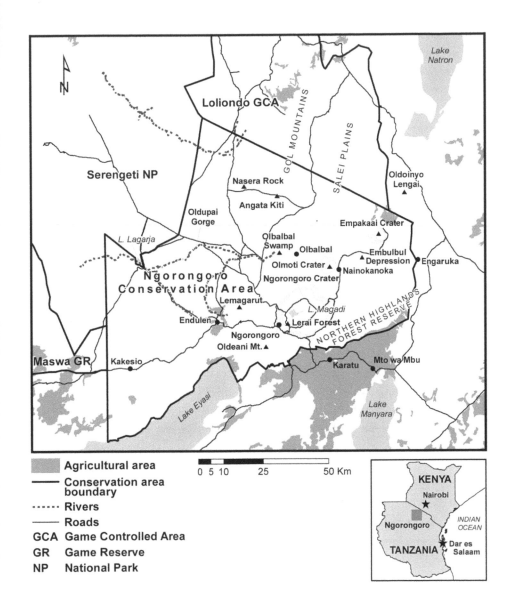

MAP 17.
Ngorongoro Conservation Area and environs in Tanzania. (Map
adapted from DMA 1992 and WPDA 2009 by Shem Kifugo and
Russell L. Kruska.)

grassland, swamps, and wetlands to bushland, woodland, montane heath, and wet highland forest. Globally, this variation is unusual: compared to twenty-one other savannas and steppes in Africa, Asia, North America, and Australia, the Ngorongoro Conservation Area has very productive, diverse, and reliably green vegetation, partly because of its large changes in elevation over short distances (see Map 9).[6]

This topographical variation allows some wildlife to remain resident in the Ngorongoro Highlands year-round and also provides important grazing lands for the large influx of migratory wildlife on the short-grass plains in the wet season. About half of the massive Serengeti wildebeest population spends some time in the short-grass plains that spill into the NCA, though this mass of animals does not migrate into and out of Ngorongoro Crater itself. Most wildlife resident in the crater, however, stay in or near the crater: only about half its zebras and a fifth of its wildebeest (in an average year) climb out of the crater to graze with the Serengeti herds on the short-grass plains. When the Serengeti herds head north toward Kenya's Mara as the short-grass plains dry, most of the Ngorongoro animals turn east and return to the crater. Elands, buffaloes, rhinos, and elephants have a different schedule: when some of the zebras and wildebeest leave the crater in the wet season, they remain in the crater; then, when the grasses dry, some retreat to the Crater Highlands, while others remain in the crater's Lerai Forest.[7]

In 1954, five years before the Ngorongoro Conservation Area was established, about 10,000 people (perhaps more) lived in the region that was to become the NCA. By 2002, the Tanzanian government reported that about 61,000 people lived in sixteen villages inside the borders of the NCA, including Maasai herders, Tatog agro-pastoralists, Hadza hunter-gatherers, and a smattering of people from all over Tanzania working for the Ngorongoro Conservation Area Authority (NCAA), which manages the NCA, or tourism companies. Clearly, human populations grew quickly here in the 1990s, though according to interviews conducted by ecologist Victor Runyoro, only 5 percent of residents were born outside the conservation area, and only 10 percent of the growth was caused by in-migration. (These figures need to be treated with caution, however, because the recurring threat of resettlement of immigrants may affect how the interviewees reported their origins.) The upshot, in any case, is that population growth is a major challenge for the NCA management.[8]

By both Tanzanian and global standards, even the richest Maasai families in Ngorongoro are poor. The average Tanzanian makes five times more income than the average Maasai in Ngorongoro. In the late 1990s, 35 percent of the adults and over half the children in Ngorongoro were under- or malnourished. Almost half of the Maasai children who are of school age do not attend school, and only 1 percent of those who do attend school go beyond primary school. As elsewhere in Maasailand, livestock are still the mainstay (73 percent) of family incomes, supplemented by crops (20 percent), but only 2 percent comes from tourism—chiefly handicrafts. Over time, livestock numbers have not kept pace with the growing number of people, so livestock products have made up less and less of each family's foodbasket.[9]

As in many pastoral lands in east Africa, not only are there fewer livestock per person in Ngorongoro, but families need more cash to buy medicines, cover school fees, and pay taxes. Typically, families are compelled to sell livestock to meet these needs, and thus gradually further erode the cornerstone of their wealth. Human ecologist Terry McCabe and his colleagues asked Maasai why they were selling reproductive cows, sacrificing their long-term food security for immediate cash needs. "What else could I do?" came the response. "My children were hungry." Without enough livestock to support their families, Maasai then increasingly turn to crop farming to bolster food stocks and generate more cash, which is a traditional way for them to deal with hard times like drought.[10]

In the wet season, herders prefer to graze their cattle on the highly nutritious short-grass forage of the western and northern Ngorongoro. But the million-strong wildebeest migration grazes here too, partly to obtain nutrients critical for gestation and lactation and also to bear their calves where predators are highly visible. Wildebeest carry the viral disease malignant catarrhal fever (MCF), which is secreted from the noses and eyes of wildebeest calves for their first three months of life. MCF is deadly to cattle, but not to wildebeest—which means that for the three months of the year when wildebeest are calving, the Maasai must avoid these lush short-grass plains.[11]

Contraction of MCF was not a problem for Maasai between the time rinderpest was introduced in the 1890s until it was controlled in the 1960s. During this period, wildebeest populations—which were also susceptible to rinderpest—were low and Maasai could use the plains freely. With control of the disease, the wildebeest populations grew rapidly, presumably back to near pre-1890s levels. By 1976, the Serengeti wildebeest population again covered the short-grass plains in the wet season, making the entire area off limits to Maasai and their cattle for several months. Ecologist Randall Boone and his colleagues estimate that Maasai could increase the size of their cattle herds by about 20 percent if they had access to the short-grass plains when the wildebeest calve.[12]

Avoiding MCF has another cost for herders. While a three-month hiatus during wildebeest calving may seem like a minor inconvenience for the Maasai, those who graze on the poorer pastures of the wet Ngorongoro Highlands put their animals at risk of contracting another deadly disease, tick-borne East Coast fever (ECF)—and just at a time when the adult ticks are most abundant. In the past, ECF killed 30–90 percent of first-year Maasai calves, though recent veterinary efforts have reduced the risk for ECF considerably, helping to revitalize pastoralism in Ngorongoro.[13]

Do the restrictions required by conservation policy on how Maasai can use the land also affect their livelihoods? Anthropologist Kathleen Galvin and ecologist James Ellis and their students set out to answer this question by comparing the livestock holdings, cropped area, income, and nutritional status of families in Ngorongoro with those in Loliondo, which is just north of Ngorongoro but does not have the same restrictions on grazing and cultivation. Their team found that Ngorongoro Maasai are poorer, owning only a third the livestock and with farms half the size as the Loliondo Maasai. Infants, children, and adults weigh less in Ngorongoro compared to Loliondo—1.5 kg (3.3 lb) less on average for two-year-olds, 3.5 kg (7.7 lb) for their mothers. They conclude that Ngorongoro's land use policy, which originally aimed to both conserve wildlife and promote human development, has failed the Ngorongoro Maasai.[14]

One way conservation policy can positively affect pastoral residents is through employment and sharing profits. However, only 12 percent of the permanent employees of the NCAA are local residents, compared to the nearly all-Maasai staff in Kenya's Mara. In terms of revenue, the NCA receives about 50 percent of the game-viewing fees for the entire country, which from 2003 to 2008 was an estimated $8–11 million annually. Some Maasai claim, however, that official revenue figures are underreported, with half of the profits lost to corruption. Indeed, many of the profits are thought to bankroll one of Tanzania's political parties, whose senior representatives have repeatedly said that they intend to evict pastoral residents from the NCA altogether. Of the revenues actually received, Maasai see about 10 percent in the form of improved social services. In terms of direct family income, tourism (the selling of crafts) boosts wages by about 2 percent, or $1.20 per person per year. Thus, Ngorongoro families live with the costs of wildlife and conservation policy, see huge profits garnered by government and the tourism industry from wildlife, but see basically no increase in their household budgets to lift them out of poverty.[15]

It is ironic that the conservation policy that restricts Maasai land use in order to protect wildlife more or less forces Maasai to plant more crops to meet their needs, which, if farming expands, will remove wildlife habitat. Farmers cultivate maize, beans, potatoes, and vegetables in the NCA and have been farming in the Ngorongoro Highlands since before the park was established. Although cultivation was banned in the NCA in 1975 for fear farming would harm wildlife, the ban was temporarily lifted in 1992 until another way could be found to meet the food needs of the Ngorongoro residents. The expansion of farming had the intended results: Maasai households became significantly more food secure. Since 1992, the ban on cultivation was temporarily reinstated in 2001 and again in 2009 (remaining in effect as of 2011).[16]

But the Maasai were not the only ones farming in Ngorongoro. According to ecologist Alan Kijazi, in 1993, shortly after the first cultivation ban was lifted, about half of the farmed plots were more than 20 acres (8 ha) in size and were cultivated by non-Maasai farmers, attracted to the NCA now that farming was again allowed. With the rescission of the ban, Randall Boone, Victor Runyoro, and colleagues estimated that the area of farms rose from 5,000 acres (2023 ha, 20 km^2, or 8 mi^2) in 1992 to 9,800 acres (3966 ha, 39.6 km^2, or 15 mi^2) in 2000, and to 13,600 acres (5503 ha, 55 km^2, or 21 mi^2) in 2004. Even at 13,600 acres, this is a minuscule 0.66 percent of the total NCA land area. When Boone examined satellite imagery in 2010, he found that the large farming fields were largely gone, thanks to efforts by the NCAA to discourage large-scale farming to minimize threats to wildlife.[17]

How is the Ngorongoro doing as a model to protect wildlife? Is settlement, cultivation, grazing, or burning causing major wildlife declines, or is a moderate level of use, as in the Mara (Chapter 8), attractive to some wildlife species? As we have seen elsewhere, farming is the most wildlife-*in*compatible way that Maasai use the land. Ecologist Randall Boone and his colleagues used computer simulation modeling to estimate how expanded farming would affect wildlife and found that 0.5 percent cultivation (the 2000 level in Ngorongoro) had little impact on wildlife. They then calculated what would happen if cultivation increased by five times to 2.5 percent coverage, expanding adjacent to places where people already farmed. They concluded that after ten years Ngorongoro would see a 15 percent increase in browsing antelopes, and the following declines: 33 percent of resident zebras, 10 percent of elephants, 8 percent of grazing antelopes, and 3 percent of buffaloes. Ecologist Patricia Moehlman thinks a 33 percent reduction in the zebra population would have a serious effect on population viability. And impacts on wildlife could be higher if Maasai farmed in the places most important to wildlife.[18]

In 1974, the NCA management evicted the Maasai who lived in the Ngorongoro Crater, leading to long-term impacts on wildlife. After that year, some Maasai were allowed to bring their cattle into the crater for water, salt, and daytime pasturage, but they could not graze extensively, burn the grassland, or build settlements. After the Maasai moved out of the crater and stopped burning the crater grasslands, small to medium-sized grazers began to disappear and large grazers became more common. Wildebeest and gazelles, which thrive on short, nutritious grass, were replaced by Cape buffaloes, which can thrive on tall, higher-fiber grass. Buffalo populations in turn grew from about 200 in 1970 to over 2,500 in 2005. Despite these changes in the populations of various species, the biomass, or total weight of wildlife, in the crater has remained remarkably constant.[19]

Victor Runyoro and colleagues think removal of the Maasai and stopping their grassland management was one reason small wildlife declined while buffaloes became more common in the crater after 1974. I think there may be additional impacts of Maasai living and grazing their livestock on the crater grasslands that affect wildlife. As we saw in Chapters 2, 6 and 8, pastoralists likely enrich savannas by creating patches of more nutritious forage in burn scars and in short, grazed grass around settlements. This short grass around settlements is not only nutritious but may also be safer for grazers, since predators are more visible in the closely cropped grass—an effect with potentially special weight in Ngorongoro, which historically supported more predators per hectare than most savannas in Africa. In addition, cattle compete with buffaloes for forage, especially during the dry season, which means that buffaloes may have been repelled by the Maasai when they lived in the crater.[20]

Maasai may inadvertently protect lion cubs but hurt lionesses as well. Biologist Bernard Kissui and colleagues saw more lion cubs (per km^2) survive their first year if their pride lived closer to Maasai daytime grazing areas in the crater. This inadvertent protection may come from repelling Cape buffalo (which can trample cubs) and hyenas. At the same time, however, the opposite occurred with adult lionesses. More died in prides that were closer to Maasai and their herds; Ngorongoro Maasai are known to spear lions on occasion.[21]

For lions at least, there are pluses and minuses to living around Maasai. What this and Runyoro's data suggest is that the Maasai have had significant effects on both herbivores and carnivores in the crater, not all negative. It also means that some of the interactions among the rich wildlife herds that the NCA was designed to protect may be human-enhanced, though some may not. I think this argues that the crater's former human residents should have a stronger hand in managing wildlife there, and may even argue that the crater would benefit if the Maasai were allowed to reside there again. The latter, however, is probably politically impossible.

The Maasai may have also helped the Lerai Forest, the main forest on the floor of the crater, full of yellow fever acacia trees (*Acacia xanthophloea*), to become established and flourish. Ecologist Antoni Milewski suggests that the Lerai Forest may have started in a tree plantation in an abandoned livestock corral, as we saw with *Acacia tortilis* in Chapter 6. He also suggests that Maasai and their livestock may have reduced elephant damage to the forest by their presence, thus encouraging the trees to thrive. These ideas are difficult to confirm or deny, but it would be intriguing to try.[22]

Many of the trees in the Lerai Forest are now dying. Few young trees grow in the main part of the forest, although I have seen many small seedlings in open grassland at its edges. One culprit may be high salt concentrations in the soils of the forest. Biologist Anthony Mills analyzed the soils of the dying Lerai Forest and those of a healthy forest. Lerai Forest soils were more saline, down to a depth of 40 cm (16 in). Why have these soils become so salty? Mills suggests that salt accumulates when there is less fresh water flowing into the crater, as occurs when tourist lodges that dot the crater rim withdraw ever larger amounts of water from incoming streams. With less flow, trees may die from both drought stress and increasing salinity. Another culprit may be a die-off of old, even-aged trees, as we saw for Amboseli in Chapter 9.[23]

One final question to ask is whether Maasai livestock are overgrazing Ngorongoro grasslands. Surveys of soil erosion in the NCA compared to other grazing lands nearby suggest that erosion is low in Ngorongoro, except along livestock trails or where soils are particularly unstable. Another indicator of decline or degradation is the spread of unpalatable grasses, which grazing can contribute to by removing other, competing grasses. Anthropologist Katherine Homewood and ecologist Alan Rodgers outline the ongoing debate in the NCA about the possible spread of the tall, highly unpalatable elephant grass (*Eleusine jaegeri* Pilg.)—a species that has been abundant in Ngorongoro for a long time, and so it is unclear if it is spreading or not. Maasai traditionally burn tough elephant grass to remove it and encourage tender new growth. But experiments at Empakaai Crater to control *Eleusine* and encourage other palatable species showed that digging up the grass by the roots works even better than burning it. Overall, range scientists Winston and Lynne Trollope conclude this about Maasai management of Ngorongoro: "It is . . . interesting to note that all the surveys conducted in the areas inhabited by the Maasai pastoralists had very high forage and fuel production potentials indicating that their management of the rangelands is very sustainable . . . . The range does not appear to be overgrazed and there are no bare patches or signs of any form of accelerated soil erosion."[24]

A HISTORICAL PERSPECTIVE

Our picture of Ngorongoro remains incomplete without an understanding of major changes this savanna and its peoples have undergone over time. As described in Chapter 3, our oldest evidence of australopithecines walking upright comes from the 3.7-million-year-old Laetoli footprints, which archeologists found in the southwestern part of the Ngorongoro Conservation Area in the late 1970s. Australopithecines and then ancient humans progressively learned to engineer their environment, first by scavenging, then by burning the savanna and hunting ever more proficiently. About 2,000–2,500 years ago, the first pastoral people (southern Cushites), who also probably grew dryland crops, reached the Ngorongoro at about the same time that Bantu farmers settled around Lake Victoria to the northwest of Ngorongoro and Serengeti. It is not clear when farmers first cultivated in the Ngorongoro Highlands; if not the earlier southern Cushites, it may have been the Tatog herder-farmers more than 500 years ago, perhaps creating its current mix of forest, shrubland, and grassland. A large village of about 10,000 people (as large as Mombasa or Kilwa of the time), possibly occupied by the Tatog, ran sophisticated irrigated sorghum farms northeast of the present-day NCA at what is now Engaruka from the late 1400s to the early 1800s. When the Maasai arrived in Ngorongoro about three hundred years ago in the 1700s, they pushed the Tatog herder-farmers out of the Ngorongoro Highlands and then lived and grazed in this area. As elsewhere, the Maasai in Ngorongoro were hard hit by the rinderpest. In 1892, Oskar Baumann traveled through the Ngorongoro and described the Maasai living in their *boma* (thorn-enclosed settlement) in Ngorongoro Crater as being in an advanced state of starvation, desperate for food, because of the ravages of the disease. By the early 1900s, the Maasai began to recover with newly rebuilt herds and shared the pasture in the crater with two German colonists, the brothers Friedrich and Adolph Siedentopf, who lived and farmed there from about 1899 to 1916.[25]

Part of the reason the colonial government created Ngorongoro as a multiple-use area was to compensate the Maasai for the loss of Serengeti National Park, which was established in 1951. Between 1928 and 1959, the colonial government restricted hunting and agriculture in Ngorongoro, though it allowed local people to continue their customary land use. In 1959, the colonial governor of Tanganyika established Ngorongoro Conservation Area (NCA) and addressed the Maasai Federal Council with this promise: "I should like to make it clear to you all that it is the intention of the government to develop the [Ngorongoro] Crater in the interests of the people who use it. At the same time, the Government intends to protect the game animals in the area, but should there be any conflict between the interests of the game and the human inhabitants, those of the latter must take precedence."[26]

Some suggest that the Maasai were forced out of the Serengeti by international interests and the colonial government and that violence was involved. The Maasai gave up access to Serengeti's permanent springs at Moru and Ngare Nanyuki but were promised new dams and boreholes in Ngorongoro. The dams, however, "silted or breached within the first two years after construction. The boreholes, too, proved a disappointment due to poor maintenance, while that at Kakesio delivered water too saline for livestock, leave alone for human consumption. Worse still, these inadequacies were never remedied nor were the dams ever repaired"— until recently.[27]

Since its establishment, Ngorongoro has struggled with its dual mandate, restricting and then loosening rules about how indigenous Ngorongoro residents (Maasai, Tatog, and Hadza) can use the land and other natural resources. From 1959 to 1974, people could settle anywhere in the conservation area. But starting in 1974, conservation authorities created a core protected area surrounded by a buffer zone, prohibiting settlement in Ngorongoro Crater (but not the rest of the NCA). Further restriction ensued the next year with the banning of cultivation across the whole NCA. In return, the government promised to improve the supply of grain to resident families by creating a regional trading company, the aim being to improve livestock sales. In 1979, Ngorongoro was internationally recognized for its unique character when UNESCO declared the NCA a World Heritage Site, one of the few with both significant cultural and natural aspects. In 1992, the government lifted the ban on cultivation temporarily to ease the food shortage among the Maasai. In 1996, the new Ngorongoro General Management Plan, developed by a wide range of stakeholders, including Maasai, laid out ways to reduce the number of people living in Ngorongoro, in part by moving all people who worked for the NCAA and tourist lodges out of the conservation area. Recently, the NCAA built a school, dispensary, police station, and a road from the conservation area to Oldoinyo Sambu, about 70 km (42 mi) north of the NCA, and offered 2 acres of land per person there to encourage immigrant farmers to move from Ngorongoro to Oldoinyo Sambu. By 2007, about 15 percent of the 1,725 immigrant farmers had relocated voluntarily out of Ngorongoro.[28]

Gradually, Maasai started to have more of a voice in how the NCA was managed. Established in 1994, the NCA Pastoral Council provides a forum for discussion between resident pastoralists and the NCA Authority. Until the council was formed, Ngorongoro residents resented the NCAA management, feeling powerless. Today, locals select eighteen of the forty-one representatives on the NCA Pastoral Council; the other twenty-three hold other political offices. Since 1998–1999 the council has received about 10 percent of the total revenues of the NCA, which it then disburses for education and training, providing scholarships for Ngorongoro students to attend secondary school and colleges or universities. In the end, though, neither the members of the council nor other Ngorongoro residents believe the council can effectively represent the community. Lawyers Issa Shivji and Wilbert Kapinga explain why: "First, the Council is not statutory; second it is not an executive body whose decisions are binding; third it has limited powers and no say over the Authority's legislative activity. What is more, its composition is dominated by ward councilors and village chairmen who, although elected, were elected with an altogether different mandate."[29]

Part of the reason Maasai lack power in the NCA is that they have no title to land. In rural areas of Tanzania, most land is village land (rather than private); thus, villages hold local village certificates or land titles, and the village council manages all lands within the village boundaries. Ngorongoro residents, however, do not enjoy these same land rights. Instead, the NCA Authority has statutory power over land and how it is used. This means that resident families or communities cannot, for example, make independent arrangements with tourism operators to capture more of the profits from wildlife on "their" land, because it isn't in fact theirs. If tourism operators try to make such arrangements with residents by working through the NCAA rules, the process is complex and bureaucratic, and so they go elsewhere (like to nearby Loliondo). Although Maasai and other indigenous residents do have "customary rights of occupancy" through Tanzania's Village Land Act of 1999, the NCAA dictates how land is used; this makes any potential Maasai–private sector partnerships insecure. It is not clear that private ownership of the land by Maasai would solve the problem, because the NCAA could then limit pastoral land use to the parts of the NCA not within village boundaries.[30]

One recurring debate is whether the Maasai should live in Ngorongoro at all. Without village land titles, Maasai have a real fear that they will be forced to move out of Ngorongoro altogether. Indeed, in 1979 the NCAA Board of Directors declared that pastoralists would eventually have to leave the NCA to ensure better protection of wildlife. But the Maasai disagree. When asked by Victor Runyoro what they think about resettling outside Ngorongoro, nearly half would not even answer, presumably because they were so opposed to the idea. One group representative said the Maasai would rather die in the NCA than be resettled. About a quarter of respondents said people who do not have livestock but want to cultivate beyond subsistence could be relocated. Another group said cultivation could be phased out of Ngorongoro after the livestock economy improves. Others said that it is important to maintain subsistence cultivation, even if the livestock economy improves, because pastoralists would still want to eat grains owing to changing food preferences, especially on the part of young people. These respondents said they would be willing to limit the size and location of farmed fields.[31]

WHAT'S NEXT FOR NGORONGORO?

It is clear from Ngorongoro's history that a dual mandate is difficult to manage, with many recurring challenges. The NCA Authority has made remarkable progress, however. Although it has made some significant mistakes (such as allowing major losses of rhinos to poaching, as happened elsewhere, or letting immigrant farmers in), it has also had major successes (maintaining largely healthy ecosystems and wildlife populations, for example). The challenges ahead will involve meeting its dual goals, a task recognized clearly in its 1996 and 2006 General Management Plans and by Maasai residents. Runyoro outlines these challenges as follows:[32]

- Keeping human populations and cultivation at levels that are compatible with wildlife
- Allowing pastoral residents more influence and authority in the way the NCA is run and integrating other resident tribes into NCA management
- Providing compensation to pastoralists for the restrictions on land ownership, grazing, cultivation, and settlement required to meet conservation

goals and developing new ways for residents to share in governmental and private-sector profits from wildlife

- Strengthening pastoralism as a resilient livelihood option supported by subsistence farming of limited scope
- Training pastoral residents so that they can qualify as employees ranging from walking guides to tourism company CEOs and even the head of the NCA

How could Ngorongoro residents have more say in how the NCA is managed? First, from my perspective, a full third of the seats on the NCA Board of Directors could go to indigenous residents, representing the Maasai but also the Tatog and Hadza. Continued and further representation would ensure that local voices are influential at the highest level of decision-making in the area. Ngorongoro would then become a role model for parks around the world that have restricted the rights of local, indigenous people. Second, in appointing the next conservator of the NCA, priority could be given to a local resident, all other qualities being equal. In the 1980s, Solomon Ole Saibull, a Maasai, was the second conservator of the NCAA, serving as its chief administrator, so a precedent has been set. Third, the Pastoral Council could be transformed into an independent watchdog body, with independent funding. These types of actions would profoundly reshape the relationships of the NCAA with local residents, while maintaining the current structure of the NCAA. Such actions would also require that the residents be fully accountable for maintaining a healthy Ngorongoro ecosystem with vibrant wildlife populations.[33]

Good progress is being made to reduce the number of people living in the NCA and to limit cultivation, though this issue will continue to be a challenge. All of the NCA and lodge staff, perhaps five thousand people, are being moved outside the conservation area, about 5 km (3 mi) from Lodoare Gate, to the 435-acre (176 ha) Kamyn Estate. Immigrant farmers are voluntarily moving out, which has led to a dramatic decline in larger-scale cultivation since 2003. However, even with only indigenous residents left, natural growth of their populations will eventually overcrowd the NCA unless other rules come into play. If populations rise to 150,000 residents, which would take a long time without immigration, at least a quarter of the food budget in poor families will have to come from relief or supplementary food, because livestock populations will be entirely insufficient to support this population level. Further progress in giving residents real authority in the NCA will help manage population in a fair and transparent manner.[34]

Clearly, indigenous residents could share more in the profits Ngorongoro generates. For example, new legislation could make it easier for residents to enjoy the same profit-sharing arrangements with the private sector that are now common around Maasailand in Kenya and Tanzania (see Chapter 12). And more effort could be put into education to help Maasai qualify for employment at the NCAA and in the tourism industry.

In addition, the NCAA could share more of the revenues from conservation. Most current sharing of revenues goes to people via the Pastoral Council, through social services that only indirectly help households and are weakly connected to conservation. One option is direct payments of cash to households in exchange for delivering conservation outcomes. In Kenya's Kaputiei Plains, for example, households receive payments for forgoing fencing and removing poachers' snares (see Chapter 10). In Ngorongoro, households could be compensated for forgoing use of Ngorongoro Crater for grazing (symbolic importance) or forgoing cultivation of more than one acre of land. If the NCA shared a quarter of its reported total revenue of $5–$7.5 million annually with each of the approximately 60,000 indigenous residents of the NCA, the average household income would increase by as much as 50 percent. Given that at present Ngorongoro families make only 20 percent of the average per capita income, this direct supplement could strongly reduce poverty and garner substantial support for conservation in the NCA. However, such largesse could also make it difficult to control immigration into the NCA.[35]

While it sounds attractive for the NCAA to share more of the profits from wildlife, published accounts show that revenues are currently on a par with expenditures in the NCA. If these figures are correct, then further sharing would require either cutting back on expenditures or raising new revenues. Because NCA operations likely need more rather than less support, new revenue is a more attractive option. This could be accomplished through higher fees, but more promising are entirely new sources of income like higher values for Ngorongoro livestock products (see below) or new revenue from payments for ecosystem services, like water, carbon, and biodiversity.[36]

Much progress was made in improving pastoralism as a profitable livelihood between 1998 and 2009 when the NCAA and the Ereto Ngorongoro Pastoralist Project (a bilateral project of the Tanzanian and Danish governments) restocked poor herders with livestock and strengthened traditional ways of alleviating poverty, such as social transfers of livestock, or *ewoloto*. The NCAA and the Ereto project also refurbished nonfunctional dams, wind pumps, boreholes, wells, and small pipelines. Pastoral communities say these water points allow them to spread out their grazing and revitalize herd movement so that they can let pastures recover from grazing seasonally. Clearly, if livestock populations grow and wide areas are continuously and heavily grazed, this benefit will be limited, but at least initially more extensive water may be an environmental boon. It will have to be carefully managed, however, with regular livestock movements, to give space for wildlife and avoid continuous livestock grazing.[37]

The Ereto project also provided training for community veterinarians and introduced immunizations against East Coast fever in 1999, which strongly improved calf survival. Perhaps most important was the project's effort to build local leadership and to create and strengthen local community-based organizations. Yet Maasai suggest more needs to be done to further improve the profitability of pastoralism, by raising sale prices of livestock and reducing the cost of veterinary drugs. Eventually, specialty livestock products like "Serengeti or Ngorongoro beef" may be available to sell to tourist lodges, but this would require Maasai to raise improved breeds of cattle and build better slaughter facilities to meet the quality standards of an international market.[38]

How might all these improvements in pastoralism affect wildlife and the Ngorongoro ecosystem? We don't know, but ecologists Randall Boone, Kathleen Galvin, and their colleagues set out to make some educated guesses using computer models. What would happen if farming expanded or there were more water points, or if epidemiologists found a way to control malignant catarrhal fever? As described above, even five times as much cultivation may have only a limited impact on wildlife, except zebra. But this prediction does not account for potential disruption of wildlife corridors or cultivation in sensitive areas like the crater rim or Obalbal. More water points will cause livestock to spread out, as occurred when the NCA and Ereto project repaired water sources, and this could become a problem unless properly managed. If scientists invent a cure for malignant catarrhal fever and cattle can graze the short-grass plains in the wet season, Ngorongoro could support another 20,000 cattle.[39]

And what if the NCA allowed livestock to graze in Ngorongoro Crater or let livestock populations grow? Boone and colleagues found that if the NCA opened the crater to livestock grazing, neither livestock populations nor their nutritional condition would change much. If restocking programs were to boost livestock populations by 50 percent, the ecosystem could support them only until a drought occurs, when up to half the cattle might die. If these dead livestock were continuously replaced, some classic signs of overgrazing might occur, with less palatable grass and more unpalatable grass, which would then support fewer wildlife and livestock. If restocking does continue, improved markets and sales will be important to make sure this successful poverty alleviation effort does not end up backfiring and impoverishing herders, livestock, and wildlife. While these are only good guesses, based on the best of our scientific knowledge, they do provide a catalyst for management discussions and can be tested in the future. Nothing will replace experimenting with different approaches, monitoring what happens, learning from mistakes, and improving for the future.[40]

Ngorongoro is indeed a grand experiment that would be much diminished if either people or wildlife were not part of its mandate. From one point of view, as Tanzanian law professors Issa Shivji and Wilbert Kapinga suggest, the creators of Ngorongoro violated some of the rights of indigenous residents to land, livelihood, and democratic governance in the name of wildlife conservation. It will be a long and difficult road to recover or compensate for the loss of these rights, and harder still to ensure that any restored rights are equitably distributed. It is important to recognize that compensation is not just a matter of a share in wildlife profits, but must also support indigenous residents' rights to land and other resources.[41]

From another point of view, these same Ngorongoro creators could be considered important (although indirect) supporters of pastoral land rights because creating Ngorongoro prevented the inevitable—farmers pushing herders and wildlife out of the highlands into other, more marginal lands. With more attention to giving Ngorongoro's traditional residents more say in its management and clearly linking better pastoral welfare with conservation goals, prospects are good that this experiment will thrive, if not without struggle, into the future.[42]

CHAPTER TWELVE

Savannas of Our Future

Finding Diversity in the Middle Ground

Pastoralists have always been resilient and highly adaptable under stress. . . . Despite gloomy prognostications about the future of pastoral peoples, they will continue to survive, though not necessarily in the same form. . . . Laments for the demise of pastoralism are themselves part of the historical tradition, but real pastoralists, awkwardly, have refused to die.

HISTORIANS RICHARD WALLER AND NEAL SOBANIA, *1994*[1]

Rains have declined to a big extent. During colonial time [i.e., before Tanganyika mainland's independence in 1962], by November, there would be plenty of rainfall. In the 1980s, rainfall arrived from September onwards. Big changes occurred in the late 1990s. . . . [This year, 2002] . . . November [is already a dry month]. . . . It is not easy to predict rainfall anymore.

MAASAI INFORMANT, *Tanzania, interview by ecologists Gufu Oba and Loyce Kaitira, 2002*[2]

Recent and ongoing formation of private conservancies and tourism concerns on formerly Masai group ranch land . . . in which land owners voluntarily vacate their land for wildlife in exchange for land rents, if encouraged, would promote recovery of wildlife populations in the ranches.

JOSEPH OGUTU AND COLLEAGUES, *referring to Mara conservancies, 2009*[3]

We can now start to see wildlife like our cousins. . . . We can start to drink their milk.

MAASAI *informant referring to sharing wildlife benefits, Tanzania, interview by geographer Mara Goldman, 2006*[4]

FINDING DIVERSITY IN THE
MIDDLE GROUND

Now comes the biggest question: Why is this story important? The answer is simple: This story is important because this is where our ancestors came from and savannas are part of our deepest evolutionary past. Many of us today live in urban areas, yet the basic materials for most of our basic needs (food, shelter, clothing) are created by people harvesting the bounty of rural lands; thus, it is also a story about the critical struggle to balance lifestyles, food production, and the health of the environment. If you are an east African but not a herder, this is the story of your sisters and brothers of the savannas, and their responses to the benefits and challenges of living with wildlife. If you are a pastoralist living outside east Africa, this story may connect you with other pastoralists. And if you yourself herd livestock in an east African savanna, I hope the omissions in this book annoy you enough that you decide to tell your own story more fully and more vibrantly than I could ever do here.

This tale, while broad in scope, has some clear themes that build from the past. Most startling is that current evidence suggests that our ancestors probably evolved in Africa and not elsewhere. Only here, in Africa, did the ancestors of today's savannas species—humans, large mammals, savanna plants, and other organisms—evolve, side by side, over a very long period of time. The very existence of savannas, and of wildlife, changing climates, rich soils, and topographic diversity, may indeed be key to our own evolution. The challenges and opportunities of varying climates favored some proto-humans to eventually walk upright and out into the savanna and to scavenge, gather, and hunt. When hominins evolved, the opportunity—and problem—of exploiting highly productive savanna mammals for food likely allowed those ancients humans who worked together to survive more successfully than others who did not. Hominins with larger brains and social cooperation invented ways to control their surroundings rather than be controlled by them, by using tools, fire, and other practices. Once our genus, *Homo*, appeared, humans progressively became better and better "ecosystem engineers," learning to capture ever more savanna resources to satisfy their growing needs. Thus, to fully understand our own evolution, we must view it in the context of African savannas and their large mammalian fauna.

But that is not all. Unlike elsewhere on the planet, the great Pleistocene herds of large mammals still exist in parts of eastern and southern Africa (though they are rare in western Africa)—this despite the ivory trade, colonial and native hunting, foreign diseases, and the spread of towns, cities, and farms. And they only survive in any significant numbers where people have chosen not to exterminate them: in protected areas, on commercial wildlife ranches, and in some savannas inhabited by pastoral herders. By far, the oldest way to coexist with wildlife is deeply cultural within some pastoral societies, and most protected areas and commercial ranches were carved out of these pastoral savannas in the last century. Even so, in some African pastoral societies, herders today actively kill significant numbers of wildlife; in others, like Maasailand, many herders still choose to tolerate wildlife and rarely kill them. Where this ancient tolerance continues to exist, herders appear to be attracting grazing wildlife to their settlements where close grazing by livestock keeps grass short and nutritious and where wild grazers can see potential predators at a distance in the short grass. Although in some places herders and their livestock have damaged the savannas, especially where soils and vegetation are fragile or where herders have settled permanently or water their livestock on a constant basis, it appears that this damage is limited in extent and often reversible. Rather, savannas—and consequently the wildlife that depend on them—are much more negatively impacted by being turned into maize farms, flower greenhouses, textile factories, and cities. In these cases, tolerant coexistence can turn into full-blown conflict, and when it does, people often exterminate large wildlife.

In this final chapter, I ask and propose answers to two questions. (1) What kind of life can pastoral peoples and the wildlife with whom they share the savannas of east Africa expect, if current trends continue? And (2) is it possible, using the ingenuity of the best ecosystem engineer on the planet, to reach a sustainable middle ground, where both humans and wildlife thrive and prosper? In answering these questions, I continue to weave together what others have written, but I also speculate based on logic and my own observations.

WHAT MIGHT THE FUTURE BRING
IF CURRENT TRENDS CONTINUE?

And now the impossible: a guess about the future, based on the best knowledge we have today. Of course, we cannot predict the future, and even if a prediction were right for one place it might be entirely wrong for another. This arises because each place follows a unique path created by its own history; thus, each savanna landscape and its people will respond in their own way to a complicated set of internal and external pressures. Pastoral people and savannas are now facing the challenges of climate change, globalization, growing human populations and migration, demands of growing economies, changes in land use and tenure, new technologies, new policies on conservation, and others. In the following, we will look at some generalities of what might happen if current trends continue, and then we will look at some *alternative futures* that emphasize equity, sustainability, and development.

First, consider what the currently predicted rise in greenhouse gases, with attendant climate change, will bring for pastoral peoples and savannas. We know a lot about this subject thanks to the hundreds of scientists who have contributed to the Nobel Peace Prize–winning Intergovernmental Panel on Climate Change. Although all predictions are uncertain, it is quite likely that, between 1950 and 2050, temperatures will have risen an average of 3.5°C (38.3°F) throughout Africa, a warming 50 percent greater than that predicted for the Earth as a whole. But this warming will include a lot of local variation. For example, biologist Jeanne Altmann and her baboon researchers found that between 1976 and 2000, temperatures in Amboseli, southern Kenya, had already risen more than six times faster than these continentwide predictions (by 6°C, or 42.8°F), whereas ecologist Mark Ritchie suggests that cooling may occur in the Serengeti woodlands, if recent cooling trends continue.[5]

Compared to temperature, predictions about rainfall are complex in two ways: (1) generally, scenarios differ on whether it will get wetter or drier along the equator in Kenya, Uganda, and Tanzania; and (2) most estimates agree that it will get drier far from the equator in places like Sudan and Eritrea. Future global estimates, based on general circulation models (GCMs), support a "wet equator" scenario with El Niño–like conditions. If this comes to pass, agricultural modelers Peter Jones and Phillip Thornton predict that the warmer, wetter conditions will encourage more maize production in the already wet highland forests and grasslands. In the Ngorongoro Highlands, ecologist Randall Boone and colleagues estimate that more rain will mean more grass for livestock and wildlife, but there will also be more disease-carrying ticks. In this wet-equator scenario, Ruth Doherty and colleagues suggest that wetter conditions will spill over into the lowland savannas, promoting the growth of more trees and putting more carbon in the soils.[6]

Jones and Thornton, however, suggest that most dry areas (savannas) will see lower rainfall, especially those far from the equator. This could mean real trouble for places like Somalia, eastern Ethiopia, and northern Kenya, for they may become both drier *and* hotter. Some savannas near the equator, such as in central

Uganda and Tanzania and southern Kenya, may actually receive more rain (as in the wet-equator scenario), but this may not result in improved crop and grass production because hotter temperatures will evaporate the extra moisture.[7]

In contrast to these predictions, some recent evidence suggests that it has been getting drier in east Africa since 1980, even on the equator. Climatologists Park Williams and Chris Funk believe that the global models that predict a wetter equator missed an important shift in the atmospheric Walker circulation over the Indian Ocean, which has meant less rainfall during the long rains (March–June) in eastern Africa. This drying is particularly strong on the eastern flank of the Ethiopian highlands and across central Kenya.[8]

A drier-savanna scenario accords better with what pastoral people have been observing on the ground. In 2010 and 2011 our team, led by anthropologist Kathleen Galvin, interviewed pastoral people about changes in climate across Kenya. Our informants, whether they were from Turkana, Somaliland, or Maasailand, universally agreed that rainfall has already declined. They also think droughts are becoming more severe, even in Maasailand along the equator. The severity of drought may be exacerbated by people being squeezed into less land, but their descriptions are consistent with drying. Thornton and colleagues estimate that in areas with significant drying, croplands may "flip" into rangelands by the 2090s because of the decline in rainfall, putting more importance on maintaining pastoral practices and knowledge to produce food.[9]

Ancient information from lake-sediment core samples from Lake Challa just east of Mt. Kilimanjaro suggests that climate will also be less predictable as it warms. These core samples show that, in the past, while global cooling led to a more predictable climate, the opposite occurred with warming. Thus future warming may make the climate more variable, with wetter wet periods and drier dry periods. Wetter wet periods should increase pasture and food production, but may also increase the incidence of diseases (by, for example, encouraging insect vectors). Drier dry periods may cause the kind of massive die-offs of livestock and wildlife that we saw in 2009 in Amboseli. Such a possibility means that herders and wildlife will need more ability to move to new pastures rather than less. They will also need more grass "safety nets" for hard times (protected areas for wildlife, "grass banks" for livestock). Given current trends, however, these types of opportunities seem unlikely.[10]

Not only will savanna grass production decline because of drying, but rising CO_2 and hotter temperatures will generally lower the quality of savanna grasses and other forage, making it less nutritious for livestock and wildlife alike. With

CO_2 doubling, nitrogen content in grass leaves will decrease by 10–20 percent and structural carbohydrates will rise by 10–40 percent, making forage less digestible. This means that the number of animals a typical savanna can support will be lower in the future. Livestock and wildlife will need more land to obtain the same amount of food, but this extra pasture will not be available unless people abandon croplands. Conflicts will intensify. As CO_2 rises, trees will also spread at the expense of grass, further reducing pasture, resulting in less fuel and thus fewer savanna fires, which will further encourage trees to replace grass.[11]

Globalization will continue as people and wildlife in savannas become more and more connected to people and economies elsewhere around the globe. Wet savannas, together with wetlands and river edges in dry savannas, will continue to attract farmers to grow high-value crops like the tomatoes, onions, beans, and cut flowers destined for African urban centers and cities around the world. The consumers of these products will be largely unaware of how their consumption choices affect savannas far away, with Dutch shoppers, for example, having little idea that the roses they've just bought were grown in greenhouses blocking what used to be a major wildlife corridor in Kenya. Globalization will also continue to bring more tourists to the countryside, attract nonpastoralists to buy land and develop commercial enterprises, and bring in more migrant labor. Development of tourist lodges to meet global demand will take away the very amenity they are built to exploit: wide open savanna spaces.[12]

Faster communication will come to more pastoral communities, with better access to the Internet and mobile phones, thanks to globalization, although thinly populated areas may be largely left out. These technologies will allow pastoral families to better access opportunities to profit from livestock and wildlife.

Unless these changes are carefully managed, however, globalization will simply allow the rich to get richer, while women, the poor, the sick, and the young will continue to be left out of the profits. Globalized Western values, such as environmental protection and animal rights, will continue to replace more local values having to do with sustainable use of savannas and wildlife. And with the expanding influence of global NGOs (nongovernmental organizations) and international agreements, pastoral communities may soon feel that political authority is "scaled up" out of their reach.[13]

None of the efforts to reach the middle ground between conservation and development will be successful unless pastoral families and wildlife populations are secure and nation-states are stable. In this regard, the recent past does not bode well for the future. Pastoral people cannot live better lives if they fear attacks by other tribes that kill people, raid livestock, and create a general culture of fear. Conflicts among rival pastoral groups, for example, drove herders and their livestock out of more than half of the rangeland surrounding Marsabit, northern Kenya. Guy Haro and colleagues show that good management of savannas and peace may be connected there. Because of work done by peace committees after 1999, people now cut fewer trees around settlements, fewer wildlife die at the hands of poachers, and herders now feel safe to graze in 75 percent of their rangeland. In early 2008, during the Kenyan election violence, most of my team's conservation-development work came to a halt as pastoral communities had no choice but to focus on avoiding violence and asserting their rights. Conflict can also create thinly populated areas that remain in wildlife habitat as communities focus on fighting adversaries elsewhere. The conflict in South Sudan, for example, inadvertently protected one such place in Boma National Park, which now boasts the continent's second largest wildlife migration of perhaps 800,000 white-eared kobs, mongalla gazelles, tiangs (topis), and other species, despite two decades of war. But in the future, the more variable rainfall that is predicted to accompany climate change will likely intensify social conflicts across Africa, if the past is any measure: from 1990 to 2008, conflicts were a third higher than average when rainfall was either unusually low or unusually high.[14]

Even where governments are stable, political corruption (the illegal use of public office for private ends) will make reaching the middle ground difficult. Today across Africa, countries with less corruption have more healthy elephant and black rhino populations than those with more corruption. Corruption appears to affect wildlife populations more than human population density, GDP, or national spending. Pastoral populations are negatively impacted by corruption as well. In Uganda, for example, tourism revenues earmarked for local communities often go to park officials instead, while in southern Kenya, rich pastoral elites keep more than their fair share of revenues. If continued, misappropriation of funds will exacerbate local conflicts, sour attitudes of herders toward wildlife, and remove income that could help to lift households out of poverty.[15]

Continued growth of human populations in savannas, added to more consumption of savanna resources, will put more pressure on savannas and make it difficult to keep landscapes open for herders, livestock, and wildlife. Russ Kruska and I estimate that between 2010 and 2050, human populations will double in both wet and dry savannas of east Africa. This growth will double the pressure on resources, making pastoral livelihoods difficult to maintain and wildlife existence more precarious. This includes many species within parks, which depend for their survival on outside lands—lands that are increasingly being used in ways that are incompatible with wildlife.[16]

Our estimate of human population growth assumes that rural-urban migration will continue at the same rate as in the recent past. If migration from rural to urban areas is stepped up, however, there might eventually be fewer people in savannas, as ecologist Truman Young suggests. Fewer people may make it easier to keep the land open and expand wildlife habitat and pastoral grazing areas.[17]

Because wetter forested lands are already farmed, farming is expanding into savannas in part to feed growing human populations. Geographer Jenny Olson and her colleagues have mapped the amount of farmland that will be needed to feed much of east Africa by the year 2040. They concluded that there will be big changes in Uganda, where only part of Mbarara and Karamojong will remain in open savanna; Burundi, which will be almost entirely under cultivation; the western half of Rwanda; central Tanzania and around Arusha; and central southern Kenya. Southern Ethiopia, Somalia, and northern Kenya will remain open savanna and desert. Maasailand will be entirely surrounded by farms, with a thick band separating Tarangire in Tanzania from the rest of Maasailand to the north and east. Nonpastoral farmers will run many of these farms, but herders will also become farmers. A Maasai elder from near Tarangire described the desire of pastoral families to expand into farming this way: "You cannot expect us to remain in history. Many projects come and recommend we remain pastoralists, but we have discovered the new foods. Now we want to grow crops and keep our cows." Note that this elder does not suggest they will give up herding; rather, they want to add in farming. Olson and colleagues estimate, however, that plowing the savanna up into farms will cause the climate to dry further in the region, exacerbating the effects of climate change caused by elevated greenhouse gases.[18]

With more people and farming, the soft, fluid boundaries of open savannas will give way to harder, more stationary boundaries, especially in well-watered places (Map 18). Using our human population projections (but not accounting for the expansion of farming), Russ Kruska and I estimate that hard-boundary savannas will at least double in extent—from 6 to 12 percent of current savanna area—between 2010 and 2050. This doubling occurs at about the same rate in both the drier (climate-driven) savannas and wetter (disturbance-driven) savannas (see Map 7). Soft-boundary savannas, meanwhile, will fall from 62 to 47 percent coverage on average (from 49 to 38 percent in wet savannas, 75 to 56 percent in dry savannas).

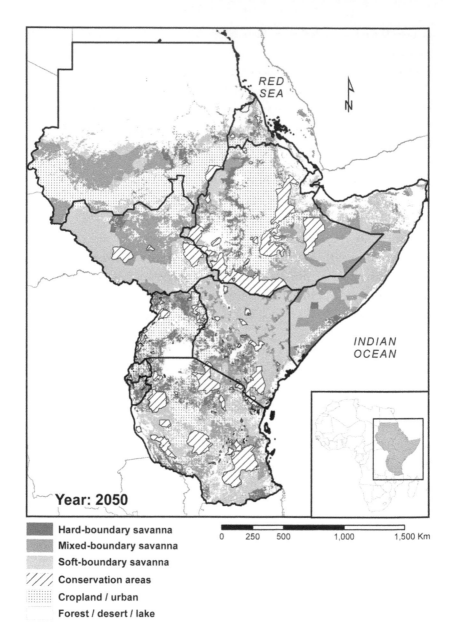

RED
SEA

INDIAN
OCEAN

Year: 2050

Hard-boundary savanna
Mixed-boundary savanna
Soft-boundary savanna
/// Conservation areas
Cropland / urban
Forest / desert / lake

| 0 | 250 | 500 | 1,000 | 1,500 Km |

MAP 18.
Possible distribution of savannas with hard, mixed, and soft
boundaries in east Africa in the year 2050. (Map adapted from
GLC 2003 and GRUMP 2004 by Russell L. Kruska.)

Bounded savannas will spread as people build more stationary boundaries, like fences, which will slow or stop the movement of wildlife and livestock. Only if herders gain access to new sources of feed and water that allow livestock to survive through drought will animal numbers remain high, leading to heavy and continuous grazing. It is unlikely that these new sources will include new pastures, unless Young's depopulation idea is right and farmers abandon cropland, freeing up land for grazing.

More likely, as expanding farms and towns fragment landscapes into smaller and smaller patches, fewer livestock (perhaps half as many) will be supported, and much less wildlife, bringing less income for herding families. Farming itself will remove savanna vegetation, and heavy grazing around farms will cause the productivity of soils, plants, and animals to fall. Where livestock populations are high, only a few types of wildlife will remain, such as plains zebras and Grant's gazelles, which currently exist in heavily stocked places like Laikipia, Kenya. With the larger wildlife gone, people will complain about damage done to their farms by smaller wildlife species like bush-pigs, warthogs, monkeys, and rats. Plowing land and growing crops will also release up to half of the carbon that is currently stored in savanna grasses. These changes will happen first in wetter savannas, areas that naturally support more wildlife; thus, the decline in wildlife will be disproportionately large in these areas, unless major new initiatives are launched.[19]

With all this change, it is logical to assume that pastoralism will die out altogether, something that observers have foretold for more than a century. Certainly, people have been abandoning pastoralism: in the wetter savannas of east Africa, herding families are becoming farmers who build permanent homes and grow crops instead of herding. In these areas, nomadic pastoralism will no doubt become increasingly rare. Today, most pastoral families find that while livestock are still the cornerstone of their livelihoods, they must add other sources of income—farming, business, teaching—to be able to feed their families and meet growing cash needs. But that does not mean that wet savannas will become wall-to-wall farms. In the future, herder-farmers will still hold cattle but will shift to herding more sheep and goats as open land diminishes and families need to sell animals more often for cash. In drier (climate-driven) savannas, in the absence of crop breeds that resist drought, I think many families will still find herding to be the best way to turn sunlight into food. And if some croplands flip to rangelands or if rural areas depopulate, rangelands may expand. According to anthropologist Peter Little and economist John McPeak, who work in dry Kenyan savannas,

"While serious development and food security problems still confront pastoral communities of northern Kenya, the pursuit of non-pastoral, sedentary-like activities does not forecast an end to pastoralism. Indeed . . . non-pastoral ('supplemental') activities . . . may be what will allow mobile pastoralism to continue in the area for the foreseeable future."[20]

If current government policy and development assumptions that promote crop farming over pastoralism and wildlife continue, finding a middle ground will be very difficult. With policies in place that award larger agricultural subsidies to crop farmers than to livestock herders, farmers will continue to migrate to savannas, herders will increasingly turn to farming, and savanna wildlife will disappear. Many policymakers, especially those with little experience in drylands, assume that adoption of crop farming and permanent settlement of herding families is the best way to develop pastoral lands, even though crop cultivation often fails in these areas. Instead, success and sustainability of production, especially in the face of climate change, may depend more on keeping land open so both livestock and wildlife can move as droughts become more frequent. Obviously, pastoralism will be more compatible with wildlife than will crop farming or settlement. Even so, crop farming is important, especially to feed growing populations, with sustainable intensification needed wherever possible to release pressure on pastoral lands (though this will be difficult to accomplish).[21]

Conservation policy supporting continued expansion of protected areas into pastoral lands will be opposed by advocates of pastoral human and land rights, and championed by those who see a continued decline in wildlife. Anthropologist Daniel Brockington and his colleagues paint a bleak picture of the global expansion of protected areas, the violation of local use rights, and the role of international capitalist economies in driving "industrial conservation." Yet it is not as if all east African countries are lagging behind other countries in their investments in parks: Tanzania has more land in protected areas than any country in the Western world, and the national government of Kenya spends more on protected areas per square kilometer than the United States does. If current trends continue, I predict that concerns over human and land rights will slow establishment of new protected areas in this region.[22]

Will conservation policy devolve the ownership, management, and use of wildlife to local communities in the future? Currently, devolution of wildlife management to local communities takes somewhat different forms for parks and reserves, on the one hand, and for lands owned communally or privately by pastoralists, on the other. Although pastoralists have managed some protected areas like the Mara and Samburu national reserves in Kenya for decades through local county governments, it is rare for pastoral communities to have any say in what goes on in nationally managed parks and reserves. Unless there are major policy changes, this will likely continue. On communal land, where groups of herding families manage the land together, the last decades brought much effort to devolve wildlife use, management, and profits to those groups through initiatives in community-based conservation or community-based natural resource management. Increasingly, elected politicians are putting pressure on wildlife authorities to account for the costs of wildlife to local communities, such as injury or killing of people or livestock by wildlife. But there is still a long way to go. National representatives remain hesitant to transfer much profit to the local level; conservation and livelihoods goals are not often met; many community-conservation initiatives are not "downwardly accountable" to local herding families; benefit sharing is often far from equitable; and communities have little say in the development of national policy. As land held in common becomes privatized, full responsibility for management of wildlife (and livestock) habitat comes under the control of individual pastoral families, although ownership of the wildlife itself is often still in government hands. I expect the future to bring increased conflict between those who believe national and international goals of conservation cannot be led by local communities and those who believe that devolution of responsibility to local communities is the most desirable end goal.[23]

Historically, most of the benefits from wildlife flowed outside savannas to the state while all the costs were borne by local communities, but this has changed and will continue to change in the future. In Tanzania, for example, according to governance scholars Fred Nelson and Arun Agrawal, "under uniform state ownership, local people have few incentives to value wildlife or invest in conservation. Instead, they remained saddled with the costs of living with wildlife, while the benefits of the resource are captured by the state and private sector." Yet governments do recognize the clear need for local communities to benefit from wildlife and, sometimes, to be compensated for losses caused by wildlife. Today, the Kenya Wildlife Service, Kenyan local county councils, and the Tanzanian National Parks Authority all share some of the proceeds from tourist fees with local communities. Many private tourism businesses pay local communities fees for use of campsites, wildlife viewing, or other sharing of benefits. In Tanzania, a new policy to create village-controlled wildlife management areas on village land is being implemented, which would eventually put most of the wildlife existing outside the state-run protected areas under local control. Still, difficulties abound. I expect continued experimentation with new forms of responsibility for wildlife but slow progress on transferring authority and profits to the local level, especially in Tanzania.[24]

Even when profits from wildlife go to local communities, they are often small compared with other more profitable ways to use the land, especially where rainfall is abundant. Such poor profits provide little incentive for pastoral families to sustain wildlife on their land. Ecological economist Michael Norton-Griffiths shows this clearly when he compares Kenyan families' end-of-season "cash in the pocket" depending on whether they farm, herd livestock, or profit from wildlife. Farming is always more profitable than either pastoralism or wildlife, except where the climate is so dry that crops simply cannot grow. The only pastoralists who earn more than farmers are those who receive high profits from wildlife, which supplement profits from livestock. At present, pastoralists capture but a tiny fraction of the profits that wildlife afford, so with major changes, profits could rise substantially. Because very few of Kenya's savannas have the open vistas and vegetation that are ideal for high-value photographic tourism, Norton-Griffiths points out that if hunting were allowed, as in Tanzania and elsewhere, people in much broader parts of the country could profit from the presence of wildlife.[25]

Where people do not profit significantly from wildlife, I think wildlife will disappear from open rangelands in the future, and this will cause wildlife to decline, somewhat, in nearby parks. We can see this process taking place today outside Nairobi in the Kaputiei Plains: as land prices rise, payments to families to keep land open in the leasing program must also rise to keep people from selling their land for development. Wildlife populations both inside and outside Nairobi National Park are declining, and we see similar declines across Kenya. As herders settle, they perceive more conflict with wildlife because they adopt a lifestyle less compatible with wildlife. In 1977 near Amboseli, for example, 32 percent of herders reported conflicts with wildlife; by 1996 that number had risen to 75 percent, partly because herders had taken up farming and complained about damage to crops. Lack of benefits may put the future of parks into question. When John Akama and colleagues asked Maasai herders and Kikuyu farmers what they thought of Nairobi National Park, many responded that parks do not benefit them and that the land should be devoted to agricultural production. Thus, the middle ground may be viable only in places where people perceive strong benefits from wildlife.[26]

Finally, what of the positive interactions we see between people and wildlife today, a legacy of tolerance, principally in Maasailand? All of the changes described above will make these positive interactions difficult to maintain, unless conservation efforts build on this local wisdom in the future. If they don't, I see the balance tipping against coexistence of pastoralists with wildlife, from compatibility to conflict.

IMAGINING A MIDDLE GROUND

The scenario depicted above looks like the end of both pastoralism and wildlife. Will future savannas have only hard boundaries, with pastoralism restricted to the driest places, remote from towns, and with wildlife confined to parks? Will there be no place where wildlife and livestock mix? Or are there alternative futures that provide new opportunities for the middle ground?

Before proceeding, I want to be clear that for some issues it will be hard to find middle ground. When an elephant kills a herder, there is perhaps no human-centric solution that does not include the death of the elephant. From an animal rights point of view, in contrast, the spearing of elephants is unacceptable at all costs. In such cases, middle ground is nowhere in sight. In a related vein, however, it is important to acknowledge that the extremes have significant and useful contributions to make—the middle-ground approach may not be the only one. For example, all who enjoy, profit from, and strive to conserve savanna wildlife in east Africa have to ask: Would there be any wildlife left in the savannas if it weren't for pastoral herders? And without herder-conserved wildlife, how full would the national treasury be? Those supporting pastoral development have to ask: Even though parks were carved from pastoral lands, wouldn't their rich soils be farmed for wheat or tomatoes or beans today if the parks did not exist? Would there be a Serengeti wildebeest migration to profit from today if Serengeti National Park had not been created, with its wildlife-only strategy?

The middle ground does not hold all the solutions to the tough problems associated with pastoral savannas, but it does encourage us to face our most essential challenges in all their complexity. In so doing, it may present us with new, less explored options that are inherently more sustainable and equitable. The middle ground will no doubt suggest some unique strategies, such as uniting the goals of conservation and pastoral development into a more powerful common cause. The middle ground, in the end, is for those who seek an elusive win-win, where not only do herders have more resources to feed their families, educate their children, and access better medical care, but wildlife can also continue to thrive.

So what might this elusive middle ground look like? In fact, I believe it offers a glimpse not merely of a win-win, but of a triple win: local empowerment of pastoral people, improved pastoral welfare, and better conservation of wildlife, which may also benefit livestock. The middle ground might include various innovative actions. For example, what if national parks established a locally appointed or elected advisory committee made up of diverse members (including men and women, young and old) from all of the communities within 10 km (6 mi) of the edge of the park? And then gave this committee real power to determine who bears the costs and receives the benefits from wildlife? The committee might also negotiate agreements with pastoral herders to report all poaching activities, collect poaching snares, and rest pastures at the edge of the park seasonally, along with the resources and authority to implement these decisions. Or what if profits from tourism were entirely transparent? Imagine a scene in which government, private sector, and local community representatives (and not just the leading elites) jointly determine what all the profits from tourism, gate fees, and other concessions are to be used for. How about pastoral communities that are fully accountable for maintaining and restoring healthy wildlife populations, clean water, and sustainable grasslands? Or think about research that is designed with and for and carried out through a collaboration of park managers and local community members.

A first step toward this middle ground is perhaps the most difficult: the development, led by Africans, of what ecologist Fikret Berkes calls a "pluralistic, cross-cultural conservation ethic," one that broadens and redefines the meaning of conservation to fit an African context. As I hope this book makes clear, conservation in savannas without people, a Western cultural notion, does not fit landscapes that were partly created and maintained by people in east Africa over many millennia. In my experience, an African conservation ethic for pastoral savannas would be an ethic that promotes conservation-with-development and development-with-conservation. This ethic would have sustainable use as a centerpiece, requiring conservation to be integrated with the health and welfare (and development) of pastoral households and communities. With an overarching goal of community development, good stewardship would be one means of getting there, rather than the other way around, thereby harnessing largely untapped knowledge, ideas, and people within communities around parks. It would mean that people value natural resources not just for the profits they represent, but also for their cultural and spiritual values. One of the early attempts to implement this new ethic was the well-known CAMPFIRE program (Communal Areas Management Programme for Indigenous Resources) in Zimbabwe starting in the mid-1980s, but overall, implementation has been difficult (particularly devolution of power over wildlife from central government to local communities). Leadership by Africans, especially those from rural areas, has been rare at best. Conservation-with-development and development-with-conservation are difficult and often fail, which could suggest that the notion be abandoned altogether. But since when are the most daunting problems easy to solve quickly? Only a redoubling of effort, led by Africans for Africans, will promote the sustainability of savannas and the people that depend on them.[27]

If there were a new African-centric (even pastoral-centric) conservation-with-development ethic, what might some of its main principles be? I would suggest five: empowerment (which would include devolution of power from the central government to local communities and locally defined goals and leadership), benefits (redirected to pastoral households and communities), equity (and transparency), adaptive stewardship and management (including sustainable use, adaptability, and clearly defined conservation outcomes), and collaboration (partnerships harnessing the power of diverse perspectives and resources).[28]

EMPOWERMENT

The first step in empowerment is a paradigm shift, on the part of all concerned, to recognize that many of the traditional practices of African pastoralists are highly appropriate for managing savanna environments. Even though the idea of widespread degradation of savannas has been challenged for decades, it still persists. As we saw in Chapter 7, our best evidence suggests that while savanna decline and difficult-to-reverse degradation do occur, there is little strong evidence to suggest that this is happening over wide areas, perhaps because savannas are more resilient than we tend to think. The larger concerns are about the loss of savannas altogether to farmland, towns, and other uses.

With this paradigm shift, it becomes clear that much of the leadership for management of savannas should lie with those most ignored: local people who live with wildlife—their costs as well as benefits—in pastoral savannas. The men might be the first to lead, but my male pastoral colleagues would be the first to say that this ethic needs leadership from women, as well as the young and the poor. With leadership that is accountable to local communities in all their variety, it will ensure that local voices and concerns are at the forefront of decisions about conservation.

Full empowerment requires that part of the authority over wildlife must reside with local people. Currently, most wildlife in east Africa are under the control of central governments, even if particular parcels of land (including wildlife habitat) are owned in common by groups or privately by individuals. By contrast, in southern Africa wildlife are sometimes owned and managed by communities or individual land owners. It is logical that conflicts between people and wildlife will only lessen, and management of wildlife will only improve, if local communities have some rights to own, manage, use, and profit from wildlife. Here we have the "subsidiarity principle," that people should solve problems locally as much as possible with only as much government oversight as necessary.[29]

Several challenges stand in the way of a devolution or transference of some of the power over wildlife to local communities: unwillingness of national government (often in partnership with powerful donors and NGOs) to share that power; lack of accountable leaders and institutions at all levels of government; and lack of strong local institutions to manage wildlife and the profits from them. During Zimbabwe's CAMPFIRE program, successful devolution has built new confidence and strength in rural governance, increased flows of profits to local districts (although not enough to villages), and maintained elephant and buffalo populations. In the future, devolution may even take the form of co-management of protected areas by a consortium of national and local governments, local NGOs, and community members and local businesses.[30]

Even more important than authority over wildlife, a middle-ground future would ensure that pastoral communities have secure access to land and its resources, with ancestral rights to land honored and upheld at the highest levels of government. In this future, wealthy foreigners or local people would not be able to "grab" pastoral land or have special access to it—as in a 1993 case when the Tanzanian government made a special arrangement with the defense minister of the United Arab Emirates, working through the Ortelo Business Corporation, to allow hunting and live animal capture rights on Maasai land near Ololosokwan village in the Loliondo Game Controlled Area. Local Maasai complained to journalist Alex Renton in 2009 that they were not consulted when the original deal was made and stated that they are often beaten or their settlements burned by Tanzanian police when they try to access their traditional lands. Renton describes other purchases of land and evictions of Maasai from Loliondo and Ngorongoro as well. A new future would stop these kinds of deals.[31]

While empowerment sounds good in the abstract, it is difficult to achieve in practice because it requires that those who have power relinquish some of it. Power influences not only who controls and has access to resources, but also how success is measured and who shares the risk of failure. In this case, central governments would have to relinquish some authority over wildlife and some of the profits from wildlife. This did occur in one case in Zimbabwe in the 1980s, but it is rare. I have seen young pastoral leaders confront elected officials in person, at political demonstrations and through the power of the media, to push them to loosen central control. Other pastoral initiatives gain power more subtly, by working within the system, achieving successes, and then sharing the credit for that success with elected officials. In the end, though, very few local organizations or leaders today have the right to plan for, manage, use, and profit from wildlife and other natural resources in pastoral savannas of east Africa. And power sharing cannot stop with the government: NGOs and donors should let local needs drive funding and program priorities; foreign scientists should support the priorities of local scientists and communities; and the private sector, particularly the tourism industry, should pay a fair price for access to land and wildlife.[32]

BENEFIT SHARING

A second principle of the conservation-with-development ethic is that the people who live with the costs of wildlife must capture a larger share of the benefits of wildlife than they do today. Most pastoralists value wildlife for their cultural, social, and political benefits, with wildlife being a major part of their mythology and stories, appreciated for their aesthetics, and used, I think, as a political tool to keep land open. But they must capture financial returns as well—a critical leg of the elusive triple-win situation of pastoral empowerment, improved pastoral welfare, and stronger conservation of wildlife. The key is to weaken the incentives for farmers and pastoralists to turn savannas into maize fields by making wildlife more profitable, particularly in places like Kenya where lucrative hunting is not allowed. This means that all who now profit from wildlife (governments, the private sector, NGOs, the pastoral elite) will have to do two things: (1) share a bigger slice of the pie of profits with local communities and (2) help make the profit pie bigger for all (though it is not clear just how much larger the pie can get).[33]

It is difficult to know what a fair share of the existing pie is—how much, for example, local communities can charge tourism operators for access to their land (and the wildlife on it) and still attract that business. This figure will vary from place to place and over time. Another way to get a fair share is to cut out the middle men (often the district government or wealthy pastoral leaders) that siphon off huge amounts of photographic or hunting tourism profits; this can be done by setting up community-private partnerships that make direct payments to community members.[34]

To make the pie larger, pastoralists may need to become tour operators themselves. Failure to do so, according to ecological economist Michael Norton Griffiths, will mean that only 10–15 percent of the pie will remain on their plate. Communities can start by taking over transportation and then accommodation, eventually replacing foreign owners of tourism businesses. Even more powerful may be business partnerships that couple the power of local and international institutions. A different future would include these new ways of making the profit pie bigger: (1) tourists paying more to experience the wildlife of pastoral savannas; (2) adopting hunting in areas where it is now prohibited and where photographic tourism is not possible; (3) increasing the access fees in places where species or landscapes are unique; (4) building new programs (through NGOs, government, or local businesses) to benefit from carbon sequestration, clean water, healthy wildlife populations, and other ecosystem enhancements; and (5) promoting biodiversity-related goods, and related price premiums, through certification schemes. Success of benefit sharing depends, according to geographers Karen Archabald and Lisa Naughton-Treves, on "long-term institutional support, appropriate identification of the target community and project type, transparency and accountability, and adequate funding."[35]

One major way to increase profits, particularly in bushy and tsetse-infested savannas where photographic tourism is not possible, is through hunting. This is applicable mostly to Kenya, where hunting is currently prohibited. Such habitats may represent most of northern Kenya. The profits to communities from the CAMPFIRE program in Zimbabwe were almost solely from hunting. Unlike photographic tourism, hunting provides very consistent streams of income, year in and year out, at least in southern Africa. Although animal rights advocates are strongly opposed to hunting on moral grounds, the reality of maintaining other lands open for wildlife may require hunting in remote areas of Kenya in the future. That said, hunting is of course not possible in places where the focus of conservation is on endangered species like the great apes or charismatic animal populations like the Amboseli elephants.[36]

EQUITY

A third principle of a viable conservation-with-development ethic is equitable, and transparent, sharing of benefits, or profits, from wildlife conservation. Public counting of funds is often a first step in this direction, but because existing power structures both outside and inside pastoral society are resistant to efforts to share more equitably, this can be very hard to accomplish. When communities do achieve equity, those who share in benefits have a better attitude toward wildlife (and conversely, the misappropriation of funds derived from wildlife can turn pastoral community members against wildlife). Any change that transfers power downward toward the village or community level tends to open the possibility of more equitable sharing. New communication technologies and social media could be used to report and highlight inequitable practices. In Kenya, group ranch committees or the local country councils often siphon off profits from wildlife. When group ranches subdivide into many and smaller, privately owned parcels, however, benefits are more likely to reach individual families, as happens in the conservancies in southern Narok District of Kenya. Of course, who does and who does not get a private land parcel in the first place can be very inequitable. In the future, according to economist Omar Azfar and colleagues, "corruption should decline if regulatory reform reduces the monopoly and discretionary power of officials, if greater transparency increases the probability of being caught, and if higher pay or better working conditions for officials increase the pecuniary benefits of being honest."[37]

ADAPTIVE STEWARDSHIP AND MANAGEMENT

A fourth principle is the need for conservation-with-development, which requires maintaining or restoring healthy lands and wildlife through good stewardship to reap the benefits of income. All too often, however, I see pastoral communities that want to share in the profits from wildlife but are unwilling to reinvest some of those profits in the ecosystem to ensure a long-term return. To maintain wildlife populations, pastoral managers need to experiment and adapt the way they use the land. Accomplishing this requires that they track the health of wildlife populations and their habitats through clear, repeatable, accountable monitoring. Even here, devolution of responsibility is important: if local community members participate and lead in project design, data collection, and the reporting of findings from monitoring, they may also agree more often about the hard land-use decisions they may have to make. Participatory monitoring creates a shared understanding of the consequences of different actions and allows communities to adapt as conditions change. This can be inside-outside monitoring as well, with pastoral PhDs leading these efforts and bringing in outside help as needed.

Conservation-with-development sometimes requires hard choices because conservation can hurt development and development can hurt conservation. When the new conservation area of Mkomazi was established in Tanzania in 1988, for example, pasturage that hundreds of pastoral families relied on to grow their livestock was placed off-limits. Today, improvement of a road to better connect trade between Kenya and Tanzania has cut the Kaputiei Plains wildlife and live-stock corridor in half. The road thus constitutes an obstacle to herders, but it provides benefits as well—a double-edged sword. Nothing about this is easy; the most important thing is to pay attention to the trade-offs, so that there is some hope of balancing concessions with desirable outcomes.

COLLABORATION

Collaboration is the final principle of this new ethic. Experiences around the world suggest that collaboration, if inclusive of diverse perspectives, can sometimes create enduring solutions to problems that previously were intractable. Collaboration is most necessary when the natural resources we are trying to conserve cross ownership and management boundaries on broad landscapes, from private land to public land and back. These are "common pool," boundary-crossing resources and processes, which in east African savannas include water, moving wildlife, pest infestations, weed invasions, fires, and air. While collaborative partnerships often take considerable time and effort, their strength derives from the power of the new middle they create and the innovation that that power promotes.[38]

Developing this ethic will not happen without effort. It will take long (and sometimes hard) listening, negotiating, and integrating the ideas and actions of people with diverse interests. It will also take individual and daring action by business people, policymakers, scientists, and community members. To negotiate complexity will require great care as individuals seek to think holistically across different ways of knowing and perceiving the world—sociocultural, economic, political, scientific. It will be important to work collaboratively, construct networks for sharing and learning, adopt experimental and experiential learning approaches, and accommodate diverse views. In this, traditional pastoral approaches to dialogue where all participants have a voice may prove more promising than rapid (and impatient) Western approaches.[39]

CURRENT EXAMPLES

Are any of these principles in place now? And if so, how might they evolve further? Examples indeed abound of different approaches extending toward the middle ground, from former wildlife-only parks that now share revenues with local communities, to pastoralist–private sector joint business ventures, to tourism companies entirely owned and operated by pastoral community members (Figure 23). Even though the models differ in how much power they give to local peoples, more devolution is possible in every case.

	Description	Examples
High	Pastoralist tourism business owners	Selenkay Safari Camp, Riverside Camp, Ol Choro Oiroua
	Pastoralist/private sector joint tourism business ventures on private or communal pastoral land	Olare Orok Conservancy, Il Ngwesi, Shompole Community Trust
	Private landowners, paying for use of community land and sharing profits for development	Maasailand Preservation Trust, Porini Camps, Basecamp, Dorobo Safaris, Klein's Camp
	NGO/private landowners, providing resources for nearby communal development	Tanzania hunting leases (by law), Cullman and Hurt, Lewa Conservancy, some Laikipia ranches
	Local government land ownership or management, sharing of revenues	Kenyan county council–managed reserves, Tanzania Wildlife Management Areas
	National government land ownership, sharing of revenues	Kenya Wildlife Service, Tanzania National Parks, Uganda Wildlife Authority
Low	Government/private land ownership, no revenue sharing	Some Tanzanian hunting/tourism concessions, some Kenyan commercial ranches

FIGURE 23.
Degree of influence that pastoral communities have over
the profits from wildlife on land used currently or historically
for grazing (NGO = nongovernmental organization).
(Sources: Archabald and Naughton-Treves 2001, Carter et al.
2008, Schroeder 2008, Reid pers. obs.)

While movement of all conservation efforts to the top of Figure 23 might seem the most desirable, I think a diversity of approaches will serve integrated conservation and development goals better. Diversity improves the ability of all partners in conservation and development to adapt more quickly to challenges such as climate change, globalization, and changes in security. As mentioned earlier, national and local government policies have supported the sharing of park gate fees and other revenues with nearby communities for decades, to fund development projects like schools, clinics, and wells and to pay for school fees (bursaries). In some cases, individuals (mostly nonpastoralists) own large ranches (Laikipia, Machakos) or have rights to hunt wildlife through the national government (Tanzania). These individuals often contribute to local development, and some are deeply involved in humanitarian assistance. True recognition of local rights to wildlife profits occur only when private companies and pastoral communities begin joint ventures, with the private companies paying for access to land or other resources. In some places these have evolved into true pastoral–private sector partnerships with joint decision-making. And more recently, some pastoral entrepreneurs have started their own ecotourism businesses so that they capture all of the benefits (and burdens) of this growing market sector. There are other bright spots in the form of NGOs lending active support to pastoral interests, developing training schools for pastoral ecotour-ism guides and businesses, and so forth. But such far-reaching experiments as co-management of protected areas by governments and pastoral communities or full transfer of the ownership of wildlife to pastoral communities remain for the future.[40]

WHO DOES WHAT TO REALIZE THIS NEW ETHIC?

What might local pastoral community members do to build on these beginnings? All the pastoral people I know agree that holding on to land is a first priority if pastoral opportunities are to be furthered. Just above this bottom line is organiz-ing to gain more political power and making noise to slow or stop changes that are not in the interest of pastoral communities. Also of utmost importance is recog-nizing that pastoral communities have some crippling weaknesses that must be overcome if they are to succeed. Ecologist David Western identifies one as the "diffuse governance structure of pastoral societies. These work slowly and surely in traditional disputes, but hamper the Maasai when it comes to establish-ing landowner associations, producer cooperatives and the political advocacy needed to counter their marginalized position."[41]

Stronger pastoral leaders and grass-roots institutions, committed to transpar-ency and equity, will make sure that conservation and development are more locally driven and profits stay in local communities. Close attention to the leader-ship role of women is strongly needed, partly because women are "keener to form cooperatives and mobilise themselves as a group to share responsibilities, provide support, and even to initiate change," according to gender scholar Fiona Flintan. And leadership by youth is needed to create intergenerational coalitions of power.

These stronger and new institutions must fight to stop damaging pastoral practices like lion poisoning. And they will also need to blend old ways with new—conservation easements, land purchases, conservation restoration credits, ecosystem service payments—to keep land open.[42]

What would policy look like if it supported these principles? First, it would require policymakers to shift their thinking and embrace the following ideas: (1) pastoral people are not undeveloped and primitive; (2) pastoral land is productively used and is *not* wasteland; (3) as a rule, livestock grazing is more sustainable than farming; (4) the mobility of pastoral herds supports sustainable livestock and pasture production; (5) pastoral communities possess critically needed knowledge about how to manage savanna ecosystems and wildlife; and (6) common land ownership is a legitimate and appropriate way to approach land use.

This paradigm shift leads to at least seven specific policy options for the future. First, policymakers could reverse agricultural subsidies that favor farming over pastoralism. Second, policymakers could give local communities more responsibility and authority over wildlife, encouraging them to innovate and build on pastoral cultural rules and norms to develop coalitions with other local groups to share what they learn. This devolution should include resources for long-term building of local leadership and institutions. Third, pastoral communities need new tenure policies that secure pastoral land access and ownership. These would include land-use planning (zoning for pastoral use), reallocation of land taken from pastoral communities in the past, and support of new pastoral-led land trusts for purchase of land to keep it open for livestock and wildlife. This policy must compensate for the fact that even with individual land titles, as David Western points out, pastoral communities are disadvantaged in a "free market economy where wealthier ethnic groups can secure loans to buy land from the poorer ones."[43]

Fourth, pastoral communities need policies that support keeping land open and encourage herd mobility, including common pastures, communal grazing on private land, and agreements to trade pasture rights seasonally. Mobile populations may also need mobile services such as schools, clinics, and markets to be brought to them so that they do not have to graze their herds close to towns to access these services. Fifth, national and international economic policy needs to encourage payment (by government, civil society, and the private sector) for the "ecosystem services"—carbon replenishment, clean water, clean air, biodiversity, and soil fertility—that pastoral land managers now provide at no cost, and whose benefits are enjoyed by people far from pastoral savannas. Such policy could also support efforts to restore land, carbon, water, and biodiversity. Sixth, markets, livestock breeds, and other productive assets need to be improved to keep pastoralism profitable, so that pastoral families can stay in pastoralism. This would include support to pastoral communities to develop grass banks, timely livestock sales during drought, and new opportunities to diversify household economies, like high-value conservation beef, renewable energy, and niche-market camel milk.[44]

Finally, a policy is needed that allows for co-management of protected areas, starting with local advisory committees and moving toward sharing of responsibility and authority over wildlife and savanna ecosystems with authorities at the national level. This would recognize the former sacred and cultural areas inside parks as well as areas of special ecological or biological significance. This co-management would include park outreach programs to forge direct connections between pastoral development and conservation of wildlife, with funds going directly to communities who live with wildlife, not ones farther away. And while I am at it, how about full consultation with pastoral communities whenever new policy is developed that affects their lands?[45]

What can NGOs, donors, and the private sector do to support these principles, more than they already do? International conservation NGOs and donors play a critical role in supporting and empowering local voices. Very often they have far more influence over how conservation gets done in east Africa than any local community group does. This asymmetry of power is not a problem if these NGOs are aligned with local interests, but such is not always the case. Western-funded conservation NGOs often promote scientific knowledge over indigenous knowledge and international conservation priorities over local needs; sometimes, too, they view local communities as tools for reaching specific conservation goals rather than as collaborative partners in land stewardship.[46]

Even local pastoral NGOs do not necessarily serve the human and land rights of local society, particularly when they get deflected by the priorities of donors. Recently, new partnerships with the private sector have supported pastoral ownership of land, especially in Tanzania. But some private-sector conservation businesses still privilege wildlife over people.[47]

Some of these initiatives could be improved by recognizing that communities deserve to be compensated for the ecosystem services that good stewardship provides and that they have other rights, such as those to land and wildlife, that need to be taken into account. Private commercial ranches or farms, whether managed by individuals or NGOs, need to partner more with local communities, given that they are on land taken from pastoral people by colonial and postcolonial governments. And if private businesses are involved in illegal activities silently condoned by government, such as poaching, this must stop.[48]

What can scientists do to support the principles I have outlined above? In my opinion, science needs to be reoriented so it better serves local needs. This will require changing scientific incentives both within and outside Africa. It is important, too, to recognize that scientists wield the power of expertise, which they need not only to acknowledge but also to use very carefully in a community context. Scientists can empower local voices by integrating indigenous ways of knowing with scientific ways of knowing to co-create a new hybrid knowledge base. This is more than a citizen science that encourages local communities to collect objective data, which scientists then analyze. Rather, it goes to the heart of the scientific model, co-creating a new hybrid model that unites different but equally valid ways of understanding the world. This would mean negotiating and co-creating what information should be collected, how questions should be stated, and what the information means once it is collected. One example of creating this hybrid knowledge is setting habitat conservation priorities based both on local, long-term observations of the best habitats for individual wildlife species and on scientific aerial surveys of the distributions of those species. This could include a united assessment by landscape and season of what species are and are not compatible with livestock grazing and what other factors are affecting wildlife populations. On the way to implementing this new model, scientists can build capacity at the local level to collect information in a repeatable and reliable way, thus empowering local voices. Our team has experimented with this model and have found that the gains for people (pastoralists and scientists alike) and wildlife can be large.[49]

In terms of formal education, there is a need for more African scientists, trained in reputable African universities, who lead African research and publish that research in recognized journals. Key to this African-led research is creating well-paying employment opportunities for African scientists to do science when they are done with their educations and governments that provide grants for those researchers. For too long, bright African scientists have had no other choice but to take better-paying jobs outside of Africa, contributing to African "brain drain." Also for too long, foreigners have come to Africa and built their careers, as I did, on information given freely by local communities. It is fair to ask what the foregone opportunities to African society are of having foreign researchers, pulling in handsome salaries, determining what science knows about east African problems. Science education in Africa must be more relevant, focusing on issues of importance to local communities and on ways to communicate scientific information so that it is useful to and used by local society. There is a need for a whole cadre of scientists who know how to work with policymakers to promote evidence-supported policy development. And we need lots of people who can work closely with stewards of the land, be they herders, wildlife managers, or others, to keep track of change so that owners and managers can adapt flexibly to the change that is all around us.[50]

IMAGINE A FUTURE WHERE . . .

It's about giving space, giving
a chance to both livestock and wildlife.

DAVID NKEDIANYE, *director of the
Reto-o-Reto Foundation, 2007*[51]

Let us imagine, together, a world in which

- *Pastoral lands are secure and conflicts are rare,* as are disease and deaths of people, livestock, and wildlife due to starvation during droughts.
- *Pastoral families live above the poverty line and have access to quality health care and education.* This can partly be accomplished by improving profits from pastoralism through better veterinary care, improved breeds, and better markets and from wildlife by sharing the existing pie and making that pie bigger in innovative ways.
- *A "new" pastoralism arises where pastoralists are not only healthy, educated, and wealthy but also major leaders in land stewardship*—and admired and supported locally and globally for their ability to conserve biodiversity and share benefits equitably and transparently.[52]
- *Pastoral leaders eschew corruption and refuse to benefit personally from wildlife profits,* preferring the political support that community receipt of profits brings. These leaders also support protected areas and their managers, are highly effective in the halls of power, and are effective agents of policy change at the local and national levels.
- *Strong local organizations link open land for livestock with rights, conservation, and welfare,* including human rights, land rights, and pastoral (cultural, spiritual, social, economic) welfare.

- *Pastoral society recognizes and supports pastoral women*—their brilliance, energy, and leadership as critical innovators for the future.
- *New sources of profit are created* as east African nations recognize and reward those who live in savannas for the ecosystem services they deliver to their region and the globe.
- *Wildlife populations are healthy and rebounding* from recent declines, with reintroduction of wildlife in areas where they have disappeared and reintroduction of rare wildlife species.
- *Savanna landscapes have fewer hard boundaries*—instead incorporating more soft boundaries and allowing for fluid movement of livestock and wildlife. Landscapes with wetlands, riverine areas, and swamps, so important to pastoral society and wildlife, receive special protection.
- *Currently farmed landscapes are sustainably intensified to provide enough food for growing national populations*—by building soil fertility, managing water efficiently, and using renewable energy.
- *Development practitioners and conservationists find middle-ground solutions,* led by pastoral people, that fully include the needs of both people and wildlife.
- *Policymakers argue for pastoralism, pastoral land rights and mobility, common land tenure, and mobile pastoral services.* These policymakers point to pastoral people as leaders in the search for national sustainability, fight hard to stop corruption, and support the interests of the marginalized in society.
- *The tourism and hunting industries transfer profit to local communities.* As much profit as possible, in a transparent manner, is transferred to the locals who live with the costs of wildlife. Not only are locals trained to guide, but they run every part of tourism businesses; rather than being servants, they are now leaders in the tourism sector.
- *Protected-area managers train and promote pastoralists* for management positions and experiment with co-management initiatives that link conservation with pastoral production.
- *Ministries of livestock and wildlife are more often run and staffed by pastoralists*—the people with the real-life experience to run them.
- *Conservation NGOs are driven less by Western values and more by local values.* Western values of conservation give ground to local values of sustainable use and poverty reduction.

- *Development NGOs adapt their priorities,* giving the same priority to long-term sustainability of pastoral societies and their environments as they do to short-term development fixes.

- *Scientists respond to local communities,* basing much of their science on what local communities need, and provide policymakers with clear and workable recommendations.

- *Human population growth and consumption rates fall everywhere,* with higher quality of life for all through equity and efficiency as we continue to question the assumption that economic growth is sustainable over the long term.

- *Citizens around the world pay local communities to maintain healthy savanna ecosystems and wildlife populations* through, for example, a small, global ecosystem services tax. This income from wildlife becomes the safety net for herders who still live with wildlife: it gets families through droughts, sends children to school, and provides business building opportunities for pastoral entrepreneurs. (These same citizens demand transparency from businesses so that they are aware of the consequences of their choices in the marketplace and do not buy flowers that may have been grown in greenhouses in wildlife corridors.)

Our species rose to dominate the Earth in part because of skills we learned long ago in savannas. Today, savanna environments and savanna peoples hold genetic, behavioral, and cultural information that point to new ways for humanity to continue to adapt. It is time for the world to give herding societies their proper respect and to recognize their wisdom, rather than viewing them as backward and in desperate need of externally driven "development." And it is time that herding societies take more pride in their environmentally winning ways, take responsibility for their herding excesses, and more assertively project their voice into the world.

NOTES

1. SEARCHING FOR THE MIDDLE GROUND

1. Parkipuny 1991a, 29 (referring to indigenous herders, but his comment applies to hunter gatherers too).

2. Prins 1992, 121.

3. Kangwana and Ole Mako 2001, 156.

4. I use the spelling *savanna* without the ending *h* because it is closer to the spelling of the Caribbean Spanish word *ʒavana* from which it derives. See Chapter 4 for details and references on evolution, but note that it is not clear if the genus *Homo* itself evolved in Africa or Asia (Wood 2011).

5. *New conservation model:* MacKenzie 1987; *Western philosophy:* Peterson 2001; *9 percent park coverage:* analysis by Russ Kruska of the area in savanna parks compared to the total savanna area in Map 7.

6. *Wildlife outside parks:* Western 1989; *savanna area outside parks:* Kruska's analysis (see n. 5); *pastoral poverty example:* Okwi et al. 2007.

7. *Borana:* Desta and Coppock 2004, Gemedo-Dalle et al. 2006; *Maasai views:* Mapinduzi et al. 2003, Nkedianye 2004, Oba and Kaitira 2006, Goldman 2007, Kaelo 2007. *Creation of Maps 3–6:* The naming of language and ethnic groups is highly complex and is still debated. Russ Kruska and I adapted these maps from linguist Joseph Greenberg's 1963 map of the four language phyla in the region and anthropologist George Murdock's 1959 map of the diversity of the peoples of east Africa and their historical locations more than fifty years ago. We then added Murdock's livelihood categories. We deviated from these sources twice: following Heine and Mohlig 1980, we put Murdock's Somali and Oromo (= Galla) under Cushites in the Afro-Asiatic

language phylum and grouped the Kalenjin with the Nilo-Saharan language phylum. We updated the language names using Gordon 2005; see www.ethnologue.com.

8. *Displacement, land rights:* Parkipuny 1991a, Mwaikusa 1993, Geisler and de Sousa 2001, Brockington 2002, Neumann 2002, Igoe 2003, Brockington and Igoe 2006; *Ngorongoro:* McCabe et al. 1992, Thompson 1997; *bearing costs:* Thompson and Homewood 2002, Roe and Elliot 2004; *limiting economic growth:* Norton-Griffiths and Southey 1995.

9. *Baseline:* Arcese and Sinclair 1997; *Kenya:* Ottichilo et al. 2000b, Said 2003; *Amboseli elephants:* Moss 2001; *Uganda:* Lamprey et al. 2003; *Tanzania:* Stoner et al. 2006.

10. *Human-centric, ecocentric perspectives:* Thompson and Barton 1994, Korten-kamp and Moore 2001, Peterson 2001; *human- and ecocentric view in African societies:* Callicott 1994; *conservation more important than people:* Rolston 2003.

11. *Quotes:* Kelbessa 2005, 22; Turton 1987, 182. *Maasai stories:* Kipury 1983; *school-children:* Ali and Maskill 2004.

12. *Middle ground:* Western 1989, Borgerhoff-Mulder and Coppolillo 2005, Brock-ington et al. 2006, Homewood 2008.

13. *Marching Sahara:* Lamprey 1983; *greening Sahara:* Olsson et al. 2005; *irrational pastoralists:* Herskovits 1926; *rational strategy:* Ellis and Swift 1988; *bush invasion:* Pratt and Gwynne 1977; CO_2: Bond and Midgley 2000; *pastoral measures of degradation:* Oba and Kaitira 2006; *pastoralists damaging grazing lands:* Hardin 1968; *rules to sustain rangelands:* Ostrom 1990, Bromley 1991; *pastoral leaders:* Desta and Coppock 2002; *violation of human rights:* Brockington et al. 2006; *pastoralists destroying biodiversity:* Prins 1992; *pastoralists saving biodiversity:* Western 1989.

14. *Condition improving:* MEA 2005a, HDR 2006; *literacy, health:* HDR 2006; *democratic elections:* Modelski and Perry 2002, Levy et al. 2005.

15. *Poverty rates:* World Bank Poverty Net, www.worldbank.org/poverty; *in-comes, children dying, doctors, women:* HDR 2006; *CEO salaries:* my calculations based on $30,000/day from www.aflcio.org/corporatewatch/paywatch/, compared to $1/day earnings.

16. *Riches into food and fiber:* Vitousek et al. 1986, MEA 2005a; *spectacular success:* Adams et al. 2004; *sustainable development:* WCED 1987.

17. *Ecosystem services:* MEA 2005a.

18. *Value of ecosystem services:* Costanza et al. 1997, MEA 2005a; *biodiversity as foundation:* Roe and Elliot 2004, MEA 2005a; *intact rain forest:* Balmford et al. 2002.

19. *Human domination:* Vitousek et al. 1997; *extinctions:* Pimm et al. 1995, Mace et al. 2005; *land conversion:* Tilman et al. 2001; *invasive species:* Vitousek et al. 1997, Mooney and Hobbs 2000, Zavaleta and Hulvey 2004; *diseases:* Daszak 2001; *climate change:* Thomas et al. 2004.

20. *Dryland GDP, infant mortality:* Safriel et al. 2005; *dryland human populations:* MEA 2005a; *water scarcity:* MEA 2005b.

2. SAVANNAS OF OUR BIRTH

1. I have changed the names of all my Turkana and Maasai informants to protect their anonymity in the stories that begin each chapter, except for Chapter 10. References to why animals migrated come later in this chapter.

2. *Savanna definition:* Bourlière and Hadley 1983, Cole 1986, Scholes and Walker 1993, Scholes and Archer 1997. *Creating Map 7 of African savannas:* My colleague Russell Kruska reclassified the Global Land Cover map, GLC 2003, into cropland, desert, savanna, and forest for the tropics and subtropics (between 35°N and 35°S). Cropland included all GLC categories for cropland and cropland mosaics; deserts included all bare areas and any additional areas below 100 mm (3.9 in) rainfall; savannas included all areas above 100 mm rainfall covered mostly by either short grasses, herbs, or shrubs or had an open tree canopy (tree cover of 15–40 percent); forest included all areas above 100 mm rainfall with a closed tree canopy (tree cover of 41–100 percent). This may underestimate the extent of savannas because some of the closed tree–cover areas may be savanna woodlands (for example, 42 percent tree cover with 58 percent grass cover). We then divided the savannas into two types: those receiving 100–650 mm (3.9–25.6 in) rainfall, called "rainfall-determined savannas"; and those receiving more than 650 mm rainfall per year, called "disturbance-driven savannas."

3. *Two types of savannas:* adapted from Sankaran et al. 2005.

4. Note that some scholars equate savannas with rangelands, but I do not use that term because it implies use only by livestock, which is inaccurate, particularly in wetter savannas today (see Grice and Hodgkinson 2002, Huntley 1982). I stick to the word *savanna,* broadly defined, without reference to how people use the land. Also note that the distinction between open grassy savanna dotted with trees and a tree-covered woodland is important to scholars of human evolution; these closed woodlands are called savanna woodlands in this book.

5. *Global mammal distributions:* Bigalke 1978, Maglio and Cooke 1978; *antiquity of extant African mammals:* Kingdon 1989.

6. *Global extent of pastoralism:* FAO 1999, Reid et al. 2004b; *global distribution of pastoral peoples:* Sandford 1983, Galaty 1994, Blench 2000; *proportion of developed countries that are pastoral:* Leneman and Reid 2001. Kruska overlaid a map of world population on a map of arid and semiarid lands to calculate the total maximum number of pastoralists who live in drylands worldwide. *Milk and meat production in African pastoral lands:* recalculated from de Haan et al. 1997; *marginalization of pastoralists:* Niamir-Fuller 1999a.

7. *Fluid boundaries between pastoralists and farmers:* Galaty 1993a; *people's view of selves:* Smith 1992.

8. *Quote:* Blench 2000, 10. *Pastoral diets:* Galvin 1992; *tracking forage:* Western 1982b, Niamir-Fuller 1999b; *subsistence and commercial herding, subsistence use of livestock products:* Behnke 1983, 1985.

9. These herders are what Niamir Fuller (199b) calls "full transhumants," but I think this terminology is confusing for east Africa, since transhumance "refers to regular seasonal movements of livestock between well-defined pasture areas: dry to wet season, or low to highland" (Niamir-Fuller 1999b, 1). Movements of pastoral people in the driest savannas can be quite irregular and pastures ill-defined.

10. Pastoral ecosystems are rarely "open access"; more often, access is communal, with resource access rules (Bromley 1991). *Nomads in Mauritania, Kenya, Namibia:* Behnke 1999, Zeidane 1999, McPeak and Little 2005.

11. *Fulani movements:* Turner 1999b; *transhumants, Somalia:* Little 2003.

12. *Five classes of herder:* adapted from Fratkin et al. 1994, Niamir-Fuller 1999c, Fratkin and Roth 2005, Baxter 1975, Blench 2000, Homewood 2008. Although such classifications may be "intellectually sterile exercises" (Dyson-Hudson and Dyson-Hudson 1980, 18), here it is a heuristic tool. *Settling herders:* McPeak and Little 2005; *cultivating herders:* Turton 1991, Coppock 1994, Fratkin et al. 1994, Lane 1996b, Niamir-Fuller 1999a, Little 1992; *herder farmers:* Fratkin et al. 1994.

13. *Map analysis:* percentage area of each polygon on map 7.

14. *ITCZ:* Farmer 1986; *bimodal rainfall, first food production:* Marshall 1990b.

15. *Map 8 source:* Peter G. Jones, unpublished map.

16. *Radar data:* Jarvis et al. 2008. *Map analysis:* The value given to each of the 30 million grid cells is the average difference in elevation (m) between the focal grid cell and each of the surrounding 1×1 km grid cells within 5 km of the focal cell.

17. Sankaran et al. 2005.

18. *Soil nutrients, different savannas:* Huntley 1982, Medina 1987, Mistry 2000; *soil nutrients, grass nutrients, herbivores:* Bell 1982, Ben-Shahar and Coe 1992, Fritz and Duncan 1994 (but note that Ben-Shahar does not account for soil texture). The estimate of two to three times greater biomass is based on my calculations of the equations in Fritz and Duncan 1994. *Rich, base-saturated soils, southern Africa:* Huntley 1982. *Nutrient cycling, rich and poor soils:* Scholes 1990; *Kenyan poverty maps:* CBS 2003, Okwi et al. 2007.

19. *Data for Map 11:* IEA 1998.

20. See cautionary note in Chapter 1 about the ethnic groups included on Maps 4–6 and their spelling. Globally, just less than half of all pastoralists live in six east African countries: Ethiopia, Kenya, Somalia, South Sudan, Sudan, and Tanzania; see Galaty 1994, Blench 2000.

21. For references on human evolution, see Chapter 4.

22. *General savanna function:* Scholes and Archer 1997; *savanna woody cover:* Sankaran et al. 2005.

23. *Grass removal, heavy soils:* Knoop and Walker 1985; *grass removel, lighter soils:* Riginos 2009.

24. *Fire, wetter savannas without people:* Sankaran et al. 2005; *more fire, wetter savannas:* Higgins et al. 2000; *African savannas burnt annually:* Hao and Liu 1994, Bond and Van Wilgen 1996; *fire effects on plants:* Trapnell 1959, Trollope 1982.

25. *Elephants vs. small browsers:* Western and Maitumo 2004.

26. *Elephants' effect on mature trees:* Eltringham 1980; *elephant impacts on tree cover:* van Wijngaarden 1985; *elephant damage, Kenya:* Waithaka 1994.

27. *Elephants, fire, vegetation:* Dublin 1995; *other key papers on the subject of elephants, fire, and vegation:* Buechner and Dawkins 1961, Laws 1970, Croze 1974, Caughley 1976, Norton-Griffiths 1979, Weyerhauser 1985.

28. *Serengeti impala study:* Belsky 1984; *Manyara impala:* Prins and Vanderjeugd 1993; *rodents:* Maclean et al. 2011.

29. *Fire causes, savannas:* Bond and Van Wilgen 1996; *Maasai clearing land for grazing:* Talbot 1972; *San burning for hunting:* Lee and DeVore 1976; *burning by farmers:* Schüle 1990.

30. *Forest from savanna in Guinea:* Fairhead and Leach 1996.

31. *Acacia trees in Turkana corrals:* Reid and Ellis 1995; *Turkana mobility:* McCabe 2004.

32. *Grasses, grazing:* Owen and Wiegert 1981, Stebbins 1981; *grasses adapted to fire:* Bond and Van Wilgen 1996; *grasses adapted to drought:* Coughenour 1985.

33. *Plurals on wildlife species names:* I am purposely making wildlife names plural to recognize individuals (except wildebeest and hartebeest), as we do with livestock; *digestive physiology:* Hofmann and Stewart 1972, Janis 1976, Cooper and Owen-Smith 1985, Demment and Van Soest 1985, Hofmann 1989.

34. *Dung beetles:* Anderson and Coe 1974, Cambefort 1991, Hanski and Cambefort 1991.

35. *Termite consumption, Lamto:* Konaté et al. 2003; *termite consumption elsewhere:* Watson 1967, Darlington 1985, Sugimoto et al. 2000; *termite abundances:* Bignell and Eggleton 2000, Ferrar 1982; *termites, earthworms:* Scholes and Walker 1993.

36. *Soil mixing, nutrient concentration by termites:* Wood 1988; *termite soil used for farming:* Carter and Murwira 1995; *small mammals, termite mounds:* Keesing 1998; *mound distribution:* Pringle et al. 2010.

37. *Biomass belowground:* Bourlière and Hadley 1983; *consumption and biomass of coleopterans and large mammals:* Scholes and Walker 1993.

38. Darwin's quote (1909–14, 43) was based on a conversation with his colleague Andrew Smith. *Serengeti-Mara biomass:* Broten and Said 1995; *biomass, composition of forests, savannas:* Bourlière and Hadley 1983.

39. *Animal diets:* Talbot and Talbot 1962, Field 1972.

40. *Feeding specialization:* Lamprey 1963, Leuthold and Leuthold 1976.

41. *Grazers sticking together:* Jarman and Sinclair 1979, McNaughton 1984, Sinclair 1985, McNaughton and Georgiadis 1986, McNaughton 1988, Rainy and Rainy 1989, De Boer and Prins 1990, Fryxell 1995, Fryxell et al. 2007.

42. *Large grazers affecting small animals:* Keesing 2000, Young et al. 2005, Mc-Cauley et al. 2006, McCauley et al. 2008.

43. *Push of scarce or salty water:* Gereta and Wolanski 1998, Wolanski et al. 1999. Not all members of a migrating species migrate (such as wildebeest) and some species don't migrate at all (such as hartebeest and buffaloes).

44. *Mud underfoot:* Anderson and Talbot 1965. *Nutrients:* McNaughton 1988, 1990; Murray 1995; Augustine 2003. *Protein/energy as plant parts age:* Bell 1971.

45. *Amboseli:* Western 1973.

46. *Predation vs. food:* Sinclair 1977, 1985; Sinclair et al. 2003, 2007.

47. *Elephant impacts on other species:* Cumming et al. 1997, Ogada et al. 2008.

48. *Fast cycling, nutrient-rich savannas:* Scholes 1990; *acceleration/deceleration of nutrient cycling:* Ritchie et al. 1998, Singer and Schoenecker 2003; *nitrogen availability, Yellowstone:* Frank and Groffman 1998; *grazers increasing grassland productivity, nitrogen availability:* Coppock et al. 1983, McNaughton et al. 1997, Ritchie et al. 1998, Frank et al. 2002, Singer and Schoenecker 2003; *decomposition speed:* Seagle et al. 1993; *preferential grazing on previously grazed patches:* McNaughton 1984, Day and Detling 1990; *grazers enhancing grassland productivity disputed:* Belsky 1986.

3. PASTORAL PEOPLE, LIVESTOCK, AND WILDLIFE

1. *Percent of savannas inside parks:* analysis by Russ Kruska of Map 7; *percentage of wildlife outside parks, Kenya:* Western 1989, Said 2003.

2. *Pastoralists publishing research:* Pasha 1986, Parkipuny 1991a, Oba et al. 2000b, Ntiati 2002, Ole-Miaron 2003, Nkedianye 2004, Kaelo 2007.

3. *Bimodal rainfall and pastoralism:* Ellis and Galvin 1994; Marshall 1990b.

4. *Pastoral people using livestock:* Dyson-Hudson and Dyson-Hudson 1982, Sandford 1983, Bonte and Galaty 1991, Fratkin et al. 1994; *Barabaig:* Klima 1970, Lane 1996b; *Bahima, Rwanda:* Bonte 1991; *bridewealth:* Gulliver 1955; *Gabra:* Tablino 1999; *cattle quote:* Arhem 1985, 17; *camel quote:* Little 2003, 36.

5. *Muslims forbidding bleeding:* Field 2005; *bleeding techniques:* Galvin and Little 1999, Reid pers. obs.; *amount of blood taken:* McCabe 1984.

6. *Challenges facing pastoralists:* Gulliver 1975, Little 2003; *nutritional stress:* Galvin 1992; *Somalia drought:* Little 2003; *floods, Nile:* Johnson 1991; *floods, Omo:* Turton 1991; *El Niño, Somalia:* Little 2003; *evil three quote:* Little 2003, 65.

7. *Rainfall variability:* Conrad 1941, Western 1982b, Farmer 1986, Ellis and Galvin 1994.

8. *Movement reasons, frequency:* Gulliver 1975; Dyson-Hudson and Dyson-Hudson 1980; McCabe 1987, 1994; McCabe et al. 1999.

9. *Livestock herding globally:* Blench 2000.

10. *Distance traveled, livestock:* Little 2003; *watering frequency:* Heady and Child 1994, Holecheck et al. 2004, Coughenour 2008; *shoat, cattle uses:* Field 2005; *livestock reproduction rates:* Dyson-Hudson and Dyson-Hudson 1980.

11. *Cattle types, coat colors:* Western 1973; *opposite day grazing patterns:* Western and Finch 1986, Homewood and Rodgers 1991.

12. *Camel milk production, uses:* Field 2005; *energy flow, camels:* Coughenour et al. 1985a.

13. *Herd splitting:* Baxter 1975, Lewis 1975.

14. *Key resources:* Scoones 1991; *quote:* Little 2003, 22; *Baringo drought:* Homewood and Lewis 1987.

15. *Irrational herders:* Herskovits 1926; *large herds, more left:* McPeak and Little 2005.

16. *Ranking cattle lowest:* McCabe 2004.

17. *10,000 Turkana killed: Economist* article, cited in McCabe 2004.

18. *Sudan raiding:* Johnson 1991; *Daasanach (spelled Dassanetch by Turton):* Turton 1991, 153; *Kuria quote:* Fleisher 2000, 2.

19. *Herder access needs:* Dyson-Hudson and Dyson-Hudson 1980; *reasons for raiding:* McCabe 2004, Johnson 1991; *political support of raiders:* Hendrickson et al. 1998; *quote:* Sobania 1991, 139.

20. *Predator attacks:* Mizutani et al. 2003, Ogada et al. 2003, Reid unpublished interviews; *mortality by disease, predators:* Mizutani et al. 2003; *killing rates, Laikipia ranches:* Ogada et al. 2003.

21. *Elephants injuring, killing:* Kaelo 2007 (quote, p. 53); *Cape buffalo:* Kangwana and ole Mako 2001, Akama et al. 1995, Reid unpublished interviews; *crop destruction:* Kangwana and ole Mako 2001.

22. *Quote:* Baxter 1975, 213.

23. *Daasanach (also spelled Dassanetch):* Sobania 1991; *types of social ties:* Galaty 1991.

24. *Female bonds:* Ryan et al. 2000; *quote:* Baxter 1975, 212.

25. *Property rights models:* McCarthy et al. 1999, Swallow and Kamara 1999; *general property rights:* Ostrom 1990, Ostrom 2003, Bromley 1991; *quote, reciprocal grazing rights:* Little 2003, 31; *customary rules of access:* McCabe 1990; *pastoral paradox*: Fernández-Giménez 2002; *quote, judicial systems:* Niamir-Fuller 1999b, 2.

26. *Somali safe passage:* Little 2003; P. Little, pers. comm.

27. *Overhunting example:* Peterson 2003; *little wildlife, Kambaland:* Said and Reid 2007, Reid unpublished interviews of Kamba.

28. *Small wildlife left on farms:* Newmark et al. 1993; *farmers in compact villages:* Kjekshus 1996.

29. *Pastoralists' use of wildlife parts:* Aboud 1989; *herders eating wild meat:* Dyson-Hudson and Dyson-Hudson 1970, Western 1973, Galaty 1991, Prins 1992; *wildlife as "second cattle":* Western 1982b, 202.

30. *Facilitation:* Prins 2000, Arsenault and Owen-Smith 2002; *cultural, economic value of wildlife to herders:* Emerton 2001, Kangwana and Ole Mako 2001, Prins 2000 (for Bigalke's data); *African wild ass:* Moehlman 2005.

31. *Domestic animal–wildlife disease transmission:* Grootenhuis 2000, Butler et al. 2004, Cleaveland et al. 2008; *malignant catarrhal fever:* Cleaveland et al. 2008.

32. *Competition:* Prins 2000, Odadi et al. 2011; *elephants destroying crops:* Aboud 1989, Cochrane et al. 2005.

33. *Limiting wildlife access to water:* Williams 1998, De Leeuw et al. 2001; *afraid to fetch water:* Aboud 1989; *livestock impacts, water points:* Andrew 1988, Barker et al. 1989, Georgiadis and McNaughton 1990, James et al. 1999.

34. *Human effects on wildlife:* Eltringham 1990, Happold 1995; *elephants, human populations:* Parker and Graham 1989a, Parker and Graham 1989b, Hoare and du Toit 1999; *Mara people, wildlife:* Reid et al. 2003; *lower conflicts:* Newmark 1996.

35. *Negative human impact on carnivores:* Woodroffe and Ginsberg 1998; *Ghana:* Brashares et al. 2001; *parks, human population growth:* Wittemyer et al. 2008.

36. *Dogs and disease:* Cleaveland et al. 2001; *chasing carnivores:* Ogada et al. 2003; *crop raiding:* Naughton-Treves 1998, Sitati et al. 2005; *farming area, wildlife:* Fritz et al. 2003, Norton-Griffiths et al. 2008.

37. *Effect of human population density on wildlife:* Parker and Graham 1989b, Barnes et al. 1991, Newmark et al. 1993, Hoare and du Toit 1999, Reid et al. 2003.

4. MOVING CONTINENTS, VARYING CLIMATE, AND ABUNDANT WILDLIFE

1. *Laetoli:* Leakey and Hay 1979; *geologic time scale used here:* Walker and Geissman 2009. Hominins are defined as all humanlike species after the split with chimps, 6–7 Ma (Kingdon 2003).

2. *List of firsts:* Johanson and Edey 1981, Waters and Odero 1986, DiMichele and Hook 1993, Reader 1997, Klein 1999; *Earth warping:* Waters and Odero 1986.

3. *Dates and ecosystem descriptions:* DiMichele and Hook 1993, 208–224; *quote:* ibid., 205.

4. *Early savannas, climate, plant types, fires:* Scott and Jones 1994, Falcon-Lang 2000, Scott 2000; *location of continents, carboniferous animals:* DiMichele and Hook 1993; *oxygen and burning:* Clark and Robinson 1993, Berner et al. 2003; *basidiomycetes:* Robinson 1991.

5. *First herbivory:* DiMichele and Hook 1993, Sues and Reisz 1998.

6. *Positions, continents:* Wing and Sues 1993; *Senegal next to Florida:* Klein 1999; *Jurassic vegetation:* van der Hammen 1983, Rees et al. 2004.

7. *Jurassic African tropical climate, Tendaguru dinosaurs, migrations:* Rees et al. 2004; *Brachiosaurus:* Russell et al. 1980; *dinosaur herbivory, fire:* Schüle 1990; *evidence of beetle-dinosaur associations inconclusive:* Davis et al. 2002; *long-term climate:* Wing and Sues 1993.

8. *Megazostrodon:* Durand 2005; *early mammal size, habits, warm-blooded:* Crompton and Jenkins 1978.

9. *Evolution of angiosperms:* Barrett and Willis 2001; *insect explosion:* Grimaldi 1999; *new angiosperm food, mammals:* Klein 1999; *high-energy food plants, human population growth:* Murdock 1959, Diamond 2002; *angiosperm abundance:* Jacobs 2004.

10. *Asteroid, dinosaur extinctions:* Alvarez et al. 1980; *Chicxulub crater:* Hildebrand et al. 1991; *dating debate:* Keller et al. 2004; *large group supporting Mexican crater:* Schulte et al. 2010; *volcanism:* Keller 2005; *fungal diseases:* Casadevall 2005; *no extinction:* Sarjeant and Currie 2001; *opportunity for mammals, evolution of primates:* Klein 1999.

11. *Isolation of Africa, anthropoid apes:* Maglio and Cooke 1978, Kingdon 1989; *Oligocene faunas:* Maglio 1978, Potts and Behrensmeyer 1993, Turner and Anton 2004; *Hyainailouros sulzeri:* Turner and Anton 2004.

12. *Summary, savanna origins, grass pollen:* Cerling 1992, Cerling et al. 1997, Jacobs et al. 1999, Jacobs 2004; *molecular clock:* Janssen and Bremer 2004.

13. *Combretoxylon:* Boureau et al. 1983; *drying phase:* Klein 1999; *first acacia:* Herendeen and Jacobs 2000; *fossils of vegetation in Africa are very rare for 65–40 Ma, making this educated guesswork:* Jacobs 2004.

14. *Coevolution of grasses, grazers:* Stebbins 1981, Prasad et al. 2005; *silica, grazers:* McNaughton et al. 1985; *dinosaur coprolites, insect grazing:* Prasad et al. 2005; *other explanations of apparent coevolution:* Stromberg 2011.

15. *Rift Valley formation:* Waters and Odero 1986; *volcano ages:* Baker et al. 1971; *volcanic history, Ol Doinyo Lengai:* www.volcano.si.edu

16. *African mammal migration:* Kingdon 1989, Kappelman et al. 2003; *first records, extinctions:* Coppens et al. 1978, Maglio 1978; *hypsodont teeth new to Africa:* Schüle 1990; *evolution of ruminant digestion*: Van Soest 1994; *ungulate diversity:* Prothero and Schoch 2002; *early ungulates, Uzbekistan, Asian steppe:* Archibald 1996.

17. *Mammals into and out of Africa, extinctions:* Coppens et al. 1978, Maglio 1978; *hyraxes:* Maglio 1978, Kappelman et al. 2003.

18. *Carnivores:* Maglio 1978, Savage 1978, Turner and Anton 2004.

19. C_3 and C_4 plants are so named for the number of carbon atoms (3 or 4) in the first molecule they produce as they turn CO_2 in the atmosphere into carbohydrate during photosynthesis. *C_4 grass in east Africa*: Uno et al. 2011; *C_4 biology*: Ehleringer et al. 1997; *C_4 plants, tooth enamel*: Cerling 1992, Cerling et al. 1997; *CO_2 concentrations*: Tripati et al. 2009; *low CO_2 favoring C_4 plants*: Cerling et al. 1997, Ehleringer et al. 1997, Bond and Midgley 2000; *fire, CO_2, and C_4 plants*: Bond et al. 2003; *grassland expansion accelerating cooling and drying:* Retallack et al. 2002; *savanna expansion:* Cerling 1992, Segalen et al. 2007.

20. It is possible that anthropoid apes traveled from Africa to Asia and back again between 18 Ma and 10.5 Ma; see Kingdon 2003. *Genetic similarities, humans and apes:* Chen and Li 2001; *divergence times, common ancestor population size:* Brunet et al. 2002, Mikkelsen et al. 2005; *junk DNA:* Veitia and Bottani 2009.

21. *Messinian Salinity Crisis:* Brain 1981; *straight-line descent:* Asfaw 1999, White and al. 2006; *timing of appearance, different species:* Klein 1999; *bushy model, quote:* Wood 2002, 134; *other bushy model:* Strait and Wood 1999a, Leakey et al. 2001.

22. *Oldest hominin:* Senut et al. 2001, Brunet et al. 2002, 2005; *chimps and humans are Homo:* Curnoe and Thorne 2003.

23. *Early bipedalism:* Pickford et al. 2002; *tools:* Schick and Toth 1993, Ambrose 2001; *brain sizes:* Klein 1999; *movement, Homo erectus:* Finlayson 2005; *Indonesian Homo erectus descendant:* Brown et al. 2004, Argue et al. 2009; *multiple evidence of African roots for Homo sapiens:* Cavalli-Sforza 2000; *eastern African origin of humans:* Strait and Wood 1999b, Gonder et al. 2007; *also southern Africa:* Behar et al. 2008; *oldest African Homo sapiens:* McDougall et al. 2005; *australopithecines, out of Africa:* Foley and Gamble 2009. *Homo* fossils from Dmanisi, Georgia, have some characters older than *Homo erectus;* thus, which *Homo* was first to walk out of Africa is under debate (see Rightmire et al. 2006), or even if *Homo* walked *into* Africa (see Wood 2011).

24. *Ardipithecus, tree climber:* Lovejoy et al. 2009; *wooded savanna, Australopithecus:* Cerling et al. 2011; *discovery, description of Lucy:* Johanson and Edey 1981; *brain sizes:* Klein 1999.

25. *Walking, climbing, australopithecenes:* Leakey et al. 1995, Pickford and Senut 2001; *reconstruction of Hadar climate:* Bonnefille et al. 2004; *mammals alive at 3 Ma:* Kingdon 1989; *extinct mammals:* Maglio 1978; *reconstructions, ancient savannas:* Turner and Anton 2004. Regarding early descriptions of Laetoli vegetation, Leakey and Harris (1987) suggested a savanna grassland, but more recent work (Kingston and Harrison 2007) suggests extensive woodland habitat as well.

26. *Why humans evolved*: Vrba et al. 1995, Bobe and Behrensmeyer 2004.

27. *Quote, variability hypothesis:* Potts 1996, 922; *use of termite mounds:* Ambrose 2001.

28. *Climate variability periods, marine evidence and human evolution:* DeMenocal 1995, 2004; *lake filling and drying:* Trauth et al. 2005, 2009; *increase in climate variability:* Potts 1996; *faunal turnover:* Bobe and Behrensmeyer 2004.

29. *Subsistence of first hominins:* Plummer 2004; *woodland vs savanna abundances:* Leonard and Robertson 1997.

30. *Body size and longevity:* Wood 1999; *brain energetics, meat vs. USO debate:* Plummer 2004; *USO abundance, edibility, consumption:* Wrangham et al. 1999, Laden and Wrangham 2005; *carrion avoidance:* Ragir et al. 2000; *Homo habilis, H. erectus, and tools:* Schick and Toth 1993, Ambrose 2001; *meat and social organization:* Rose and Marshall 1996.

31. *Eutrophic savannas:* Bell 1982; *more mammal biomass on rich soils*: Fritz and Duncan 1994; *grazed grass more productive*: McNaughton 1976, 1979.

32. Blumenschine 1987, 1995; *first access, carcasses:* Bunn and Ezzo 1993, Dominguez-Rodrigo 2002.

33. *Fire evidence:* Gowlett et al. 1981, Brain and Sillen 1988, Bellomo 1994, Goren-Inbar et al. 2004.

34. *Burning in Plio-Pleistocene:* Bird and Calio 1998; *contemporary hunters, fire:* Laris 2002, author's observations in the Ituri Forest, author's interviews with Okiek.

35. *Technology development:* Schick and Toth 1993, McBrearty and Brooks 2000; *oldest spears:* Thieme 1997; *hunting large animals:* Klein 1999, McBrearty and Brooks 2000, Marean 1997, Ambrose 2001, Ambrose 2010; *superpredators:* S. Ambrose, pers. comm.

36. *Overkill timing, genera lost:* Barnosky et al. 2004; *sizes of animals lost, effects of megafaunal loss:* Owen-Smith 1987; *extinction causes:* Brook and Bowman 2002, Burney and Flannery 2005; *habitat change, burning:* Caldararo 2002; *lack of North America evidence:* Grayson and Meltzer 2003.

37. *Genera lost:* Maglio 1978, Barnosky et al. 2004; *earlier extinction, Africa:* Martin and Klein 1984, Owen-Smith 1987, Surovell et al. 2005; *early hunting proficiency:* McBrearty and Brooks 2000, Barnosky et al. 2004.

5. ECOSYSTEM ENGINEERS COME OF AGE

1. Waller 1988, 76.

2. Lugard 1893, quoted in Ford 1971, 140.

3. When reporting dates, I use a base year of 2000; for example, 12,000 years ago is 12,000 BP (before present) or 10,000 BC.

4. *Nine locations:* Diamond 2002; *cradle of agriculture:* Lev-Yadun et al. 2000; *Nabta Playa, Egypt:* Wendorf and Schild 1984; *domestication, Africa's first cattle:* Hanotte et al. 2002, Marshall and Hildebrand 2002; *donkey domestication:* Marshall 2007, Rossel et al. 2008; *camels:* Field 2005.

5. *Khoisan hunter-gatherers:* Murdock 1959; *absorbed by herders moving south:* Ambrose 1984b; *Khoisan genetic age:* Zhivotovsky et al. 2003; *disease constraints:* Gifford-Gonzalez 1998.

6. *Settlement near Khartoum, wild sorghum:* Haaland 1995, Iliffe 1995; *human movements, crops:* Newman 1995.

7. *Southern Cushites, cultivation:* Ambrose 1984a, Ehret 1984; *Bantu farmers:* Newman 1995; *Maasai stole land:* Shetler 2007.

8. *Human movements, languages, quote:* Newman 1995, 166.

9. *Spread of iron:* Miller and van der Merwe 1994, Iliffe 1995, Schmidt 1997; *malaria, spread of Bantu peoples:* Webb 2005.

10. *Bantu farmers adopt sorghum, cattle:* Iliffe 1995.

11. Marshall 1990b, Marshall and Hildebrand 2002; *start of biomodal rainfall:* Richardson and Richardson 1972, Marshall, 1990b.

12. *Quote:* Smith 1992, 9; *hunters, herders:* Smith 2005; *Kalahari:* Cashdan 1984.

13. *Quote:* Collett 1987, 136; *pastoral road to extinction:* Prins 1992.

14. *Quote:* Hoare 1999, 689; *obstacles to farming:* Parker and Graham 1989a, Barnes 1996; *well-defended villages:* Laws et al. 1975; *deforestation:* Chapman and Chapman 1996; *erosion:* Eriksson et al. 2000.

15. *Slave trade, quote:* Lovejoy 2000, 24. I estimated the total number of slaves traded based on the Lovejoy's summary of data from many sources.

16. *Ivory trade history:* Iliffe 1995, Hakansson 2004; *Aksum:* Newman 1995.

17. *Ivory demand, Tabora porters, female tusk quote:* Beachey 1967, 275; *causes quote:* Hakansson 2004, 569; *early coastal ivory trade:* Iliffe 1995.

18. *African elephant hunters:* Steinhart 2000, Hakansson 2004; *Dorobo, ivory:* Berntsen 1976, Dyson-Hudson and Dyson-Hudson 1980.

19. An estimate of Africa's elephant carrying capacity in 1814 is 26.9 million, but the actual population may have been 19–20 million; see Milner-Gulland and Beddington 1993. *Ivory exports:* Hakansson 2004; *elephant distribution:* Spinage 1973; *no Amboseli elephants seen in 1883:* Thomson 1885. *No Serengeti elephants in 1900:* Dublin 1991.

20. *1987 population estimate:* Douglas-Hamilton 1988; *2002 population estimate:* Blanc et al. 2003; *percent of pre-1814 populations:* my calculations.

21. *Quotes:* Koponen 1988, 637 and 652; *Tanzanian firearms:* Iliffe 1995.

22. *Social memory quote:* Koponen 1988, 661; *colonialism, disease:* Ford 1971, Diamond 1997, Fenn 2001. *Smallpox, cholera—in Tanzania:* Koponen 1988; *in Somalia:* Mohamed 1999; *in Sudan:* Hartwig 1981; *in Ethiopia:* Pankhurst 1965. Thanks to Holly Reid for encouraging me to investigate cholera.

23. *Source of rinderpest:* Pankhurst and Johnson 1988; *quote:* Ford 1971, 123; *animals affected, quote:* Plowright 1982, 4; *rinderpest control:* Atang and Plowright 1969.

24. *Karamojong:* Barber 1968, in Dyson-Hudson and Dyson Hudson 1980; *Mursi:* Turton 1988; *Samburu, Rendille:* Sobania 1991; *highest highlands:* Pankhurst and Johnson 1988; *Il Chamus:* Anderson 2002; *Jie and quote:* Lamphear 1976, 224; *Turkana:* Dyson-Hudson 1999.

25. *Soup kettles quote:* Waller 1988, 79; *vulture quote:* Plowright 1982, 3 (quoting Branagan and Hammond 1965); *jigger fleas:* Ford 1971.

26. *Tree death from old age:* Young and Lindsay 1988, Prins and Vanderjeugd 1993. For the debate on the cause of tree decline in Amboseli, see Chapter 9.

27. *Rinderpest, expansion of woodlands:* Ford 1971.

28. *Quote:* Dublin 1991, 173; *burning woodlands:* Lamprey and Waller 1990.

29. *Fire frequency after rinderpest:* Norton-Griffiths 1979, Dublin et al. 1990; *Tsavo:* Gillson 2006; *why people burned:* Waller 1988; *predator loss:* Sinclair 1977; *food web effects:* McNaughton 1992.

30. *History of colonialism:* Tignor 1976, Davidson 1991, Illife 1995.

31. *Territorial history, Maasai and other Maa speakers:* Jacobs 1975, Robertshaw 1991, Sutton 1993, Hughes 2002. Note that colonialism started in about the 1890s in

many countries of east Africa, although some countries, like Ethiopia were never really colonized.

32. *Quote:* Hughes 2002, 125.

33. *Land loss, Tanzania*: Brockington 2002.

34. *Quote:* Waller and Sobania 1994, 45–46.

35. *Dinka/Nuer:* Johnson 1991; *Rendille/Samburu:* Sobania 1991; *new conflicts over scarce resources, Kenya:* Oba 2011; *quarantines:* Waller 2004; *reducing drought coping:* Waller and Sobania 1994.

36. *Quote:* Anderson and Grove 1987, 4; *access to Nile, settlers:* Tignor 1976; *hunters:* MacKenzie 1987, Beinart 2000.

37. *Quote:* Hoben 1995, 1004; *degradation narratives:* McCann 1999, Beinart 2000; *Baringo quote:* Anderson 2002, 156; *white hunters quote:* MacKenzie 1987, 50.

38. *Narratives overstating degradation:* Homewood and Brockington 1999, McCann 1999, Brockington 2002; *elephant loss, ivory trade:* Spinage 1973; *hunting and wildlife decline:* MacKenzie 1987; *quagga:* Reed 1970; *Cape forests:* Grove 1987; *Machakos:* Tiffen et al. 1994.

39. *Mammal range loss:* Ceballos and Ehrlich 2002; *soil erosion, Machakos:* Tiffen et al. 1994.

40. *Elite, conservation:* MacKenzie 1987, Mwaikusa 1993; *Kenya hunting regulation, protected areas:* Lado 1996; *early Tanzania game laws*: Leader-Williams 2000.

41. *Quote:* Neumann 1997, 568; *irrational Maasai, conservation area income:* Collett 1987; *when parks established:* Brockington 2002, Turton 1987.

42. *Anstey quotes and Mkomazi history:* Brockington 2002, 38.

43. *Protected areas established on key resources*: Reid, pers. obs.

44. *Community-based conservation, selected references:* Western and Wright 1994, Barrett and Arcese 1995, Meffe et al. 2002, Goldman 2003, Berkes 2004, Borgerhoff-Mulder and Coppolillo 2005, Berkes 2007, Igoe and Croucher 2007, Brockington et al. 2008, Homewood et al. 2009, Child and Barnes 2010, Dressler et al. 2010, Nelson et al. 2010, Dukes et al. 2012.

45. *Movement into, out of pastoralism:* Galaty 1993a, McPeak and Little 2005; *speed of change, nomadism in pastoral society:* Roth and Fratkin 2005; *livelihood diversification:* Little et al. 2001, McCabe 2003, Homewood 2008, Homewood et al. 2009.

46. *Borana:* Kamara et al. 2004; *farmers dominating rangelands:* Campbell et al. 2000.

47. *Wet savannas:* Said 2003; *Borana leaders:* Huqqaa n.d., Desta and Coppock 2004.

48. *Privatization, pastoral land:* Galaty 1992, Rutten 1992; *quote, who gets private land:* Kirk 1999, 35; *wealthy pastoral elite, land:* Rutten 1992, Mwangi 2005.

49. *Maasai group ranches, privatization:* Pasha 1986, Rutten 1992, Galaty 1994, Reid et al. 2008b, Nkedianye et al. 2009.

50. *Reinforcement of prejudices:* Horowitz and Little 1987; *excision, pastoral land for parks:* Homewood and Rodgers 1984, Brockington 2002; *Tanzanian resettlement:* Moses Ole Neselle, pers. comm.

51. *Tradition breaking down:* Little and Brokensha 1987.

52. *Sedentarization:* Salzman 1980, Hogg 1986, Mayer et al. 1986, Kamara et al. 2004, Fratkin and Roth 2005; *quote:* Blench 2000, iii.

53. *Settled and mobile lifestyles:* McPeak and Little 2005.

54. *What is cultivated first:* Galaty 1994; *base of Mt. Kilimanjaro:* Campbell et al. 2003b; *Nguruman:* Reid pers. obs.; *Pianupe:* R. Lamprey, pers. comm.

55. *Wildlife only areas:* Brockington 2002, Neumann 2002; *settlements, fenced areas:* Williams 1998, De Leeuw et al. 2001.

56. *Kenya:* Said 2003; *Tanzania:* Caro et al. 1998.

57. *Sudan:* Wilson 1980; *South Sudan:* Fryxell 1983, www.wcs.org/saving-wildlife/hoofed-mammals/kob.aspx; *Ethiopia, hunting:* Stephens et al. 2001; *wolves, rabies:* Marino et al. 2006. I could find no current published information on wildlife populations in other parts of Sudan, Eritrea, Djibouti, Somalia, Ethiopia, and Burundi, nor information on population trends, nor studies of how land use affects wildlife.

58. *Uganda:* Lamprey and Mitchelmore 1996, Lamprey et al. 2003; *Rwanda:* Kanyamibwa 1998.

59. *Problems with the segregation of pastoralists and wildlife:* Western and Gichohi 1993.

6. CAN PASTORAL PEOPLE AND LIVESTOCK ENRICH SAVANNA LANDSCAPES?

1. *Quote:* Loreau et al. 2001, 804; *Serengeti:* McNaughton 1994.

2. *Intermediate disturbance hypothesis:* Grime 1973, Connell 1978, Huston 1979, Huston 1994, Wilkinson 1999; *intermediate disturbance response not widespread:* e.g., Mackey and Currie 2001; *animal responses to grazing:* Joubert and Ryan 1999, Seymour and Dean 1999.

3. *Savannas decline if not grazed:* Oba et al. 2000b; *grazing and plant diversity worldwide:* Milchunas et al. 1988, Milchunas and Laurenroth 1993.

4. *More species = more biodiversity?:* Belsky et al. 1999; *weeds brought in by livestock:* Milton 2004. The effects of grazing on weeds in African savannas is poorly understood.

5. *History of grazing:* Mack and Thompson 1982, Milchunas and Laurenroth 1993; *grazing, Asian savannas:* Perevolotsky and Seligman 1998, Fernández-Giménez and Allen-Diaz 1999.

6. *Grazing impacts, U.S. west (selected):* Laurenroth et al. 1999, Milchunas et al. 1998, Neff et al. 2005.

7. *400,000 years of fires:* Bird and Calio 1998.

8. Herskovits 1926; *tragedy of the commons, quotes:* Hardin 1968, 1244; *Sahel overgrazing:* Lamprey 1983, Sinclair and Fryxell 1985.

9. *Review, equilibrium/nonequlibrium systems:* Vetter 2005, Homewood 2008; *climate vs. overgrazing:* Behnke et al. 1993, Ellis and Swift 1988, Homewood and Rodgers 1987; *continuum:* Briske et al. 2003; *thresholds:* Friedel 1991, Stringham et al. 2003.

10. *Review papers:* Milchunas et al. 1988, Bakker 1989, Milchunas and Laurenroth 1993, Hodgson and Illius 1996, Olff and Ritchie 1998, Proulx and Mazumder 1998, Harrison et al. 2003; *marine snails:* Lubchenco 1978; *Great Plains:* Hartnett et al. 1996; *Serengeti plants:* Belsky 1992.

11. *Grazing stimulating plant production:* Frank et al. 2002; McNaughton 1976, 1979, 1985; Milchunas and Laurenroth 1993; Tansley and Adamson 1925. *Yellowstone:* Frank and McNaughton 1993, Frank and Groffman 1998; *increased carrying capacity:* McNaughton et al. 1997; *counterarguments:* Belsky 1986.

12. *Grasses:* Coughenour et al. 1985b, 1985c; *sedge:* McNaughton et al. 1983; *shrubs:* Coughenour et al. 1990b; *modeling:* Coughenour 1991.

13. *Grazing lawns:* McNaughton 1984; *repeat grazing benefits:* McNaughton 1985; *intermediate biomass:* Demment and Van Soest 1985, Fryxell 1991, Hobbs and Swift 1988, McNaughton 1984, Owen-Smith and Novellie 1982, Reid et al., in prep.

14. *Grazing lawns only created by wildlife:* McNaughton 1986; *largest herbivores:* Owen-Smith 1988, du Toit and Cumming 1999.

15. *Heavier livestock grazing than wildlife:* Skarpe 1991, du Toit and Cumming 1999.

16. *Kaputiei Plains:* Gichohi 1996; *Kenya and Niger:* Oba et al. 2001, Hiernaux 1998; *Somalia:* Thurow and Hussein 1989; *Namibia, South Africa:* Ward et al. 1998, Todd and Hoffman 1999.

17. *Grazing effects on plant types*: Milchunas et al. 1988, Hobbs and Huenneke 1992.

18. *Grazing responses, unproductive savannas—Kenya*: Oba et al. 2001; *Niger:* Hiernaux 1998; *Somalia:* Thurow and Hussein 1989.

19. The effects of livestock on productivity in east Africa are hard to judge because most studies on grazing do not separate plant production from grazing offtake. *No response—Kenya:* Ellis and Swift 1988, Coughenour et al. 1990a; *Namibia:* Sullivan and Rohde 2005; *Zimbabwe:* Kelly and Walker 1977. *Humped response—Somalia:* Thurow and Hussein 1989; *Kenya:* Mworia et al. 1997; *Mali:* Hiernaux 1993. *Indigofera:* Oba 1994.

20. *Bison, Great Plains:* Milchunas et al. 1998.

21. *Maasai observations:* Goldman 2006.

22. *Fire increasing rangeland diversity:* Fuhlendorf and Engle 2001; *extensive fires reducing diversity, Australia*: Yibarbuk et al. 2001.

23. *Fire effects on grass:* Reid pers. obs.; *nitrogen return through rainfall:* Frost and Robertson 1987; *Zambia burning experiment:* Trapnell et al. 1976; *increased nutrients after burns:* van de Vijver et al. 1999, McNaughton 1985.

24. *Fire and species numbers:* Parsons and Stohlgren 1989, Meyer and Schiffman 1999, d'Antonio 2000, Harrison et al. 2003, Leach and Givinish 1996; *change in Serengeti species after burns:* Belsky 1992; *burning, type of species:* McNaughton 1985.

25. *Easier to digest:* Mentis and Tainton 1984, Hobbs and Spowart 1984, Frost and Robertson 1987, Hobbs et al. 1991; *smaller grazers:* Sensenig et al. 2010; *weight gain:* Anderson et al. 1970, Svejcar 1989; *calmer grazing:* Sutherst 1987.

26. *Lions, vegetation hiding:* Hopcraft et al. 2005.

27. *Aboriginal burning practices/effects:* Russell-Smith et al. 1997, Yibarbuk et al. 2001; *Native Americans burning California shrublands:* Keeley 2002, Anderson 2005.

28. *African herders diversify savannas by burning*: Reid pers. obs.

29. *Ngorongoro burning effects:* Runyoro et al. 1995.

30. Ancient cattle in the Mara were larger with smaller horns than cattle in east Africa today, and were probably of indigenous, taurine genetic stock; see Marshall 1990a.

31. *Turkana homestead locations:* Reid 1992, Reid and Ellis 1995; *Amboseli Maasai homestead locations:* Western and Dunne 1979; *Mara:* Reid pers. obs.

32. *Bomas visible on satellite imagery:* Reid pers. obs.

33. *Nutrients, boma soils—in Laikipia*: Augustine 2003, Veblen and Young 2010; *in Amboseli*: Muchiru 1992, Muchiru et al. 2008.

34. *Last longer, dry savannas, dung piles:* Reid pers. obs.

35. *Boma grass nutrients:* Augustine 2003; *Turkana:* Reid 1992; *Maasai, "grass is too fat":* Reid unpublished interviews.

36. *Plant, animal succession in century-old settlements:* Muchiru et al. 2008; *cyanide:* Georgiadis and McNaughton 1988, 1990.

37. *Wildlife, livestock preference for old bomas:* Young et al. 1995; *wildlife maintenance of bomas*: Veblen and Young 2010; *birds:* Morris et al. 2008, Söderström and Reid 2010.

38. *Turkana settlements promoting acacia trees:* Reid 1992, Reid and Ellis 1995.

39. *Large trees as small ecosystems:* Dean et al. 1999.

40. *Old settlements in South Africa:* Blackmore et al. 1990; *in central Kenya:* Young et al. 1995, Augustine 2003.

41. *Mining of nutrients:* Augustine 2003.

42. *Nutrient-poor African grasses:* Pratt and Gwynne 1977; *lactating females:* McDowell 1985; *boma nutrients, impala:* Augustine 2003, 2005.

43. *Serengeti nutrient hotspots:* McNaughton and Banyikwa 1995; *nutrient hotspots in North America:* Coppock et al. 1983, Jaramillo and Detling 1988, Day and Detling 1990.

44. This work was initiated by Michael Rainy, Cathleen Wilson, Eoin Harris, Russ Kruska, Pakuo Lesorogol, Sauna Lemiruni, Lerale Lesorogol and myself, followed by work with Joe Ogutu, Tom Hobbs, Dickson Kaelo, Jeff Worden, Clare Bedelian, Sandra Van Dijk, John Rakwa, Meole Sananka, Joseph Temut, James

Kaigil, Charles Matankory, Daniel Naurori, John Siololo, and other excellent team members.

45. *Wildlife cluster, intermediate distances:* Reid et al. 2003, Ogutu et al. 2010.

46. *Why Serengeti wildlife aggregate:* McNaughton 1984, Sinclair 1985, McNaughton 1988, Fryxell 1995, Fryxell et al. 2007; *multiple species associations:* Rainy and Rainy 1989.

47. *Settlement effects on wildlife:* Reid et al. 2001, 2003.

48. *Wildlife near bomas at night:* Sandra Van Dijk, Dickson Kaelo, Kari Veblen, and Robin Reid, unpublished interviews; Goldman 2009. *Lions near bomas:* Goldman 2006.

49. *Wildlife livestock experiments:* Young et al. 2005; *facilitation:* Odadi et al. 2011.

50. *Grazing succession:* Vesey-Fitzgerald 1960, Bell 1971; *in Amboseli:* Western 1973.

51. *Hyenas:* Boydston et al. 2003; *Ethiopian wolves:* Sillero-Zubiri and Gotteli 1995.

52. *Species that tolerate herders, livestock:* Reid et al. 2003.

53. *Diversity of mixed livestock-wildlife systems:* Western 1989; *abundance at park edges:* Western and Gichohi 1993, Reid et al. 2003.

7. WHEN COEXISTENCE TURNS INTO CONFLICT

1. Quoted in Goldman 2006, 417.

2. *Pastoral hunting:* Prins 1992; *"second cattle":* Western 1982b, 202; *emergency food quote:* Western 1982b, 192; *no longer warriors:* Nkedianye 2004; *Tarangire:* Kangwana and Ole Mako 2001.

3. *Northern Kenya:* Said 2003; *Karamojong:* Lamprey et al. 2003, 24; *Rwanda:* Kanyamibwa 1998; *Maasai boys:* Reid unpublished interviews; *dogs killing wildlife:* Stephens et al. 2001; *Cushitic herders, wild meat:* Galaty 1991.

4. *Carnivores attacking livestock:* Mizutani et al. 2003, Ogada et al. 2003; *resentment of lions:* Hazzah et al. 2009.

5. *Impact of elephants:* Western 1982a, Aboud 1989, Hoare 2000, Moss 2001, Sitati et al. 2003, Kaelo 2007; *buffalo:* Kangwana and Ole Mako 2001.

6. *Amboseli elephants speared for killing Maasai:* Moss 2001; *poisoning predators:* Maclennan et al. 2009; *wild dogs:* Nkwame 2009.

7. *Trypanosomosis, East Coast fever:* Ford 1971; *MCF:* Reid et al. 1984; *rabies, canine distemper, hybridizing:* Sillero-Zubiri et al. 1996, Packer et al. 1999, Cleaveland et al. 2000.

8. *Review of competition:* Prins 2000; *pastoral observations:* Kangwana and Ole Mako 2001; *Mara, Kaputiei competition:* Reid unpublished interviews; *competition when food limiting:* Voeten and Prins 1999, Odadi et al. 2011; *diet overlap:* Field et al. 1972, Ego et al. 2003; *elk and cattle:* Hobbs et al. 1996a, 1996b.

9. *People excluding wildlife:* van Wijngaarden 1985, Williams 1998; *wildlife size at water:* Ayeni 1975; *Maasai chasing wildlife:* Gichohi 1996; *northern Kenya waterpoints:* de Leeuw et al. 2001; *Kalahari water:* Perkins and Thomas 1993, Verlinden 1997.

Water point effects—in Africa: Georgiadis 1987, Tolsma et al. 1987; *in Australia:* James et al. 1999, Landsberg et al. 2003.

10. *Fires, Australia:* Russell-Smith et al. 1997, Yibarbuk et al. 2001; *fires and woody plants:* Pratt and Gwynne 1977, Archer 1989, van de Vijver et al. 1999a, Roques et al. 2001, van Langevelde et al. 2003.

11. Huqqaa n.d., quoted in Desta and Coppock 2004.

12. *Overgrazing controversy, evidence not supporting overgrazing:* Homewood and Rodgers 1987, Brockington 2002, Sullivan and Rohde 2002; *evidence supporting overgrazing:* Lamprey 1983, Sinclair and Fryxell 1985; *review of overgrazing:* Vetter 2005.

13. *Quote:* Behnke and Scoones 1991, 16; *pastoralists using livestock as an indicator of degradation:* Oba and Kaitira 2006.

14. *In-depth herder measurements of rangeland:* Oba et al. 2000a; *differences between herder and scientist measures:* Oba and Kotile 2001; *Ariaal herder measures:* Roba and Oba 2009; *Tanzanian Maasai observations:* Oba and Kaitira 2006.

15. *Woody plants spread:* Coppock 1994, Oba et al. 2000a; *herder quote, 1960s landscape:* Tiki et al. 2011, 73; *grass, more milk, calves:* Roba and Oba 2009; *reasons for loss of grass:* Oba et al. 2000a, Tiki et al. 2011.

16. *Middle Awash:* Abule et al. 2005; *northern Tanzania, more woody plants:* Oba and Kaitira 2006; *Pokot:* Bollig and Schulte 1999; *Turkana:* Reid unpublished interviews.

17. *Grazing and woody plant spread:* Asner et al. 2004; *bush encroachment in equilibrial savannas:* Coppock 1993; *Kalahari sacrifice zones, boreholes:* Perkins and Thomas 1993, Moleele and Perkins 1998, Thomas et al. 2000.

18. *Soils, woody savannas:* Asner et al. 2004; *slowing wind erosion:* Perkins and Thomas 1993.

19. *Heavy grazing, shifting perennials to annual plants:* Milton et al. 1994; *in Niger:* Hiernaux 1998; *in Mali:* Turner 1999a; *in Somalia:* Thurow and Hussein 1989; *in South Africa:* Parsons et al. 1997, Fynn and O'Connor 2000; *in Namaqualand:* Todd and Hoffman 1999. *Heavy wildlife grazing:* van Wijngaarden 1985; *many perennials, heavy grazing:* Perkins and Thomas 1993; *Namibia, no impacts:* Ward et al. 1998.

20. *Annual plants, more protein:* Kelly and Walker 1977; *quote:* Bollig and Schulte 1999, 512; *ephermal plants die quickly:* O'Connor 1991; *degradation review:* Milton et al. 1994.

21. *Palatable plants disappearing—Borana:* Oba et al. 2000a; *Kenya:* Mworia et al. 1997; *Mali:* Breman and Cissé 1977; *Somalia:* Thurow and Hussein 1989; *South Africa:* O'Connor 1995, Todd and Hoffman 1999. *Toxic, spiny, weeds:* Milton 1994; *weeds, South Africa:* Todd 2006; *weeds, Tanzania:* Tobler et al. 2003.

22. *Sahelian soils, grazing:* Hiernaux et al. 1999; *soil crusts, fertility:* Belnap and Lange 2001, Allsopp 1999; *Namib Desert:* Ward et al. 1998.

23. *Borana, increased erosion some places:* Coppock 1993, 1994; *but not others:* Oba et al. 2000a. *South Africa:* Parsons et al. 1997; *400 years:* Biot 1993, Behnke et al. 1993.

24. *Zimbabwe:* Scoones 1992; *southern Kenya:* Lamprey and Reid 2004; *Namibia:* Sullivan 1999; *Sahel:* Mortimore and Turner 2005; *no effect, South Africa:* Fynn and O'Connor 2000; *Cape Province*: Dean and Macdonald 1994.

25. *Quotes* (here and next paragraph): Ash et al. 2002, 111.

26. *Maasai, degradable savannas*: Oba and Kaitira 2006; *West Africa:* Hiernaux 1993; *4–25-year recovery:* Harrison and Shackleton 1999; *Karoo:* Wiegand and Milton 1996; *"stuck" bushlands:* Westoby 1979–80, Coppock 1993, Anderies et al. 2002.

27. *Global assessments:* Mabbutt 1984, GLASOD 1990; *woody plant expansion:* Asner et al. 2004; *Sahel vegetation degradation, remote sensing:* Tucker et al. 1991, Nicholson et al. 1998, Prince et al. 1998, Hein and de Ridder 2006, Heumann et al. 2007, Hein et al. 2011; *degradation poorly measured:* Dodd 1994; *Nicholson findings disputed:* Hein and De Ridder 2006; *not overstocked:* Ellis et al. 1999.

28. *Water points:* De Leeuw et al. 2001; *Mara:* Reid et al. 2003, Lamprey and Reid 2004, Ogutu et al. 2009; *elephant extinction:* Parker and Graham 1989b.

29. *Fragmentation of east African savannas*: Boone and Hobbs 2004; Reid et al. 2004a, 2008b.

30. *Rising conflicts:* Campbell et al. 2000; *crop attractants:* Eltringham 1990; *conflict summary:* Naughton-Treves and Treves 2005.

31. *Field size and river access:* Fritz et al. 2003; *elephant population decline:* Hoare and du Toit 1999; *Mara wildlife, farming:* Norton-Griffiths et al. 2008.

32. *Sensitivity of wildlife:* Happold 1995, Woodroffe and Ginsberg 1998, Cardillo et al. 2005; *more smaller species on farms:* Prendini et al. 1996, Cumming et al. 1997, Reid et al. 1997, Wilson et al. 1997, Shackleton 2000, Gardiner et al. 2005; *Serengeti:* Sinclair et al. 2002.

8. THE SERENGETI–MARA

1. *Grandeur of the migration:* Sinclair 1995; *benchmark:* Sinclair et al. 2002; *financial returns:* Honey 2008, Norton-Griffiths et al. 2008; *Mau forest, rivers:* Gereta et al. 2002, Akotsi et al. 2006; *wheat farms:* Ottichilo et al. 2001, Serneels and Lambin 2001; *million people:* Campbell and Hofer 1995; *poaching offtake:* Campbell and Hofer 1995, Hofer et al. 1996, Mduma et al. 1999; *Loita wildebeest:* Ottichilo et al. 2001, Serneels and Lambin 2001; *most studied ecosystem:* Sinclair and Norton-Griffiths 1979, Sinclair and Arcese 1995, Sinclair et al. 2008b.

2. *Peoples, Serengeti-Mara:* Lamprey and Reid 2004, Shetler 2007; *hominins:* Leakey and Hay 1979; *Serengeti history:* Neumann 1995, Shetler 2007; *conservation benefits:* Thompson and Homewood 2002, Norton-Griffiths et al. 2008.

3. *Parks as benchmarks:* Sinclair et al. 2002; *shaped by humans:* Collett 1987.

4. *Global migrant loss:* Harris et al. 2009; *Serengeti migration:* Sinclair et al. 2008a.

5. Sinclair et al. 2007, 586.

6. *Ecosystem size:* Sinclair 1995, WDPA 2009.

7. *Ecosystem definition*: Sinclair and Norton-Griffiths 1979, Thirgood et al. 2004; *rules for resource use:* Polasky et al. 2008, Thirgood et al. 2008.

8. *Outsized importance:* Sinclair and Norton-Griffiths 1979.

9. *Male animal weights:* Estes 1993; *Serengeti species numbers:* Mduma and Hopcraft 2008; *large mammal populations:* my calculations combining Serengeti (Mduma and Hopcraft 2008) and Mara (Ottichilo et al. 2000a); *bird, frog, insect species numbers:* Sinclair et al. 2008a.

10. *Migration causes:* Wolanski and Gereta 2001, Boone et al. 2006b, Holdo et al. 2009a; *annual rainfall:* Norton-Griffiths et al. 1975, Ogutu et al. 2011; *migration cycle:* Maddock 1979.

11. *Wildebeest mating, survival, running:* Estes 1966, 1976; *birth synchrony:* Sinclair et al. 2000; *Maasai:* Goldman 2007.

12. *Number, poverty of people west of Serengeti*: Campbell and Hofer 1995; *2007 poverty, Republic of Tanzania*: http://www.nbs.go.tz.

13. *Maswa boundary:* Reed et al. 2009; *people attracted to parks:* Wittemyer et al. 2008; *Mara settlement growth:* Norton-Griffiths et al. 2008; *grazing in the reserve:* Butt et al. 2009.

14. *Mara wildlife loss:* Lamprey and Reid 2004, Ogutu et al. 2009, 2011; *settlement growth*: Norton-Griffiths et al. 2008.

15. *Leasing land, plowing best calving ground:* Ottichilo et al. 2001; *wildebeest lost:* Serneels and Lambin 2001; *poachers:* R. Lamprey pers. comm.

16. *2001–2004 poaching:* Ogutu et al. 2009; *arrests, late 2011:* Mara-Conservancy 2011.

17. *Migration, village land:* Thirgood et al. 2004, Nyahongo et al. 2009; *50,000 poachers, hunt for food and cash, third of income:* Loibooki et al. 2002; *40,000 wildebeest, sustainable:* Mduma et al. 1999; *160,000 of all species:* Hofer et al. 1996; *hunt when crops fail:* Barrett and Arcese 1998; *eat meat daily:* Nyahongo et al. 2009; *prices fall by half:* Holmern et al. 2002; *impalas:* Setsaas et al. 2007.

18. *Drying of Mara:* Wolanski and Gereta 2001; *68 percent drop:* Gereta et al. 2009; *irrigation withdrawals, pumps:* Thompson et al. 2009a; *34 percent forest loss:* Mnaya et al. forthcoming; *annual 1 percent loss:* Akotsi et al. 2006; *floods and dries faster:* Gereta et al. 2002, Mati et al. 2008, Mango et al. 2011; *blocking access routes:* Peters et al. 2008, Sinclair et al. 2008a.

19. *Rabies, canine distemper:* Roelke-Parker et al. 1996, Cleaveland et al. 2008; *malignant catarrhal fever:* Homewood and Rodgers 1991, McCabe 1997, Cleaveland et al. 2008.

20. *Elephants:* Kaelo 2007; *fifth of cash income:* Holmern et al. 2007; *hyenas:* Kolowski and Holekamp 2006.

21. Ogutu et al. 2005; *Mara Count:* Reid et al. 2003; *hyena behavior:* Boydston et al. 2003, Pangle and Holekamp 2010; *close proximity:* Kolowski and Holekamp 2009; *wild dogs, cheetah, competition from lions:* Ogutu and Dublin 1998, Maddox 2003, Durant et al. 2007.

22. *Maasai lion poisoning, Amboseli:* Hazzah et al. 2009; *Maasai poisoning of Mara hyena clans:* Holekamp et al. 1993; *Mara lions poisoned*: Reid unpublished interviews; *wild dogs poisoned, Loliondo:* Nkwame 2009; *Mara vultures poisoned:* Virani et al. 2011; *Tarangire hyenas poisoned:* Kissui 2008.

23. *Mara:* Reid et al. 2003; *Tanzania:* Maddox 2003.

24. *Moderate human population, maximum wildlife, Mara*: Reid et al. 2003.

25. *Parks sharing benefits:* Lamprey and Reid 2004, Schroeder 2008, Thirgood et al. 2008; *wealthy Maasai benefit:* Thompson and Homewood 2002; *hunters sharing profits:* Nelson et al. 2009.

26. *Mara conservancies:* Thompson et al. 2009, Butt 2011.

27. *Maintaining new institutions:* Ostrom 2009; *unexpected problems with conservancies*: Butt 2011.

28. Reid unpublished key informant interviews, 2007–2010.

29. *Suspicious of WMAs:* Shetler 2007; *central government control, Loliondo CWM:* Nelson et al. 2009.

30. Parkipuny 1997, 144.

31. *Mara livestock bones*: Marshall 1990a; *Asi and Tatog, digging sticks, Maasai, dry season grazing:* Shetler 2007.

32. *Pastoralists creating the Serengeti:* Bell 1971, Jacobs 1975, McCann 1999; *water points:* Shetler 2007; *burning woodlands:* Lamprey and Waller 1990; *settlements:* Reid et al. 2003; *no general evidence:* Peters et al. 2008; *Maasai not hunting:* Parkipuny 1991a.

33. *Trade route:* Sinclair et al. 2008a; *Baumann:* Dublin 1991; *95 percent died, rinderpest control, wildebeest population growth:* Sinclair et al. 2007.

34. *Wildebeest effects on Serengeti, elephants fleeing, becoming grassland and woodland:* Sinclair et al. 2007, Holdo et al. 2009b.

35. *Maasai burning:* Lamprey and Waller 1990, Lamprey and Reid 2004; *Croton woodlands, burnt woodlands, elephants eating trees:* Dublin 1991.

36. *Segregated landscape:* Neumann 1995, Shivji and Kapinga 1998; *history, Serengeti:* Shetler 2007.

37. *Promised water development:* Thompson 1997; *forced out of the Serengeti:* Lissu 2000; *quotes:* Parkipuny 1991a, 28 and 8.

38. *Trophy hunter fee, adopt game reserves, 1960s:* Lamprey and Reid 2004; *run by county council, more say:* Parkipuny 1991a

39. *Livestock in reserve at night:* Butt et al. 2009, Butt 2011.

40. Prins 1992, 122.

41. *Chose not to extinguish:* Parkipuny 1997; *lock up in parks:* Hilborn et al. 2006.

42. *Effects of Mara River drying*: Gereta et al. 2002, 2009; *fish*: Mnaya et al. forthcoming.

43. *Monitoring and recommendations*: Nkako et al. 2005.

44. *Transboundary Plan:* Gereta 2004; *Lake Victoria access:* Mnaya et al. forthcoming.

45. *Mara droughts:* Ogutu et al. 2007; *Serengeti woodlands cooler:* Ritchie 2008.

46. *Serengeti highway:* Dobson et al. 2010, www.nytimes.com/2011/06/28/world/africa/28brief-Tanzania.html; *40 percent loss:* Holdo et al. 2011; *GPS-collared wildebeest:* www.nrel.colostate.edu/projects/gnu/tracking.php.

47. *More profits from farming:* Norton-Griffiths et al. 2008; *ecosystem service payments to Maasai, Kaputiei:* Reid et al. 2008.

48. *Poaching and development:* Barrett and Arcese 1998, Loibooki et al. 2002; *carbon:* Holdo et al. 2009.

9. AMBOSELI

1. *Chapter title, Maasai saying:* Western and Maitumo 2004; *icefield depth:* Thompson et al. 2002; *forest capturing rain, fire, treeline:* Hemp 2005.

2. *Intermarriages, wildlife, farms:* Campbell et al. 2003a, 2005; *Maasai herding and swamp farming:* BurnSilver 2009.

3. *Rainfall, average daily temperatures:* Altmann et al. 2002, Hemp 2005, Thompson et al. 2002. The name Amboseli is from a Maasai (Maa) word, *empusel,* which means "dry, parched land" (Mol 1996).

4. *Springs and swamps:* Western 1975, Worden et al. 2003. Swamp names follow Moss et al. 2011 and those on topographic maps. I define the greater Amboseli ecosystem broadly following Western 2005 and Croze and Lindsay 2011. *Closed basin:* Western 1973; *elephant distribution:* Croze and Moss 2011; *baboons:* Alberts and Altmann 2001; *Kilimanjaro wildlife corridor:* Newmark 1993, Noe 2003.

5. *Origin of water flow:* Ramsay 1965; *hydrology:* Meijerink and van Wijngaarden 1997; *cloud forests:* Agrawala et al. 2003; *fogwater:* Hemp 2005; *Nolturesh pipeline:* Campbell et al. 2005.

6. *Soils, habitat diversity:* Western 1973, Altmann 1998, Worden 2007.

7. *Wildlife diversity:* Western and Sindiyo 1972, Western 1975, Western and Manzolillo Nightingale 2004; *Maasi section:* Rutten 1992.

8. *Amboseli elephants:* Poole and Moss 1981, Moss 2001, Moss et al. 2011.

9. *Key resources:* Scoones 1991, Illius and O'Connor 1999; *elephant movements, forage nutrition:* Western and Lindsay 1984, Moss 1988, Croze and Moss 2011.

10. *Migrations, grazing succession:* Western 1973, 1975.

11. *Pipeline building and repair:* Western 1982a, Kangwana 1993; *drought frequency:* calculated from data in Altmann et al. 2002; *drought and conflict:* Moss 1988, Campbell 1999, Campbell et al. 2000; *elephant death, drought:* Lindsay 1987, Moss 2001.

12. *Elephant view of Maasai:* Kangwana 2011; *quotes:* Kangwana 1993, 21–22 (human origin) and 73 (tossing livestock); *consolation program:* Sayialel and Moss 2011; *Maasai view of elephants, wildlife:* Moss 1988, Kangwana 1993, Roque de Pinho 2009, Browne-Nuñez 2011; *trauma, aggression:* Bradshaw et al. 2005.

13. *People and farmers in Oloitokitok:* Campbell 1999, Campbell et al. 2000, 2005; *farming, profits, losses:* BurnSilver 2007, 2009; *profits from livestock:* Norton-Griffiths and Butt 2006.

14. *Maasai oral history, quote:* Western and van Praet 1973, 106.

15. *Icecap melt:* Hastenrath and Greischar 1997, Thompson et al. 2002, 2009b; *icecap area, other glaciers:* Kaser et al. 2004; *sublimation:* Mölg and Hardy 2004.

16. *Temperature rise:* Altmann et al. 2002, Hemp 2005; *drying causing melting:* Mölg et al. 2006; *melting unique, warming involved:* Thompson et al. 2009b; *melting occurs in cycles, not exceptional:* Kaser et al. 2010, Mölg et al. 2010.

17. *Plantations:* Malimbwi et al. 1992; *farm clearing:* Campbell 1999.

18. *Forest fires:* Hemp 2005, 2009; *tree survey:* Lambrechts et al. 2002.

19. *Water from icecap:* Mölg et al. 2006.

20. *Streams:* Ramsay 1965; *fogwater, runoff variability:* Hemp 2005; *rainwater to swamps:* Meijerink and van Wijngaarden 1997; *tree loss, runoff, lakes:* Calder et al. 1995, Legesse et al. 2003.

21. *Swamp water level rise:* Western 1973, Western and van Praet 1973, Lovatt Smith 1997, Meijerink and van Wijngaarden 1997; *lake levels:* Verschuren et al. 2000, Nicholson and Yin 2001.

22. *Subtle changes in rainfall, crops replacing forest:* Legesse et al. 2003; *regional rainfall:* Hay et al. 2002; *livestock numbers:* Western and Manzolillo Nightingale 2004.

23. *Vegetation change:* Western 2006, Croze and Lindsay 2011.

24. *Woodland loss, salt, elephants:* Western 1973, Western and van Praet 1973, Western and Maitumo 2004, Croze and Lindsay 2011; *woodland regeneration:* Western and Lindsay 1984.

25. *Trees died of old age:* Young and Lindsay 1988.

26. *Wildlife population change:* Western 1973, Western and van Praet 1973, Lovatt Smith 1997, Altmann 1998, Western and Manzolillo Nightingale 2004; *baboons:* Altmann et al. 1985, Altmann 1998; *vervets:* Struhsaker 1973, Isbell and Young 1993, Lee and Hauser 1998.

27. *1911 moves:* Sandford 1919; *Maasai moves, quote:* Hughes 2005, 207. *Reserve, park:* Western 1973, 1982a; Lindsay 1987.

28. *Negotiations:* Kangwana and Browne-Nuñez 2011; *rhino, lions:* Western and Sindiyo 1972, Lovatt Smith 1997, Maclennan et al. 2009; *land loss:* Collett 1987.

29. *Land loss, vulnerability:* Western and Manzolillo Nightingale 2004; *human populations:* Katampoi et al. 1990, Campbell et al. 2000.

30. *Maasai farming, swamp, expeditionary cultivation:* Campbell 1986, 1999; Campbell et al. 2005. *Esoitpus, irrigation water use:* Okello 2005a; *Kitendeni corridor:* Newmark 1993, Noe 2003; *Chyulu's and pipeline farming, diversification:* BurnSilver and Worden 2008, Homewood et al. 2009.

31. *Less movement, land privatization:* Kimani and Pickard 1998, Ntiati 2002, Mwangi 2007; *tin roofs:* Worden et al. 2003, Western and Manzolillo Nightingale 2004; *paradox of pastoral land tenure:* Fernández-Giménez 2002.

32. *Model of future effects of privatization on livestock:* Boone et al. 2005; *rangeland congestion, fragmentation:* Worden 2007, BurnSilver and Worden 2008; *loss of savanna access:* BurnSilver 2007.

33. *Farming causing wildlife declines:* Worden et al. 2003, Okello 2005a.

34. *Land privatization causing wildlife declines:* Groom 2007, Worden 2007, Western et al. 2009.

35. *Amboseli wildlife trends over time:* Western and Manzolillo Nightingale 2004, KWS 2010, Worden et al. 2010.

36. *Loss of elephants, 2009 drought:* Moss 2010; *elephant population trends:* Moss 2001, Croze and Lindsay 2011; *growth reasons:* Moss 1988, Western 1994; *consolation scheme:* Sayialel and Moss 2011.

37. *Nothing Short of a Miracle:* Lovatt Smith 1997; *segregation:* Western and Gichohi 1993; *lack of benefit sharing unfair:* Okello 2005b.

38. *Little community wildlife returns:* Western 1994, Emerton 2001, Thompson and Homewood 2002, Okello 2005a; *wildlife initiatives:* Campbell et al. 2000, Ogutu 2002, Mburu et al. 2003, Kiyiapi et al. 2005, Okello 2005a.

39. *Earthen dams, settlements away from parks:* Croze et al. 2011; *grazing associations:* BurnSilver 2007, Mwangi 2007; *ecosystem services:* Bulte et al. 2008.

10. THE KAPUTIEI PLAINS

1. *Athi-Kaputiei Plains is the correct name for this entire area;* I use Kaputiei Plains here because I focus on the northern part of these plains. The spelling Kaputiei results in spoken pronunciation in English (Ka-poo-tee-ay) closer to the Maa pronunciation of their sectional name than the commonly used Kapiti; *description of ecosystem:* Gichohi 1996, Reid et al. 2008.

2. *Maasai movements:* Jacobs 1975, Robertshaw 1991, Sutton 1993; *most spectacular concentration of wildlife:* Simons 1962, Gichohi 1996; *rinderpest:* Waller 1988; *smallpox:* Rutten 1992; *railway:* Tignor 1976.

3. *Nairobi in 1902, cattle/wildlife recovering, good hunting:* Meinertzhagen 1957; *Cowrie's description:* Anon. 1951; *wildlife, cattle ratios:* my calculations from Meinertzhagen 1957; *Kenyan government surveys:* Gichohi 1996.

4. *Maasai land losses to colonists:* Rutten 1992, Hughes 2002.

5. *1900s migrations of wildlife:* my analysis of Meinertzhagen (1957) data; *long-distance movement of wildlife:* Hillman and Hillman 1974, Nkedianye 2004.

6. *Degradation narratives:* McCann 1999, Beinart 2000; *Nairobi National Park in 1946:* Anon. 1951; *park description:* Reid pers. obs.

7. *Historical pastoral movements:* Reid unpublished interviews; *historical wildlife movements:* Foster and Coe 1968, Hillman and Hillman 1977, Gichohi 1996.

8. *Park fencing:* Foster and Coe 1968, Hillman and Hillman 1977, Reid et al. 2008b; *historical land-use changes:* Gichohi 1996.

9. *Group ranch establishement:* Rutten 1992.

10. *Group ranch privatization:* Rutten 1992; *advantages/disadvantages of privatization:* Pasha 1986.

11. *Livestock walking distances:* Gichohi 1996.

12. *Poverty levels:* Kristjanson et al. 2002; *other livelihood information:* Nkedianye 2004.

13. *Geology:* Baker 1954; *rainfall:* Gichohi 1996; *crop production reliability:* Kristjanson et al. 2002; *wildlife species, last elephants:* Stewart and Zaphiro 1963.

14. *Land use:* Nkedianye et al. 2009; *fencelines:* Reid et al. 2008b, Ogutu et al. submitted.

15. *Livestock movements, recent droughts:* Campbell 1978, Njoka 1979, Bekure et al. 1991, Kristjanson et al. 2002, Nkedianye et al. 2011.

16. *1950s mowing, burning in park:* Foster and Kearney 1967, Gichohi 1990; *zebra "spoil" the water for Maasai cattle:* Gichohi 1996, 136; *Kaputiei Plains grass protein contents:* Gichohi 1996; *more green grass, Kaputiei Plains:* Imbahale et al. 2008.

17. *Nairobi National Park* (here and next paragraph): Gichohi 1990, 1996, 2000.

18. *Shift from livestock to land:* Campbell 1993; *grass poaching in park:* Ogutu et al. submitted.

19. *Cash, poaching meat preferences:* Nkedianye 2004; *lion killing, poaching methods:* Nkedianye et al. 2009, Reid unpublished key informant interviews.

20. *Tracks of collared wildebeest:* www.nrel.colostate.edu/projects/gnu/track_nairobi2.php; *wildlife, livestock changes:* Reid et al. 2008b, Ogutu et al. submitted; *Ngorongoro:* Runyoro et al. 1995.

21. *Privatization (here and next paragraphs):* Pasha 1986, Rutten 1992, Galaty 1994, Reid et al. 2008b, Nkedianye et al. 2009.

22. *Paradox:* Fernández-Giménez 2002.

23. *Group ranch allocation, quote:* Rutten 1992, 340; Kipury 1991.

24. *Enough grass, household size declines, less leisure time:* Rutten 1992.

25. *Land ownership leading to poverty:* Rutten 1992; *Kaputiei Plains poverty rates:* Kristjanson et al. 2002; *land sales, prices:* Rutten 1992, Nkedianye 2004; *avoiding MCF:* Nkedianye 2004, Bedelian et al. 2007.

26. *Kaputiei Plains grazing rights:* Nkedianye, pers. comm.; *pastoral reciprocity arrangements, examples:* Niamir-Fuller 1999c, Reid et al. 2008a; *cattle movements:* Kshatriya 2005, Reid et al. 2008b.

27. *South African wildebeest, fencing:* Whyte and Joubert 1988; *landscape fragmentation:* Reid et al. 2004a; *Botswana fences:* Spinage 1992, Perkins 1996; *Amboseli land subdivision model:* Boone et al. 2005.

28. *Tragedy of the commons:* Hardin 1968; *rebuttals, examples:* McCabe 1990, Ostrom 1990, O'Flaherty 2003.

29. *Fencing expansion:* Reid et al. 2008b, Ogutu et al. submitted; *future land sales:* Nkedianye 2004.

30. *Kruger fencing:* Whyte and Joubert 1988; *Maasai attitudes, fencing the park:* Nkedianye 2004; *quote:* Nkedianye 2004, 21.

31. *Lease program:* Nkedianye et al. 2009; *payments less than 10 percent, endowment required:* my calculations based on data in Ogutu et al. submitted.

32. *Doubling incomes:* Kristjanson et al. 2002, Nkedianye et al. 2009.

33. *Livestock consolation program:* Ogutu et al. submitted.

34. *Recommendations for the future*: Gichohi 2000, Nkedianye 2004.

35. *Greater Kitengela Master Plan:* Nkedianye et al. 2009.

11. NGORONGORO

1. Homewood and Rodgers 1991, 1.

2. *Revenues:* Leader-Williams 2000, Thirgood et al. 2008; *conservation concerns:* Homewood and Rodgers 1991, UNESCO 2007.

3. *Impoverished by conservation*: McCabe et al. 1992, Galvin and Magennis 1999, Galvin et al. 2002; *Hadʒa hunter-gatherers:* Marlowe 2010; *Maasai concerns:* Parkipuny 1997.

4. *NCA Management Plan:* Thompson 1997.

5. *Ngorongoro description*: Herlocker and Dirschl 1972, Homewood and Rodgers 1991.

6. *Rainfall, vegetation:* Herlocker and Dirschl 1972, Homewood and Rodgers 1987; *soils:* Pratt and Gwynne 1977; *NCA compared to other savannas:* Boone et al. 2008.

7. *Wildlife movements:* Estes and Small 1981, Moehlman et al. 1997, Estes et al. 2006.

8. *Tanzanian government population figures:* Runyoro 2007; *1954 population, debated:* Homewood and Rodgers 1991; *immigrant interviews:* Kijazi et al. 1997.

9. *Maasai poverty, education, income sources:* Runyoro 2007; *malnourishment:* McCabe 2002; *incomes, Maasailand:* Homewood et al. 2009; *livestock populations:* NCAA 1999.

10. *Maasai cash needs:* Homewood et al. 1987, Potkanski 1997; *selling reproductive livestock:* McCabe et al. 1992, 1997b; *quote:* McCabe et al. 1992, 359; *Maasai crop farming:* McCabe et al. 1997a.

11. *MCF:* Plowright et al. 1960; *first three months:* Mushi et al. 1981.

12. *Rinderpest, best grazing:* Homewood et al. 1987, McCabe 1997, Cleaveland et al. 2008; *1976, wildebeest covering short grass:* Potkanski 1997; *20 percent more cattle:* Boone et al. 2002.

13. *Ticks:* Fyumagwa et al. 2007; *ECF, calves:* Field et al., 1997; *ECF control:* Kipuri and Sørensen 2008.

14. *Fewer livestock, smaller farms, poorer:* Galvin and Magennis 1999, Lynn 2000, Galvin et al. 2002; *weight differences:* Galvin et al. 2002, forthcoming.

15. *NCAA employees:* Homewood and Rodgers 1991; *sharing 10 percent NCA revenue, NCA employees, income from tourism:* Runyoro 2007; *Mara employees, 50 percent of view-*

ing fees: Leader-Williams 2000; *NCA revenues:* Runyoro 2007, Kipuri and Sørensen 2008, Thirgood et al. 2008; *underreported revenues:* Charnley 2005; *bankrolling political parties, eviction:* Homewood et al. 2004.

16. *Farmed crops:* McCabe et al. 1997a; *farming before park, 1992 ban lifted:* Perkin 1997; *more food secure:* Galvin et al. 2002, 2006; *cultivation bans in 2001, 2009:* Galvin et al. forthcoming.

17. *Large-scale farming*: Kijazi 1997; *growth in farmed area*: Boone et al. 2002, Boone et al. 2006a, Runyoro 2007; *2010 cultivation update:* Galvin et al. forthcoming.

18. *Effects of expanded farming on wildlife*: Boone et al. 2006a; *zebra population, viability loss:* Moehlman pers. comm.

19. *Gazelles replaced by buffaloes:* Runyoro et al. 1995, Estes et al. 2006; *constant biomass over time:* Moehlman et al. 1997.

20. *Stopping Maasai burning changed wildlife:* Runyoro et al. 1995.

21. *Maasai and lions:* Kissui et al. 2010; *Maasai spearing lions*: Ikanda and Packer 2008.

22. *Trees establishing in livestock corrals, Maasai reducing elephant tree damage:* A. Milewski cited in Mills 2006.

23. *Trees dying, few young trees:* Trollope et al. 2002; *salt accumulation, lodges, water:* Mills 2006.

24. *Low erosion, Eleusine, Maasai burning:* Homewood and Rodgers 1991; *Maasai burning Eleusine:* Herlocker and Dirschl 1972; *Empakaai experiments, digging better:* O'Rourke et al. 1976; *quote:* Trollope and Trollope 1995, 13–14.

25. *Synthesis of prehistory/history, Maasai push out Tatog:* Homewood and Rodgers 1991; *southern Cushites:* Ehret 1984; *Engaruka:* Westerberg et al. 2010; *Maasai arrival 300 years ago:* Galaty 1993b; *starving during rinderpest*: Shetler 2007, Homewood and Rodgers 1991; *German brothers:* Fosbrooke 1972.

26. *1928–1959 restrictions:* Charnley 2005; *governor's quote:* Parkipuny 1991a, 22.

27. *Maasai forced out of Serengeti:* Arhem 1985; *violence involved:* Lissu 2000; *spring access, quote:* Parkipuny 1997, 157.

28. *History of permitted land use*: Homewood and Rodgers 1991, Galvin et al. forthcoming; *regional trading company:* Parkipuny 1991a; *farmers encouraged to move out:* Kijazi 1997; *1996 plan:* NCAA 1996; *offered 2 acres:* UNESCO 2007.

29. *Powerless to influence the Ngorongoro management:* Kijazi 1997; *council composition:* Charnley 2005; *disbursement of funds:* Runyoro 2007, Kipuri and Sørensen 2008; *inability to effectively represent:* Odhiambo 2002; *quote:* Shivji and Kapinga 1998, 61.

30. *NCAA statutory power:* Shivji and Kapinga 1998; *no independent arrangements with tour operators, rights of occupancy:* Charnley 2005; *private ownership, no solution:* Galvin et al. 2008.

31. *Fear, forced out:* Shivji and Kapinga 1998; *NCAA Board declaration:* Parkipuny 1991b; *resettling outside the NCA, Maasai views:* Runyoro 2007.

32. *NCA plans:* NCAA 1996, 2006; *challenges recognized by planners, Maasai:* Runyoro 2007, Lane 1996a.

33. *Independent watchdog body:* Bellini 2008.

34. *Staff, immigrant farmers moving outside NCA:* UNESCO 2007; *large cultivation decline:* Galvin et al. forthcoming; *future needs for relief food:* Boone et al. 2006a, Galvin et al. 2006.

35. *Direct payments:* Ferraro and Kiss 2002; *Kaputiei Plains payments:* Reid et al. 2008b; *Maasai vs. Tanzanian income, total Ngorongoro revenues:* Runyoro 2007; *impact of shared revenue on household income*: my calculations.

36. *Ngorongoro revenues on par with expenditures*: Thirgood et al. 2008.

37. *Ewoloto committees:* Kipuri and Sørensen 2008; *Ereto project, spread out grazing:* Sørensen et al. 2005, www.ereto-npp.org.

38. *Lower-cost veterinary drugs, better-quality beef:* Runyoro 2007.

39. *Minor impacts, cultivation, except in corridors:* Boone et al. 2006; *more water points, control of MCF:* Boone et al. 2002; *Ereto:* Sørensen et al. 2005.

40. *Impacts of cultivation, water points, MCF cure on wildlife:* Boone et al. 2002.

41. *Violated rights:* Shivji and Kapinga 1998.

42. *Equitable rights:* West et al. 2006; *support of indigenous land rights:* Wanitzek and Sippel 1998, Schroeder 2008.

12. SAVANNAS OF OUR FUTURE

1. Waller and Sobania 1994, 64.

2. Oba and Kaitira 2006, 175.

3. Ogutu et al. 2009, 11.

4. Goldman 2006, 436.

5. *Warming:* Christensen et al. 2007, Altmann et al. 2002; *cooling:* Ritchie 2008.

6. *More rainfall:* Hulme et al. 2001; *GCM models:* Christensen et al. 2007; *highland maize:* Jones and Thornton 2003; *Ngorongoro Highlands:* Boone et al. 2002; *more trees and carbon:* Doherty et al. 2010.

7. *Drier dry areas:* Jones and Thornton 2003.

8. *Future rainfall:* Hulme et al. 2001, Christensen et al. 2007; *past, current rainfall:* Williams and Funk 2011; *future drought:* Li et al. 2009.

9. Based on interviews by Kathleen Galvin, David Nkedianye, Dickson Kaelo, myself, and others; *flipping:* Thornton et al. 2011.

10. *Lake cores:* Wolff et al. 2011.

11. *Lower quality grass:* Ehleringer et al. 2002, Bond et al. 2003; *trees, fire:* Bond et al. 2003.

12. *Globalization trends, rural areas:* Woods 2007.

13. *Rich getting richer:* Thompson and Homewood 2002.

14. *Peace, good management connected:* Haro et al. 2005; *kob, South Sudan:* Fryxell 1983, www.wcs.org/saving-wildlife/hoofed-mammals/kob.aspx; *rainfall, conflicts:* Hendrix and Salehyan 2010.

15. *Corruption definition:* Azfar et al. 2001; *elephants, rhinos:* Smith et al. 2003; *Uganda revenues:* Archabald and Naughton-Treves 2001; *pastoral elite:* Ogutu 2002, Groom and Harris 2008.

16. *Human populations affect park wildlife:* Woodroffe and Ginsberg 1998, Parks and Harcourt 2000, Ogutu et al. 2009.

17. *Savanna–urban migration:* Young 2006.

18. *Farmland map, more drying:* Olson et al. 2008; *Maasai elder quote:* Kangwana and Ole Mako 2001, 159; *adding farming:* McCabe et al. 2010.

19. Map 18 underestimates the extent of hard-boundary savannas because we did not include an estimate of the future expansion of farming, which will create more hard boundaries. *Fences, herbivores:* Boone and Hobbs 2004; *fragmentation, fewer livestock, income:* Thornton et al. 2006, Boone 2007; *Laikipia:* Georgiadis et al. 2007; *smaller wildlife, conflicts:* Newmark et al. 1994; *savanna carbon loss:* Reid et al. 2004b; *wetter savannas, more wildlife*: Fritz and Duncan 1994.

20. *Death of pastoralism:* Hinde and Hinde 1910; *livestock as cornerstone:* Homewood et al. 2009; *quote:* McPeak and Little 2005, 102.

21. *Agricultural subsidies, intensification:* Horowitz and Little 1987, Norton-Griffiths 1998; *crop failure:* Niamir-Fuller 1999c, Homewood et al. 2009.

22. *Industrial conservation:* Brockington et al. 2008; *Tanzania protected area, Kenyan investment:* Borgerhoff-Mulder and Coppolillo 2005.

23. *Political pressure, hesitant national politicians:* Nelson and Agrawal 2008; *downwardly accountable:* Jones and Murphree 2001; *little say in policy:* Reid et al. 2009; *goals not met:* Ribot 2003.

24. *Local people, few incentives:* Nelson and Agrawal 2008, 561; *Tanzania parks sharing:* Schroeder 2008; *private sector payments:* Leader-Williams 2000; *wildlife management areas:* Goldman 2003.

25. *Household profits from livestock, wildlife:* Homewood et al. 2009; *returns to land use:* Norton-Griffiths et al. 2008.

26. *Wildlife populations declining in parks:* Reid et al. 2008b, Western et al. 2009b, Ogutu et al. submitted; *Amboseli conflicts:* Campbell et al. 2003a; *Maasai, Kikuyu views about Nairobi National Park:* Akama et al. 1995.

27. *Cross-cultural ethic:* Berkes 2004, 628; *conservation and development, CAMPFIRE:* Hulme and Murphree 2001, Murphree 2009, Taylor 2009.

28. *Five principles, similar to Murphree's three pillars:* Murphree 2009.

29. *Local authority:* Goldman 2003, Brockington et al. 2008; *southern Africa, no strong local institutions:* Murphree 2009; *subsidiarity:* Berkes 2004.

30. *Governments unwilling to devolve:* Goldman 2003, Igoe and Croucher 2007, Nelson and Agrawal 2008; *no accountable leaders:* Ribot et al. 2006; *CAMPFIRE:* Taylor 2009.

31. *Pastoral land grabbing:* Igoe 2003, Renton 2009.

32. *Power:* Berkes 2004; *devolution, Zimbabwe:* Murphree 2009; *pastoral initiatives working in the system:* Reid et al. 2009.

33. *Nonfinancial benefits of wildlife:* Kangwana and ole Mako 2001, Berkes 2004, Okello 2005b; *need to grow profits:* Muchapondwa et al. 2008; *more profits, no hunting:* Norton-Griffiths and Butt 2006, Norton-Griffiths et al. 2008; *maximum size of pie:* Young 2006.

34. *Cutting out middle men:* Thompson and Homewood 2002, Murphree 2009, Taylor 2009; *direct payments:* Thompson et al. 2009a.

35. *Take over tourism business:* Norton-Griffiths et al. 2008; *ecosystem service payments:* Bulte et al. 2008; *success, benefit sharing:* Archabald and Naughton-Treves 2001, 135.

36. *Hunting in bushy areas:* Leader Williams et al. 1996; *95 percent of Mara land:* Norton-Griffiths et al. 2008; *CAMPFIRE and community profits:* du Toit 2002, Taylor 2009; *consistent income stream:* Murphree 2009; *no hunting, endangered species:* Archabald and Naughton-Treves 2001.

37. *Little equitable sharing:* Kellert et al. 2000, Ogutu 2002, Groom and Harris 2008; *counting of funds:* Child 1995; *misappropriation, against wildlife:* Hazzah et al. 2009; *quote:* Azfar et al. 2001, 51.

38. *Collaboration examples*: Brick et al. 2001, Meffe et al. 2002, Conley and Moote 2003, Dukes et al. 2011; *common pool resources*: Ostrom 1990, Knight and Landres 1998.

39. *Negotiation, collaboration:* Berkes 2007, Armitage et al. 2009, Reid et al. 2009; *pastoral dialogue:* Goldman 2006.

40. *Government sharing gate fees:* Archabald and Naughton-Treves 2001, Carter et al. 2008, Schroeder 2008; *joint decision-making:* Thompson et al. 2009a.

41. *Quote:* Western 2009, vii.

42. *Quote, women, cooperatives:* Flintan 2003, vi.

43. *Quote:* Western 2009, vii.

44. *Mobile herds more productive:* Boone et al. 2005; *more profitable pastoralism:* Norton-Griffiths et al. 2008.

45. *Former sacred areas in parks:* Shetler 2007; *funds to communities near wildlife:* Archabald and Naughton-Treves 2001; *profits siphoned:* Thompson and Homewood 2002, Groom and Harris 2008.

46. *Influence, international NGOs:* Goldman 2003, Brockington et al. 2008; *alignment with local interests, promotion of science:* Goldman 2003.

47. *Pastoral priorities deflected:* Igoe 2003; *Tanzania private sector:* Schroeder 2008.

48. *Illegal activities:* IUCN-WCMC 2005.

49. *New hybrid model*: Reid et al. 2009, Kristjanson et al. 2009.

50. *Science addressing local needs*: Reid et al. 2009.

51. D. Nkedianye, pers. comm.

52. *New pastoralism*: Homewood 2008.

REFERENCES

Aboud, A. 1989. The role of public involvement in wildlife-livestock conflicts: The case of Narok ranchers in Kenya. *Society and Natural Resources* 2:319–328.

Abule, E., H. A. Snyman, and G. N. Smit. 2005. Comparisons of pastoralists' perceptions about rangeland resource utilisation in the Middle Awash Valley of Ethiopia. *Journal of Environmental Management* 75:21–35.

Adams, W. M., R. Aveling, D. Brockington et al. 2004. Biodiversity conservation and the eradication of poverty. *Science* 306:1146–1149.

Agrawala, S., A. Moehner, A. Hemp et al. 2003. *Development and climate change in Tanzania: Focus on Mount Kilimanjaro.* Paris: Organisation for Economic Co-operation and Development.

Akama, J. S., C. L. Lant, and G. W. Burnett. 1995. Conflicting attitudes toward state wildlife conservation programs in Kenya. *Society and Natural Resources* 8:133–144.

Akotsi, E. F. N., M. Gachanja, and J. K. Ndirangu. 2006. Changes in forest cover in Kenya's five "water towers," 2003–2005. Report prepared for the Department of Resource Surveys and Remote Sensing and Kenya Forest Working Group, Nairobi.

Alberts, S. C., and J. Altmann. 2001. Immigration and hybridization patterns of yellow and anubis baboons in and around Amboseli, Kenya. *American Journal of Primatology* 53:139–154.

Ali, I. M., and R. Maskill. 2004. Functional wildlife parks: The views of Kenyan children who live with them. *Natural Resources Forum* 28:205–215.

Allsopp, N. 1999. Effects of grazing and cultivation on soil patterns and processes in the Paulshoek area of Namaqualand. *Plant Ecology* 142:179–187.

Altmann, J., S. C. Alberts, S. A. Altmann, and S. B. Roy. 2002. Dramatic change in local climate patterns in the Amboseli basin, Kenya. *African Journal of Ecology* 40:248–251.

Altmann, J., G. Hausfater, and S. A. Altmann. 1985. Demography of Amboseli Baboons, 1963–1983. *American Journal of Primatology* 8:113–125.

Altmann, S. 1998. *Foraging for Survival: Yearling Baboons in Africa.* University of Chicago Press, Chicago.

Alvarez, L. W., W. Alvarez, F. Asaro, and H. V. Michel. 1980. Extraterrestrial cause for the Cretaceous-Tertiary extinction: Experimental results and theoretical interpretation. *Science* 208:1095–1108.

Ambrose, S. H. 1984a. Holocene environments and human adaptation in the Central Rift Valley, Kenya. Ph.D. diss., University of California, Berkeley.

———. 1984b. The introduction of pastoral adaptations to the highlands of East Africa. In *From Hunters to Farmers: The Causes and Consequences of Food Production in Africa,* ed. J. D. Clark and S. A. Brandt, 212–239. Berkeley: University of California Press.

———. 2001. Paleolithic technology and human evolution. *Science* 291:1748–1753.

———. 2010. Coevolution of composite-tool technology, constructive memory, and language: Implications for the evolution of modern human behavior. *Current Anthropology* 51:S135–S147.

Anderies, J. M., M. A. Janssen, and B. H. Walker. 2002. Grazing management, resilience, and the dynamics of a fire-driven rangeland system. *Ecosystems* 5:23–44.

Anderson, D. M. 2002. *Eroding the Commons: The Politics of Ecology in Baringo, Kenya, 1890–1963.* Oxford: James Currey; Nairobi: EAEP; Athens: Ohio University Press.

Anderson, D. M., and R. Grove. 1987. The scramble for Eden: The past, present, and future of African conservation. In *Conservation in Africa: People, Policies, and Practice,* ed. D. M. Anderson and R. Grove, 1–12. Cambridge: Cambridge University Press.

Anderson, G. D., and L. M. Talbot. 1965. Soil factors affecting the distribution of the grassland types and their utilization by wild animals on the Serengeti Plains, Tanganyika. *Journal of Ecology* 53:33–56.

Anderson, J. M., and M. J. Coe. 1974. Decomposition of elephant dung in an arid, tropical environment. *Oecologia* 14:111–125.

Anderson, K. L., E. F. Smith, and E. Owensby. 1970. Burning bluestem range. *Journal of Range Management* 23:81–92.

Anderson, M. K. 2005. *Tending the Wild: Native American Knowledge and the Management of California's Natural Resources.* Berkeley: University of California Press.

Andrew, M.H. 1988. Grazing impact in relation to livestock watering points. *Trends in Ecology and Evolution* 3:336–339.

Angassa, A., and F. Beyene. 2003. Current range condition in southern Ethiopia in relation to traditional management strategies: The perceptions of Borana pastoralists. *Tropical Grasslands* 37:53–59.

Anon. 1951. *Report of the Trustees of the Royal National Parks of Kenya, Parts I and II, for the Years 1946–1950.* Nairobi: Royal National Parks of Kenya.

Arcese, P., and A.R.E. Sinclair. 1997. The role of protected areas as ecological baselines. *Journal of Wildlife Management* 61:587–602.

Archabald, K., and L. Naughton-Treves. 2001. Tourism revenue-sharing around national parks in Western Uganda: Early efforts to identify and reward local communities. *Environmental Conservation* 28:135–149.

Archer, S. 1989. Have southern Texas savannas been converted to woodlands in recent history? *American Naturalist* 134:545–561.

Archibald, J.D. 1996. Fossil evidence for a late Cretaceous origin for "hoofed" mammals. *Science* 272:1150–1153.

Argue, D., M.J. Morwood, T. Sutikna et al. 2009. *Homo floresiensis:* A cladistic analysis. *Journal of Human Evolution* 57:623–639.

Arhem, K. 1985. Pastoral man in the Garden of Eden: The Maasai of Ngorongoro Conservation Area, Tanzania. Uppsala Research Report in Cultural Anthropology, Uppsala, Swe.

Armitage, D.R., R. Plummer, F. Berkes et al. 2009. Adaptive co-management for social-ecological complexity. *Frontiers in Ecology and the Environment* 7:95–102.

Arsenault, R., and R. Owen-Smith. 2002. Facilitation versus competition in grazing herbivore assemblages. *Oikos* 97:313–318.

Asfaw, B. 1999. *Australopithecus garhi:* A new species of early hominid from Ethiopia. *Science* 284:1623–1623.

Ash, A.J., D.M. Stafford Smith, and N. Abel. 2002. Land degradation and secondary production in semi-arid and arid grazing systems: What is the evidence? In *Global Desertification: Do Humans Cause Deserts?*, ed. J.F. Reynolds and D.M. Stafford Smith, 111–134. Berlin: Dahlem University Press.

Asner, G.P., A.J. Elmore, L.P. Olander et al. 2004. Grazing systems, ecosystem responses, and global change. *Annual Review of Environment and Resources* 29:261–299.

Atang, P.G., and W. Plowright. 1969. Extension of the JP-15 rinderpest control campaign to Eastern Africa: The epizootiological background. *Bulletin of Epizootic Diseases of Africa* 17:161–170.

Augustine, D.J. 2003. Long-term, livestock-mediated redistribution of nitrogen and phosphorus in an East African savanna. *Journal of Applied Ecology* 40:137–149.

————. 2005. Influence of cattle management on habitat selection by impala on central Kenyan rangeland. *Journal of Wildlife Management* 68:916–923.

Ayeni, J.S.O. 1975. Utilisation of waterholes in Tsavo National Park (East). *East African Wildlife Journal* 13:305–324.

Azfar, O., Y. Lee, and A. Swamy. 2001. The causes and consequences of corruption. *Annals of the American Academy of Political and Social Science* 573:42–56.

Baker, B.H. 1954. *Geology of South Machakos District.* Nairobi: Geological Survey of Kenya.

Baker, B.H., L.A. Williams, J.A. Miller, and F.J. Fitch. 1971. Sequence and geochronology of Kenya rift volcanics. *Tectonophysics* 11:191–215.

Bakker, J.P. 1989. *Nature Management by Cutting and Grazing.* Dordrecht, Neth.: Kluwer Academic Publishers.

Balmford, A., A. Bruner, P. Cooper et al. 2002. Economic reasons for conserving wild nature. *Science* 297:950–953.

Barber, J. 1968. *Imperial Frontier: A Study of Relations between the British and the Pastoral Tribes of North East Uganda.* Nairobi: East African Publishing House.

Barker, J.R., D.J. Herlocker, and S.A. Young. 1989. Vegetal dynamics along a grazing gradient within the coastal grassland of central Somalia. *African Journal of Ecology* 27:283–289.

Barnes, R.F.W. 1996. The conflict between humans and elephants in the central African forests. *Mammal Review* 26:67–80.

Barnes, R.F.W., K.L. Barnes, M.P.T. Alers, and A. Blom. 1991. Man determines the distribution of elephants in rain forests of northeastern Gabon. *African Journal of Ecology* 29:54–63.

Barnosky, A.D., P.L. Koch, R.S. Feranec et al. 2004. Assessing the causes of Late Pleistocene extinctions on the continents. *Science* 306:70–75.

Barrett, C.B., and P. Arcese. 1995. Are integrated conservation-development projects (ICDP's) sustainable? On the conservation of large mammals in sub-Saharan Africa. *World Development* 23:1073–1084.

————. 1998. Wildlife harvest in integrated conservation and development projects: Linking harvest to household demand, agricultural production, and environmental shocks in the Serengeti. *Land Economics* 74:449–465.

Barrett, P.M., and K.J. Willis. 2001. Did dinosaurs invent flowers? Dinosaur-angiosperm coevolution revisited. *Biological Reviews* 76:411–447.

Baxter, P.T.W. 1975. Some consequences of sedentarization for social relationships. In *Pastoralism in Tropical Africa,* ed. T. Monod, 206–228. London: Oxford University Press.

Beachey, R.W. 1967. The East African ivory trade in the nineteenth century. *Journal of African History* 8:269–290.

Bedelian, C., M. Herrero, and D. Nkedianye. 2007. Maasai perception of the impact and incidence of malignant catarrhal fever (MCF) in southern Kenya. *Preventive Veterinary Medicine* 78:296–316.

Behar, D. M., R. Villems, H. Soodyall et al. 2008. The dawn of human matrilineal diversity. *American Journal of Human Genetics* 82:1130–1140.

Behnke, R. H. 1983. Production rationales: The commercialization of subsistence pastoralism. *Nomadic Peoples* 14:3–27.

————. 1985. Measuring the benefits of subsistence versus commercial livestock production in Africa. *Agricultural Systems* 16:109–135.

————. 1999. Stock-movement and range-management in a Himba community in north-western Namibia. In *Managing Mobility in African Rangelands*, ed. M. Niamir-Fuller, 184–216. London: FAO and Beijer International Institute of Ecological Economics.

Behnke, R. H., and I. Scoones. 1991. *Rethinking Range Ecology: Implications for Rangeland Management in Africa*. Woburn, U.K.: Overseas Development Institute and International Institute for Environment and Development.

Behnke, R. H., I. Scoones, and C. Kerven. 1993. *Range Ecology at Disequilibrium: New Models of Natural Variability and Pastoral Adaptation in African Savannas*. London: Overseas Development Institute.

Beinart, W. 2000. African history and environmental history. *African Affairs* 99:269–302.

Bekure, S., P. N. de Leeuw, B. E. Grandin, and P. J. H. Neate. 1991. *Maasai Herding: An Investigation of the Livestock Production System of Maasai Pastoralists in Eastern Kajiado District, Kenya*. Nairobi: International Livestock Centre for Africa.

Bell, R. H. V. 1971. A grazing ecosystem in the Serengeti. *Scientific American* 224:86–93.

————. 1982. The effect of soil nutrient availability on community structure of African ecosystems. In *Ecology of Tropical Savannas*, ed. B. J. Huntley and B. H. Walker, 191–216. Berlin: Springer-Verlag.

Bellini, J. 2008. *Ngorongoro: Broken Promises—What Price Our Heritage?* Available from Pingo's, Arusha, Tanzania.

Bellomo, R. V. 1994. Methods of determining early hominid behavioral activities associated with the controlled use of fire at FxJj 20 Main, Koobi Fora, Kenya. *Journal of Human Evolution* 27:173–195.

Belnap, J., and O. L. Lange, eds. 2001. *Biological Soil Crusts: Structure, Function, and Management*. Berlin: Springer-Verlag.

Belsky, A. J. 1984. Role of small browsing mammals in preventing woodland regeneration in the Serengeti National Park, Tanzania. *African Journal of Ecology* 22:271–279.

―――. 1986. Does herbivory benefit plants? A review of the evidence. *American Naturalist* 127:870–892.

―――. 1992. Effects of grazing, competition, disturbance and fire on species composition and diversity in grassland communities. *Journal of Vegetation Science* 3:187–200.

Belsky, A. J., A. Matzke, and S. Uselman. 1999. Survey of livestock influences on stream and riparian ecosystems in the western United States. *Journal of Soil and Water Conservation* 54:419–431.

Ben-Shahar, R., and M. J. Coe. 1992. The relationships between soil factors, grass nutrients, and the foraging behaviour of wildebeest and zebra. *Oecologia* 90:422–428.

Berkes, F. 2004. Rethinking community-based conservation. *Conservation Biology* 18:621–630.

―――. 2007. Community-based conservation in a globalized world. *Proceedings of the National Academy of Sciences* 104:15188–15193.

Berner, R. A., D. J. Beerling, R. Dudley et al. 2003. Phanerozoic atmospheric oxygen. *Annual Review of Earth Planetary Science* 31:105–134.

Berntsen, J. L. 1976. The Maasai and their neighbors: Variables of interaction. *African Economic History* 2:1–11.

Bigalke, R. C. 1978. Present-day mammals of Africa. In *Evolution of African Mammals*, ed. V. J. Magio and H. B. S. Cooke, 1–16. Cambridge, Mass.: Harvard University Press.

Bignell, D. E., and P. Eggleton. 2000. Termites in ecosystems. In *Termites: Evolution, Sociality, Symbioses, Ecology*, ed. T. Abe, D. E. Bignell, and M. Higashi, 363–387. Dordrecht, Neth.: Kluwer Academic Publishers.

Bird, M. I., and J. A. Calio. 1998. A million-year record of fire in sub-Saharan Africa. *Nature* 394:767–769.

Blackmore, A. C., M. T. Mentis, and R. J. Scholes. 1990. The origin and extent of nutrient-enriched patches within a nutrient-poor savanna in South Africa. *Journal of Biogeography* 17:463–470.

Blanc, J. J., C. R. Thouless, J. A. Hart et al. 2003. African elephant status report 2002. Occasional Paper of the IUCN Species Survival Commission No. 29. Gland, Switz.

Blench, R. 2000. Extensive pastoral livestock systems: Issues and options for the future. Paper prepared under the FAO-Japan Cooperative Project "Collective of Information on Animal Production and Health." Overseas Development Institute, London.

Blumenschine, R. J. 1987. Characteristics of an early hominid scavenging niche. *Current Anthropology* 28:383–407.

————. 1995. Percussion marks, tooth marks, and experimental determinations of the timing of hominid and carnivore access to long bones at FLK Zinjanthropus, Olduvai Gorge, Tanzania. *Journal of Human Evolution* 29:21–51.

Bobe, R., and A.K. Behrensmeyer. 2004. The expansion of grassland ecosystems in Africa in relation to mammalian evolution and the origin of the genus *Homo*. *Palaeogeography Palaeoclimatology Palaeoecology* 207:399–420.

Boitani, L., et al. 1998. A Databank for the Conservation and Management of the African Mammals, Rome, Italy, Institute of Applied Ecology. Accessible at http://gcmd.nasa.gov/records/GCMD_IAE_African Mammals.html.

Bollig, M., and A. Schulte. 1999. Environmental change and pastoral perceptions: Degradation and indigenous knowledge in two African pastoral communities. *Human Ecology* 27:493–514.

Bond, W.J., and G.F. Midgley. 2000. A proposed CO_2-controlled mechanism of woody plant invasion in grasslands and savannas. *Global Change Biology* 6:865–869.

Bond, W.J., G.F. Midgley, and F.I. Woodward. 2003. The importance of low atmospheric CO_2 and fire in promoting the spread of grasslands and savannas. *Global Change Biology* 9:973–982.

Bond, W.J., and B.W. van Wilgen. 1996. *Fire and Plants*. London: Chapman and Hall.

Bonnefille, R., R. Potts, F. Chalie et al. 2004. High-resolution vegetation and climate change associated with Pliocene *Australopithecus afarensis*. *Proceedings of the National Academy of Sciences* 101:12125–12129.

Bonte, P. 1991. "To increase cows, god created the king": The function of cattle in intralacustrine societies. In *Herders, Traders, and Warriors: Pastoralism in Africa*, ed. J.G. Galaty and P. Bonte, 62–86. Boulder, Colo.: Westview Press.

Bonte, P., and J.G. Galaty. 1991. *Herders, Warriors, and Traders: The Political Economy of African Pastoralism*. Boulder, Colo.: Westview Press.

Boone, R.B. 2007. Effects of fragmentation on cattle in African savannas under variable precipitation. *Landscape Ecology* 22:1355–1369.

Boone, R.B., S.B. BurnSilver, and R.L. Kruska. 2008. Comparing landscape and infrastructural heterogeneity within and between ecosystems. In *Fragmentation in Semi-arid and Arid Landscapes: Consequences for Human and Natural Systems*, ed. K.A. Galvin, R.S. Reid, R.H. Behnke, and N.T. Hobbs, 341–367. Dordrecht, Neth.: Springer.

Boone, R.B., S.B. BurnSilver, P.K. Thornton et al. 2005. Quantifying declines in livestock due to land subdivision. *Rangeland Ecology and Management* 58:523–532.

Boone, R.B., M.B. Coughenour, K.A. Galvin, and J.E. Ellis. 2002. Addressing management questions for Ngorongoro Conservation Area, Tanzania, using the SAVANNA modelling system. *African Journal of Ecology* 40:138–150.

Boone, R. B., K. A. Galvin, P. K. Thornton et al. 2006a. Cultivation and conservation in Ngorongoro Conservation Area, Tanzania. *Human Ecology* 34:809–828.

Boone, R. B., and N. T. Hobbs. 2004. Lines around fragments: Effects of fencing on large herbivores. *African Journal of Range and Forage Science* 21:79–90.

Boone, R. B., S. J. Thirgood, and J. G. C. Hopcraft. 2006b. Serengeti wildebeest migratory patterns modeled from rainfall and new vegetation growth. *Ecology* 87:1987–1994.

Borgerhoff-Mulder, M., and P. Coppolillo. 2005. *Conservation: Linking Ecology, Economics, and Culture*. Princeton: Princeton University Press.

Boureau, E., M. Cheboldaeff-Salard, J.-C. Koeniguer, and P. Louvet. 1983. Évolution des flores et de la végétation Tertiaires en Afrique, au nord de l'équateur. *Bothalia* 14:355–367.

Bourlière, F., and M. Hadley. 1983. Present-day savannas: An overview. In *Tropical Savannas*, vol. 13 of *Ecosystems of the World*, ed. F. Bourlière. Amsterdam: Elsevier.

Boydston, E. E., K. M. Kapheim, H. E. Watts et al. 2003. Altered behaviour in spotted hyenas associated with increased human activity. *Animal Conservation* 6:207–219.

Bradley, B. J. 2008. Reconstructing phylogenies and phenotypes: A molecular view of human evolution. *Journal of Anatomy* 212:337–353.

Bradshaw, G. A., A. N. Schore, J. L. Brown et al. 2005. Elephant breakdown. *Nature* 433:807.

Brain, C. K. 1981. *The Hunters or the Hunted?* Chicago: University of Chicago Press.

Brain, C. K., and A. Sillen. 1988. Evidence from the Swartkrans cave for the earliest use of fire. *Nature* 336:464–466.

Branagan, D., and J. Hammond. 1965. Rinderpest in Tanganyika: A review. *Bulletin of Epizootic Diseases of Africa* 13:225–246.

Brashares, J. S., P. Arcese, and M. K. Sam. 2001. Human demography and reserve size predict wildlife extinction in West Africa. *Proceedings of the Royal Society of London, Series B—Biological Sciences* 268:2473–2478.

Breman, H., and A. M. Cissé. 1977. Dynamics of Sahelian pastures in relation to drought and grazing. *Oecologia* 28:301–315.

Brick, P., D. Snow, and S. van de Wetering. 2001. *Across the Great Divide: Explorations in Collaborative Conservation and the American West*. Washington, D.C.: Island Press.

Briske, D. D., S. D. Fuhlendorf, and F. E. Smeins. 2003. Vegetation dynamics on rangelands: A critique of the current paradigms. *Journal of Applied Ecology* 40:601–614.

Brockington, D. 2002. *Fortress Conservation: The Preservation of the Mkomazi Game Reserve, Tanzania.* Oxford: International African Institute.

Brockington, D., R. Duffy, and J. Igoe. 2008. *Nature Unbound: Conservation, Capitalism, and the Future of Protected Areas.* London: Earthscan.

Brockington, D., and J. Igoe. 2006. Eviction for conservation: A global overview. *Conservation and Society* 4:424–470.

Brockington, D., J. Igoe, and K. Schmidt-Soltau. 2006. Conservation, human rights, and poverty reduction. *Conservation Biology* 20:250–252.

Bromley, D. W. 1991. *Environment and Economy: Property Rights and Public Policy.* Oxford: Blackwell.

Brook, B. W., and D. M. J. S. Bowman. 2002. Explaining the Pleistocene megafaunal extinctions: Models, chronologies, and assumptions. *Proceedings of the National Academy of Sciences* 99:14624–14627.

Broten, M. D., and M. Said. 1995. Population trends of ungulates in and around Kenya's Masai Mara Reserve. In *Serengeti: Dynamics, Management, and Conservation of an Ecosystem,* ed. A. R. E. Sinclair and P. Arcese, 169–193. Chicago: University of Chicago Press.

Brown, P., T. Sutikna, M. J. Morwood et al. 2004. A new small-bodied hominin from the Late Pleistocene of Flores, Indonesia. *Nature* 431:1055–1061.

Browne-Nuñez, C. 2011. The Maasai-elephant relationship: The evolution and influence of culture, land use, and attitudes. In *The Amboseli Elephants,* ed. C. J. Moss, H. Croze, and P. C. Lee, 291–306. Chicago: University of Chicago Press.

Brunet, M., A. Beauvilain, Y. Coppens et al. 1996. *Australopithecus bahrelghazali,* a new species of early hominid from Koro Toro region, Chad. *Comptes rendus de l'Académie des Sciences, série II, fascicule A—Sciences de la terre et des planètes* 322:907–913.

Brunet, M., F. Guy, D. Pilbeam et al. 2002. A new hominid from the Upper Miocene of Chad, central Africa. *Nature* 418:145–151.

———. 2005. New material of the earliest hominid from the Upper Miocene of Chad. *Nature* 434:752–755.

Buechner, H. K., and H. C. Dawkins. 1961. Vegetation change induced by elephants and fire in Murchison Falls National Park, Uganda. *Ecology* 42:752–766.

Bulte, E., R. B. Boone, R. Stringer, and P. K. Thornton. 2008. Elephants or onions? Paying for nature in Amboseli, Kenya. *Environment and Development Economics* 13:395–414.

Bunn, H. T., and J. A. Ezzo. 1993. Hunting and scavenging by Pliopleistocene hominids—nutritional constraints, archaeological patterns, and behavioural implications. *Journal of Archaeological Science* 20:365–398.

Burney, D. A., and T. F. Flannery. 2005. Fifty millennia of catastrophic extinctions after human contact. *Trends in Ecology and Evolution* 20:395–401.

BurnSilver, S. 2007. Critical Factors Affecting Maasai Pastoralism: The Amboseli Region, Kajiado District, Kenya. Ph.D. diss., Colorado State University, Fort Collins.

————. 2009. Pathways of continuity and change: Maasai livelihoods in Amboseli, Kajiado District, Kenya. In *Staying Maasai: Livelihoods, Conservation, and Development in East African Rangelands,* ed. K. Homewood, P. Kristjanson, and P. Trench, 161–208. New York: Springer.

BurnSilver, S., and J. Worden. 2008. Processes of fragmentation in the Amboseli Ecosystem, southern Kajiado District, Kenya. In *Fragmentation in Semi-arid and Arid Landscapes: Consequences for Human and Natural Systems,* ed. K. A. Galvin, R. S. Reid, R. H. Behnke, and N. T. Hobbs, 225–253. Dordrecht, Neth.: Springer.

Butler, J. R. A., J. T. du Toit, and J. Bingham. 2004. Free-ranging domestic dogs (*Canis familiaris*) as predators and prey in rural Zimbabwe: Threats of competition and disease to large wild carnivores. *Biological Conservation* 115:369–378.

Butt, B. 2011. Coping with uncertainty and variability: The influence of protected areas on pastoral herding strategies in East Africa. *Human Ecology* 39:289–307.

Butt, B., A. Shortridge, and A. WinklerPrins. 2009. Pastoral herd management, drought coping strategies, and cattle mobility in southern Kenya. *Annals of the Association of American Geographers* 99:309–334.

Caldararo, N. 2002. Human ecological intervention and the role of forest fires in human ecology. *Science of the Total Environment* 292:141–165.

Calder, I. R., R. L. Hall, H. G. Bastable et al. 1995. The impact of land-use change on water resources in sub-Saharan Africa: A modeling study of Lake Malawi. *Journal of Hydrology* 170:123–135.

Callicot, J. B. 1994. *Earth Insights: A Multicultural Survey of Ecological Ethics from the Mediterranean Basin to the Australian Outback.* Berkeley: University of California Press.

Cambefort, Y. 1991. Biogeography and evolution. In *Dung Beetle Ecology,* ed. I. Hanski and Y. Cambefort, 51–67. Princeton: Princeton University Press.

Campbell, D. J. 1978. *Coping with Drought in Kenya Maasailand: Pastoralists and Farmers of the Loitokitok Area, Kajiado District.* Nairobi: Institute of Development Studies.

————. 1986. The prospect for desertification in Kajiado District, Kenya. *Geographical Journal* 152:44–55.

————. 1993. Land as ours, land as mine: Economic, political, and ecological marginalization in Kajiado District. In *Being Maasai,* ed. T. Spear and R. Waller, 258–272. London: James Currey.

————. 1999. Response to drought of farmers and herders in southern Kajiado District, Kenya: A comparison of 1972–1974 to 1994–1995. *Human Ecology* 27:377–416.

Campbell, D. J., H. Gichohi, A. Mwangi, and L. Chege. 2000. Land use conflict in Kajiado District, Kenya. *Land Use Policy* 17:337–348.

Campbell, D. J., H. Gichohi, R. Reid et al. 2003a. Interactions between people and wildlife in southeast Kajiado District, Kenya. LUCID Working Paper No. 18, International Livestock Research Institute, Nairobi (www.lucideastafrica.org).

Campbell, D. J., D. P. Lusch, T. Smucker, and E. E. Wangui. 2003b. Land use change patterns and root causes in the Loitokitok area, Kajiado District, Kenya. LUCID Working Paper No. 19, International Livestock Research Institute, Nairobi.

————. 2005. Multiple methods in the study of driving forces of land use and land cover change: A case study of SE Kajiado District, Kenya. *Human Ecology* 33:763–794.

Campbell, K., and H. Hofer. 1995. People and wildlife: Spatial dynamics and zones of interaction. In *Serengeti: Dynamics, Management, and Conservation of an Ecosystem,* ed. A. R. E. Sinclair and P. Arcese, 534–570. Chicago: University of Chicago Press.

Cardillo, M., G. M. Mace, K. E. Jones et al. 2005. Multiple causes of high extinction risk in large mammal species. *Science* 309:1239–1241.

Caro, T. M., N. Pelkey, M. Borner et al. 1998. Consequences of different forms of conservation for large mammals in Tanzania: Preliminary analyses. *African Journal of Ecology* 36:303–320.

Carter, E., W. M. Adams, and J. Hutton. 2008. Private protected areas: Management regimes, tenure arrangements, and protected area categorization in East Africa. *Oryx* 42: 177–186.

Carter, S. E., and H. K. Murwira. 1995. Spatial variability in soil fertility management and crop response in Mutoko Communal Area, Zimbabwe. *Ambio* 24:77–84.

Casadevall, A. 2005. Fungal virulence, vertebrate endothermy, and dinosaur extinction: Is there a connection? *Fungal Genetics and Biology* 42:98–106.

Cashdan, E. A. 1984. The effects of food production on mobility in the central Kalahari. In *From Hunters to Farmers: The Causes and Consequences of Food Production in Africa,* ed. J. D. Clark and S. A. Brandt, 311–327. Berkeley: University of California Press.

Caughley, G. 1976. The elephant problem: An alternative hypothesis. *East African Wildlife Journal* 14:265–283.

Cavalli-Sforza, L. L. 2000. *Genes, Peoples, and Languages.* Berkeley: University of California Press.

CBS. 2003. *Geographic Dimensions of Well-Being in Kenya.* Vol. 1: *Where Are the Poor? From Districts to Locations.* Nairobi: Government of Kenya, Ministry of Planning and National Development, Central Bureau of Statistics (CBS) in collaboration with the International Livestock Research Institute (ILRI).

Ceballos, G., and P. R. Ehrlich. 2002. Mammal population losses and the extinction crisis. *Science* 296:904–907.

Cerling, T. E. 1992. Development of grasslands and savannas in East Africa during the Neogene. *Palaeogeography, Palaeoclimatology, Palaeoecology* 97:241–247.

Cerling, T. E., J. M. Harris, B. J. MacFadden et al. 1997. Global vegetation change through the Miocene/Pliocene boundary. *Nature* 389:153–158.

Cerling, T. E., J. G. Wynn, S. A. Andanje et al. 2011. Woody cover and hominin environments in the past 6 million years. *Nature* 476:51–56.

Chapman, C. A., and L. J. Chapman. 1996. Mid-elevation forests: A history of disturbance and regeneration. In *East African Ecosystems and Their Conservation,* ed. T. R. McClanahan and T. P. Young, 385–400. Oxford: Oxford University Press.

Charnley, S. 2005. From nature tourism to ecotourism? The case of the Ngorongoro Conservation Area, Tanzania. *Human Organization* 64:75–88.

Chen, F.-C., and W.-H. Li. 2001. Genomic differences between human and other hominids and effective population size of the common ancestor of humans and chimpanzees. *American Journal of Human Genetics* 68:444–456.

Child, B., and G. Barnes. 2010. The conceptual evolution and practice of community-based natural resource management in southern Africa: Past, present, and future. *Environmental Conservation* 37:283–295.

Child, G. 1995. Managing wildlife successfully in Zimbabwe. *Oryx* 29:171–177.

Christensen, J. H., B. Hewitson, A. Busuioc et al. 2007. Regional climate projections. In *Climate Change 2007: The Physical Science Basis,* ed. S. Solomon, D. Qin, M. Manning et al. , 847–940. Contribution of Working Group I to the Fourth Assessment Report of the Intergovernmental Panel on Climate Change. Cambridge and New York: Cambridge University Press.

Clark, J. S., and J. Robinson. 1993. Paleoecology of fire. In *Fire in the Environment: The Ecological, Atmospheric, and Climatic Importance of Vegetation Fires,* ed. P. J. Crutzen and J. G. Goldammer, 193–214. Chichester, U.K.: John Wiley and Sons.

Cleaveland, S., M. G. J. Appel, W. S. K. Chalmers et al. 2000. Serological and demographic evidence for domestic dogs as a source of canine distemper virus infection for Serengeti wildlife. *Veterinary Microbiology* 72:217–227.

Cleaveland, S., G. R. Hess, A. P. Dobson et al. 2001. The role of pathogens in biological conservation. In *The Ecology of Wildlife Diseases,* ed. P. J. Hudson, A. Rizzoli, B. T. Grenfell et al., 139–150. Oxford: Oxford University Press.

Cleaveland, S., C. Packer, K. Hampson et al. 2008. The multiple roles of infectious diseases in the Serengeti ecosystem. In *Serengeti III: Human Impacts on Ecosystem Dynamics,* ed. A. R. E. Sinclair, C. Packer, S. A. R. Mduma, and J. M. Fryxell, 209–239. Chicago: University of Chicago Press.

Cochrane, K., D. Nkedianye, E. Partoip et al. 2005. *Family fortunes: Analysis of changing livelihoods in Maasailand.* Final report, Project ZC0275, DFID Livestock Production Programme. Nairobi: International Livestock Research Institute.

Cole, M. M. 1986. *The Savannas: Biogeography and Geobotany.* London: Academic Press.

Collett, D. 1987. Pastoralists and wildlife: Image and reality in Kenya Maasailand. In *Conservation in Africa: People, Policies, and Practice,* ed. D. M. Anderson and R. Grove, 129–148. Cambridge: Cambridge University Press.

Conley, A., and M. A. Moote. 2003. Evaluating collaborative natural resource management. *Society and Natural Resources* 16:371–386.

Connell, J. H. 1978. Diversity in tropical rain forests and coral reefs. *Science* 199:1302–1310.

Conrad, V. 1941. The variability of precipitation. *Monthly Weather Review* 69:5–11.

Cooper, S. M., and N. Owen-Smith. 1985. Condensed tannins deter feeding by browsing ruminants in a South African savanna. *Oecologia* 67:142–146.

Coppens, Y., V. J. Maglio, C. T. Madden, and M. Beden. 1978. Proboscidea. In *Evolution of African Mammals,* ed. V. J. Maglio and H. B. S. Cooke, 336–367. Cambridge, Mass.: Harvard University Press.

Coppock, D. L. 1993. Vegetation and pastoral dynamics in the southern Ethiopian rangelands: Implications for theory and management. In *Range Ecology at Disequilibrium: New Models of Natural Variability and Pastoral Adaptation in African Savannas,* ed. R. H. Behnke, I. Scoones, and C. Kerven, 42–61. London: Overseas Development Institute.

———. 1994. *The Borana Plateau of Southern Ethiopia: Synthesis of Pastoral Research, Development, and Change, 1980–1991.* Addis Ababa: International Livestock Centre for Africa.

Coppock, D. L., J. E. Ellis, J. K. Detling, and M. I. Dyer. 1983. Plant-herbivore interactions in a North American mixed-grass prairie. II. Responses of bison to modification of vegetation by prairie dogs. *Oecologia* 56:10–15.

Costanza, R., R. d'Arge, R. de Groot et al. 1997. The value of the world's ecosystem services and natural capital. *Nature* 387:253–260.

Coughenour, M. B. 1985. Graminoid responses to grazing by large herbivores: Adaptations, exaptations, and interacting processes. *Annals of the Missouri Botanical Garden* 72:852–863.

———. 1991. Dwarf shrub and graminoid responses to clipping, nitrogen, and water: Simplified simulations of biomass and nitrogen dynamics. *Ecological Modelling* 54:81–110.

———. 2008. Causes and consequences of herbivore movement in landscape ecosystems. In *Fragmentation of Semi-arid and Arid Lands: Consequences for Human and*

Natural Systems, ed. K. A. Galvin, R. S. Reid, R. H. Behnke, and N. T. Hobbs, 45–91. Dordrecht, Neth.: Springer.

Coughenour, M. B., D. L. Coppock, and J. E. Ellis. 1990a. Herbaceous forage variability in an arid pastoral region of Kenya: Importance of topographic and rainfall gradients. *Journal of Arid Environments* 19:147–159.

Coughenour, M. B., J. K. Detling, I. F. Bamberg, and M. M. Mugambi. 1990b. Production and nitrogen responses of the African dwarf shrub *Indigofera spinosa* to water and defoliation. *Oecologia* 83:546–552.

Coughenour, M. B., J. E. Ellis, D. M. Swift et al. 1985a. Energy extraction and use in a nomadic pastoral ecosystem. *Science* 230:619–625.

Coughenour, M. B., S. J. McNaughton, and L. L. Wallace. 1985b. Adaptations of an African tallgrass (*Hyparrhenia filipendula* Stapf.) to defoliation and limitations of nutrients and water. *Oecologia* 68:50–56.

———. 1985c. Responses of an African graminoid (*Themeda triandra* Forsk.) to frequent defoliation, nitrogen, and water: A limit of adaptation to herbivory. *Oecologia* 68:105–110.

Crompton, A. W., and F. A. Jenkins. 1978. Mesozoic mammals. In *Evolution of African Mammals,* ed. V. J. Maglio and H. B. S. Cooke, 46–55. Cambridge, Mass.: Harvard University Press.

Croze, H. 1974. The Seronera bull problem: The trees. *East African Wildlife Journal* 12:29–47.

Croze, H., and K. Lindsay. 2011. Amboseli ecosystem context: Past and present. In *The Amboseli Elephants,* ed. C. J. Moss, H. Croze, and P. C. Lee, 11–28. Chicago: University of Chicago Press.

Croze, H., and C. J. Moss. 2011. Patterns of occupancy in time and space. In *The Amboseli Elephants,* ed. C. J. Moss, H. Croze, and P. C. Lee, 89–105. Chicago: University of Chicago Press.

Croze, H., C. J. Moss, and W. K. Lindsay. 2011. The future of the Amboseli elephants. In *The Amboseli Elephants,* ed. C. J. Moss, H. Croze, and P. C. Lee, 325–335. Chicago: University of Chicago Press.

Cumming, D. H., M. B. Fenton, I. L. Rautenbach et al. 1997. Elephants, woodlands, and biodiversity in southern Africa. *South African Journal of Science* 93:231–236.

Curnoe, D., and A. Thorne. 2003. Number of ancestral human species: A molecular perspective. *Homo: Journal of Comparative Human Biology* 53:201–224.

d'Antonio, C. 2000. Fire, plant invasions, and global changes. In *Invasive Species in a Changing World,* ed. H. A. Mooney and R. J. Hobbs, 65–94. Covelo, Ca.: Island Press.

Darlington, J. P. E. C. 1985. Lenticular mounds in the Kenyan highlands. *Oecologia* 66:116–121.

Darwin, C. R. 1909–1914. *The Voyage of the Beagle.* Vol. 29. The Harvard Classics. New York: P. F. Collier & Son.

Daszak, P., A. A. Cunningham, and A. D. Hyatt. 2001. Anthropogenic environmental change and the emergence of infectious diseases in wildlife. *Acta Tropica* 78:103–116.

Davidson, B. 1991. *Africa in History.* London: Phoenix Press.

Davis, A. L. V., C. H. Scholtz, and T. K. Philips. 2002. Historical biogeography of scarabaeine dung beetles. *Journal of Biogeography* 29:1217–1256.

Day, T. A., and J. K. Detling. 1990. Grassland patch dynamics and herbivore grazing preferences following urine deposition. *Ecology* 71:180–188.

Dean, W. R. J., and I. A. W. Macdonald. 1994. Historical changes in stocking rates of domestic livestock as a measure of semiarid and arid rangeland degradation in the Cape Province, South Africa. *Journal of Arid Environments* 26:281–298.

Dean, W. R. J., S. J. Milton, and F. Jeltsch. 1999. Large trees, fertile islands, and birds in arid savanna. *Journal of Arid Environments* 41:61–78.

de Boer, W. F., and H. T. T. Prins. 1990. Large herbivores that strive mightily but eat and drink as friends. *Oecologia* 82:264–274.

de Haan, C., H. Steinfeld, and H. Blackburn. 1997. *Livestock and the Environment: Finding a Balance.* Fressingfield, U.K.: WRENmedia.

de Leeuw, J., M. N. Waweru, O. O. Okello et al. 2001. Distribution and diversity of wildlife in northern Kenya in relation to livestock and permanent water points. *Biological Conservation* 100:297–306.

DeMenocal, P. B. 1995. Plio-Pleistocene African climate. *Science* 270:53–59.

———. 2004. African climate change and faunal evolution during the Pliocene-Pleistocene. *Earth and Planetary Science Letters* 220:3–24.

Demment, M. W., and P. van Soest. 1985. A nutritional explanation for body-size patterns of ruminant and non-ruminant herbivores. *American Naturalist* 125:641–672.

Desta, S., and D. L. Coppock. 2002. Cattle population dynamics in the southern Ethiopian rangelands, 1980–97. *Journal of Range Management* 55:439–451.

———. 2004. Pastoralism under pressure: Tracking system change in southern Ethiopia. *Human Ecology* 32:465–486.

Diamond, J. 1997. *Guns, Germs, and Steel.* New York: W. W. Norton.

———. 2002. Evolution, consequences, and future of plant and animal domestication. *Nature* 418:700–707.

DiMichele, W. A., and W. H. Hook. 1993. Paleozoic terrestrial ecosystems. In *Terrestrial Ecosystems through Time: Evolutionary Paleoecology of Terrestrial Plants and Animals,* A. K. Behrensmeyer, J. D. Damuth, W. A. DiMichele et al., 205–325. Chicago: University of Chicago Press.

DMA. 1992. Digital chart of the world. Defense Mapping Agency, Fairfax, Va.

Dobson, A. P., M. Borner, A. R. E. Sinclair et al. 2010. Road will ruin Serengeti. *Nature* 467:272–273.

Dodd, J. L. 1994. Desertification and degradation in Sub-Saharan Africa. *BioScience* 44:28–34.

Doherty, R. M., S. Sitch, B. Smith et al. 2010. Implications of future climate and atmospheric CO_2 content for regional biogeochemistry, biogeography, and ecosystem services across East Africa. *Global Change Biology* 16:617–640.

Dominguez-Rodrigo, M. 2002. Hunting and scavenging by early humans: The state of the debate. *Journal of World Prehistory* 16:1–54.

Douglas-Hamilton, I. 1988. *African Elephant Population Study—Phase 2*. Nairobi: WWF/UNEP.

Dressler, W., B. Buscher, M. Schoon et al. 2010. From hope to crisis and back again? A critical history of the global CBNRM narrative. *Environmental Conservation* 37:5–15.

du Toit, J. T. 2002. Wildlife harvesting guidelines for community-based wildlife management: A southern African perspective. *Biodiversity and Conservation* 11:1403–1416.

du Toit, J. T., and D. H. M. Cumming. 1999. Functional significance of ungulate diversity in African savannas and the ecological implications of the spread of pastoralism. *Biodiversity and Conservation* 8:1643–1661.

Dublin, H. 1991. Dynamics of Serengeti-Mara woodlands: An historical perspective. *Forest and Conservation History* 35:169–178.

———. 1995. Vegetation dynamics in the Serengeti-Mara ecosystem: The role of elephants, fire, and other factors. In *Serengeti II: Dynamics, Management, and Conservation of an Ecosystem*, ed. A. R. E. Sinclair and P. Arcese, 71–90. Chicago: University of Chicago Press.

Dublin, H. T., A. R. E. Sinclair, and J. McGlade. 1990. Elephants and fire as causes of multiple stable states in the Serengeti Mara woodlands. *Journal of Animal Ecology* 59:1147–1164.

Dukes, E. F., K. E. Firehock, and J. E. Birkhoff, eds. 2011. *Community-Based Collaboration: Bridging Socio-Ecological Theory and Practice*. Charlottesville: University of Virginia Press.

Durand, J. F. 2005. Major African contributions to Palaeozoic and Mesozoic vertebrate palaeontology. *Journal of African Earth Sciences* 43:53–82.

Durant, S. M., S. Bashir, T. Maddox, and M. K. Laurenson. 2007. Relating long-term studies to conservation practice: The case of the Serengeti Cheetah Project. *Conservation Biology* 21:602–611.

Dyson-Hudson, N., and R. Dyson-Hudson. 1982. The structure of East African herds and the future of East African herders. *Development and Change* 13:213–238.

Dyson-Hudson, R. 1999. Turkana in time perspective. In *Turkana Herders of the Dry Savanna*, ed. M. A. Little and P. W. Leslie, 24–40. Oxford: Oxford University Press.

Dyson-Hudson, R., and N. Dyson-Hudson. 1970. The food production system of a semi-nomadic society: The Karamojong, Uganda. In *African Food Production Systems*, ed. P. F. M. McLoughlin, 93–123. Baltimore: Johns Hopkins University Press.

————. 1980. Nomadic pastoralism. *Annual Review of Anthropology* 9:15–61.

Ego, W. K., D. M. Mbuvi, and P. F. K. Kibet. 2003. Dietary composition of wildebeest (*Connochaetes taurinus*), kongoni (*Alcephalus buselaphus*), and cattle (*Bos indicus*), grazing on a common ranch in south-central Kenya. *African Journal of Ecology* 41:83–92.

Ehleringer, J. R., T. E. Cerling, and B. R. Helliker. 1997. C_4 photosynthesis, atmospheric CO_2, and climate. *Oecologia* 112:285–299.

Ehleringer, J. R., T. E. Cerling, and M. D. Dearing. 2002. Atmospheric CO_2 as a global change driver influencing plant-animal interactions. *Integrative and Comparative Biology* 42:424–430.

Ehret, C. 1984. Historical/linguistic evidence for early African food production. In *From Hunters to Farmers: The Causes and Consequences of Food Production in Africa*, ed. J. D. Clark and S. A. Brandt, 26–35. Berkeley: University of California Press.

Ellis, J. E., and K. A. Galvin. 1994. Climate patterns and land-use practices in dry zones of Africa. *BioScience* 44:340–349.

Ellis, J., R. Reid, P. K. Thornton, and R. Kruska. 1999. Population growth and land use change among pastoral people: Local processes and continental patterns. In *Sixth International Rangeland Congress, Townsville, Australia, 19–25 July 1999*, 168–169.

Ellis, J. E., and D. M. Swift. 1988. Stability of African pastoral ecosystems: Alternative paradigms and implications for development. *Journal of Range Management* 41:450–459.

Eltringham, S. K. 1980. A quantitative assessment of range usage by large African mammals with particular reference to the effects of elephants on trees. *African Journal of Ecology* 18:53–71.

————. 1990. Wildlife carrying capacities in relation to human settlement. *Koedoe* 33:87–97.

Emerton, L. 2001. The nature of benefits and the benefits of nature. In *African Wildlife and Livelihoods*, ed. D. Hulme and M. Murphree, 208–226. Oxford: James Currey.

Eriksson, M. G., J. R. Olley, and R. W. Payton. 2000. Soil erosion history in central Tanzania based on OSL dating of colluvial and alluvial hillslope deposits. *Geomorphology* 36:107–128.

Estes, R.D. 1966. Behaviour and life history of wildebeest (*Connochaetes taurinus* Burchell). *Nature* 212:999–1000.

———. 1976. The significance of breeding synchrony in the wildebeest. *East African Wildlife Journal* 14:135–152.

———. 1993. *Safari Companion*. Harare: Tutorial Press.

Estes, R.D., J.L. Atwood, and A.B. Estes. 2006. Downward trends in Ngorongoro Crater ungulate populations, 1986–2005: Conservation concerns and the need for ecological research. *Biological Conservation* 131:106–120.

Estes, R.D., and R. Small. 1981. The large herbivore populations of Ngorongoro Crater. *African Journal of Ecology* 19:175–185.

Fairhead, J., and M. Leach. 1996. *Misreading the African Landscape*. Cambridge: Cambridge University Press.

Falcon-Lang, H.J. 2000. Fire ecology of the Carboniferous tropical zone. *Palaeogeography, Palaeoclimatology, Palaeoecology* 164:339–355.

FAO. 1999. *1998 Production Yearbook*. Rome: Food and Agriculture Organization of the United Nations.

Farmer, G. 1986. Rainfall variability in tropical Africa: some implications for policy. *Land Use Policy* 3:336–342.

Fenn, E.A. 2001. *Pox Americana: The Great Smallpox Epidemic of 1775–82*. New York: Hill and Wang.

Fernández-Giménez, M.E. 2002. Spatial and social boundaries and the paradox of pastoral land tenure: A case study from postsocialist Mongolia. *Human Ecology* 30:49–78.

Fernández-Giménez, M.E., and B. Allen-Diaz. 1999. Testing a non-equilibrium model of rangeland vegetation dynamics in Mongolia. *Journal of Applied Ecology* 6:871–885.

Ferrar, P. 1982. Termites of a southern African savanna. IV. Subterranean populations, mass determinations, and biomass estimations. *Oecologia* 52:147–151

Ferraro, P.J., and A. Kiss. 2002. Direct payments to conserve biodiversity. *Science* 298:1718–1719.

Field, C.R. 1972. The food habits of wild ungulates in Uganda by analysis of stomach contents. *East African Wildlife Journal* 10:17–42.

———. 2005. *Where There Is No Development Agency: A Manual for Pastoralists and Their Promoters*. Aylesford, Kent, U.K.: NR International.

Field, C.R., G.N. Harrington, and D. Pratchett. 1972. A comparison of grazing preferences of buffalo (*Syncerus caffer*) and Ankole cattle (*Bos indicus*) on three different pastures. *East African Wildlife Journal* 11:19–29.

Field, C.R., C. Moll, and C. Ole Sonkoi. 1997. Livestock development. In *Multiple Land Use: The Experience of the Ngorongoro Conservation Area, Tanzania,* ed. D.M. Thompson, 181–199. Gland, Switz.: IUCN

Finlayson, C. 2005. Biogeography and evolution of the genus *Homo. Trends in Ecology and Evolution* 20:457–463.

Fleisher, M.L. 2000. *Kuria Cattle Raiders.* Ann Arbor: University of Michigan Press.

Flintan, F. 2003. *Women, Gender, and ICDPs: Lessons Learnt and Ways Forward.* Vol. 1. London: International Institute for Environment and Development.

Foley, R., and C. Gamble. 2009. The ecology of social transitions in human evolution. *Philosophical Transactions of the Royal Society B—Biological Sciences* 364:3267–3279.

Ford, J. 1971. *The Role of Trypanosomiases in African Ecology.* Oxford: Clarendon Press.

Fosbrooke, H. 1972. *Ngorongoro: The Eighth Wonder.* London: Andre Deutsch.

Foster, J.B., and M.J. Coe. 1968. The biomass of game animals in Nairobi National Park (1960–1966). *Journal of Zoology, London* 155:413–425.

Foster, J.B., and D. Kearney. 1967. Nairobi National Park game census 1966. *East African Wildlife Journal* 5:112–120.

Frank, D.A., and P.M. Groffman. 1998. Ungulate vs. landscape control of soil C and N processes in grasslands of Yellowstone National Park. *Ecology* 79:2229–2241.

Frank, D.A., M.M. Kuns, and D.R. Guido. 2002. Consumer control of grassland plant production. *Ecology* 83:602–606.

Frank, D.A., and S.J. McNaughton. 1993. Evidence for the promotion of aboveground grassland production by native large herbivores in Yellowstone National Park. *Oecologia* 96:157–161.

Fratkin, E., K.A. Galvin, and E.A. Roth. 1994. *African Pastoralist Systems: An Integrated Approach.* Boulder, Colo.: Lynne Rienner Publishers.

Fratkin, E., and E.A. Roth. 2005. The setting: pastoral sedentarization in Marsabit District, Northern Kenya. In *As Pastoralists Settle: Social, Health, and Economic Consequences of Pastoral Sedentarization in Marsabit District, Kenya,* ed. E. Fratkin and E.A. Roth, 29–52. New York: Kluwer Academic Publishers.

Friedel, M.H. 1991. Range condition assessment and the concept of thresholds: A viewpoint. *Journal of Range Management* 44:422–426.

Fritz, H., and P. Duncan. 1994. On the carrying capacity for large ungulates of African savanna ecosystems. *Proceedings of the Royal Society of London, Series B—Biological Sciences* 256:77–82.

Fritz, H., S. Said, P.C. Renaud et al. 2003. The effects of agricultural fields and human settlements on the use of rivers by wildlife in the mid-Zambezi valley, Zimbabwe. *Landscape Ecology* 18:293–302.

Frost, P. G. H., and F. Robertson. 1987. The ecological effects of fire in savannas. In *Determinants of Tropical Savannas*, ed. B. H. Walker, 93–140. Oxford: IRL Press.

Fryxell, J. M. 1983. Wildlife migration in Boma National Park. *Swara* 6:12–15.

———. 1991. Forage quality and aggregation by large herbivores. *American Naturalist* 138:478–498.

———. 1995. Aggregation and migration by grazing ungulates in relation to resources and predators. In *Serengeti II: Dynamics, Management and Conservation of an Ecosystem*, ed. A. R. E. Sinclair and P. Arcese, 257–273. Chicago: University of Chicago Press.

Fryxell, J. M., A. Mosser, A. R. E. Sinclair, and C. Packer. 2007. Group formation stabilizes predator-prey dynamics. *Nature* 449:1041–U1044.

Fuhlendorf, S. D., and D. M. Engle. 2001. Restoring heterogeneity on rangelands: Ecosystem management based on evolutionary grazing patterns. *BioScience* 51:625–632.

Fynn, R. W. S., and T. G. O'Connor. 2000. Effect of stocking rate and rainfall on rangeland dynamics and cattle performance in a semi-arid savanna, South Africa. *Journal of Applied Ecology* 37:491–507.

Fyumagwa, R. D., V. Runyoro, I. G. Horak, and R. Hoare. 2007. Ecology and control of ticks as disease vectors in wildlife of the Ngorongoro Crater, Tanzania. *South African Journal of Wildlife Research* 37:79–90.

Fyumagwa, R. D., P. Simmler, B. Willi et al. 2008. Molecular detection of haemotropic *Mycoplasma* species in *Rhipicephalus sanguineus* tick species collected on lions (*Panthera leo*) from Ngorongoro Crater, Tanzania. *South African Journal of Wildlife Research* 38:117–122.

Galaty, J. G. 1991. Pastoral orbits and deadly jousts: Factors in Maasai expansion. In *Herders, Warriors, and Traders: Pastoralism in Africa*, ed. J. G. Galaty and P. Bonte, 171–198. Boulder, Colo.: Westview Press.

———. 1992. "This land is yours": Social and economic factors in privatisation, subdivision, and sale of Maasai ranches. *Nomadic Peoples* 30:26–40.

———. 1993a. "The eye that wants a person, where can it not see?": Inclusion, exclusion, and boundary shifters in Maasai identity. In *Being Maasai*, ed. T. Spear and R. Waller, 174–194. London: James Currey; Dar es Salaam: Mkuki na Nyota; Nairobi: EAEP; Athens: Ohio University Press.

———. 1993b. Maasai expansion and the new East African pastoralism. In *Being Maasai*, ed. T. Spear and R. Waller, 61–86. Oxford: James Currey; Athens: Ohio University Press.

———. 1994. Rangeland tenure and pastoralism in Africa. In *African Pastoralist Systems: An Integrated Approach*, ed. E. Fratkin, K. A. Galvin, and E. A. Roth, 185–204. Boulder, Colo.: Lynne Reiner Publishers.

Galvin, K. A. 1992. Nutritional ecology of pastoralists in dry tropical Africa. *American Journal of Human Biology* 4:209–221.

Galvin, K. A., R. B. Boone, J. T. McCabe et al. Forthcoming. Transitions in the Ngorongoro Conservation Area: The story of cultivation, human well-being, and conservation. In *Serengeti IV: Biodiversity*, ed. A. J. Sinclair, K. Metzger, and J. Fryxell. Chicago: University of Chicago Press.

Galvin, K., J. Ellis, R. B. Boone et al. 2002. Compatibility of pastoralism and conservation? A test case using integrated assessment in the Ngorongoro Conservation Area, Tanzania. In *Conservation and Mobile Indigenous Peoples: Displacement, Forced Settlement, and Sustainable Development*, ed. D. Chatty and M. Colchester, 36–60. New York: Berghan Books.

Galvin, K. A., and M. A. Little. 1999. Dietary intake and nutritional status. In *Turkana Herders of the Dry Savanna*, ed. M. A. Little and P. W. Leslie, 124–145. Oxford: Oxford University Press.

Galvin, K. A., and A. L. Magennis. 1999. Compatibility of pastoralism and conservation? A test case comparing nutritional status among two Maasai populations in Tanzania. *American Journal Of Human Biology* 11:111–112.

Galvin, K. A., P. K. Thornton, R. B. Boone, and L. M. Knapp. 2008. Ngorongoro Conservation Area, Tanzania: Fragmentation of a unique region of the greater Serengeti ecosystem. In *Fragmentation in Semi-arid and Arid Landscapes: Consequences for Human and Natural Systems*, ed. K. Galvin, R. S. Reid, R. H. Behnke, and N. T. Hobbs, 255–279. Dordrectht, Neth.: Springer.

Galvin, K. A., P. K. Thornton, J. de Pinho et al. 2006. Integrated modeling and its potential for resolving conflicts between conservation and people in the rangelands of East Africa. *Human Ecology* 34:155–183.

Gardiner, A. J., R. S. Reid, and S. Kiema. 2005. Impact of land-use on butterflies in south-western Burkina Faso. *African Entomology* 13:201–212.

Geisler, C., and R. de Sousa. 2001. From refuge to refugee: The African case. *Public Administration and Development* 21:159–170.

Gemedo-Dalle, J. Isselstein, and B. L. Maass. 2006. Indigenous ecological knowledge of Borana pastoralists in southern Ethiopia and current challenges. *International Journal of Sustainable Development and World Ecology* 13: 113–130.

Georgiadis, N. J. 1987. Responses of savanna grasslands to extreme use by pastoralist livestock (Kenya). Ph.D. diss., Syracuse University.

Georgiadis, N. J., and S. J. McNaughton. 1988. Interactions between grazers and a cyanogenic grass, *Cynodon plectostachyus*. *Oikos* 51:343–350.

———. 1990. Elemental and fibre contents of savanna grasses: Variation with grazing soil type, season, and species. *Journal of Applied Ecology* 27:623–634.

Georgiadis, N. J., J. G. N. Olwero, G. Ojwang, and S. S. Romanach. 2007. Savanna herbivore dynamics in a livestock-dominated landscape. I. Dependence on land use, rainfall, density, and time. *Biological Conservation* 137:461–472.

Gereta, E. J. 2004. Transboundary water issues threaten the Serengeti ecosystem. *Oryx* 38:14–15.

Gereta, E., E. Mwangomo, and E. Wolanski. 2009. Ecohydrology as a tool for ensuring the survival of the threatened Serengeti ecosystem. *Ecohydrology and Hydrobiology* 9:115–124.

Gereta, E., and E. Wolanski. 1998. Wildlife–water quality interactions in the Serengeti National Park, Tanzania. *African Journal of Ecology* 36:1–14.

Gereta, E., E. Wolanski, M. Borner, and S. Serneels. 2002. Use of an ecohydrology model to predict the impact on the Serengeti ecosystem of deforestation, irrigation, and the proposed Amala Weir Water Diversion Project in Kenya. *Ecohydrology and Hydrobiology* 2:135–142.

Gichohi, H. 1990. The effects of fire and grazing on grasslands of Nairobi National Park. M.Sc. thesis, University of Nairobi.

———. 1996. The ecology of a truncated ecosystem: The Athi-Kapiti Plains. Ph.D. diss., University of Leicester.

———. 2000. Functional relationships between parks and agricultural areas in east Africa: The case of Nairobi National Park. In *Wildlife Conservation and Sustainable Use*, ed. H. H. T. Prins, J. G. Grootenhuis, and T. T. Dolan, 141–168. Dordrecht, Neth.: Kluwer Academic Publishers.

Gifford-Gonzalez, D. 1998. Early pastoralists in east Africa: Ecological and social dimensions. *Journal of Anthropological Archaeology* 17:166–200.

Gillson, L. 2006. A "large infrequent disturbance" in an east African savanna. *African Journal of Ecology* 44:458–467.

GLASOD. 1990. *Global Assessment of Soil Degradation*. Wageningen, Neth.: International Soil Reference and Information Centre; Nairobi: United Nations Environment Program.

GLC. 2003. Global Land Cover 2000 Database. Joint Research Centre, European Commission, http://bioval.jrc.ec.europa.eu/products/glc2000/glc2000.php.

Goldman, M. 2003. Partitioned nature, privileged knowledge: Community-based conservation in Tanzania. *Development and Change* 34:833–862.

———. 2006. Sharing pastures, building dialogues: Maasai and wildlife conservation in northern Tanzania. Ph.D. diss., University of Wisconsin, Madison.

———. 2007. Tracking wildebeest, locating knowledge: Maasai and conservation biology understandings of wildebeest behavior in northern Tanzania. *Environment and Planning D: Society and Space* 25:307–331.

————. 2009. Constructing connectivity: Conservation corridors and conservation politics in east African rangelands. *Annals of the Association of American Geographers* 99:335–359.

Gonder, M. K., H. M. Mortensen, F. A. Reed et al. 2007. Whole-mtDNA genome sequence analysis of ancient African lineages. *Molecular Biology and Evolution* 24:757–768.

Gordon, R. G., Jr., ed. 2005. *Ethnologue: Languages of the World*. Fifteenth edition. Dallas: SIL International. Online version: www.ethnologue.com.

Goren-Inbar, N., N. Alperson, M. E. Kislev et al. 2004. Evidence of hominin control of fire at Gesher Benot Ya'aqov, Israel. *Science* 304:725–727.

Gowlett, J. A. J., J. W. K. Harris, D. Walton, and B. A. Wood. 1981. Early archaeological sites, hominid remains, and traces of fire from Chesowanja, Kenya. *Nature* 294:125–129.

Grayson, D. K., and D. J. Meltzer. 2003. A requiem for North American overkill. *Journal of Archaeological Science* 30:585–593.

Greenberg, J. H. 1963. *The Languages of Africa*. Bloomington: Indiana University Press.

Grice, A. C., and K. C. Hodgkinson. 2002. Challenges for rangeland people. In *Global Rangelands: Progress and Prospects*, ed. A. C. Grice and K. C. Hodgkinson, 1–9. Wallingford, U.K.: CAB International.

Grimaldi, D. 1999. The co-radiations of pollinating insects and angiosperms in the Cretaceous. *Annals of the Missouri Botanical Garden* 86:373–406.

Grime, J. P., 1973. Competitive exclusion in herbaceous vegetation. *Nature* 242: 344–347.

Groom, R. 2007. How to make land subdivision work: An analysis of the ecological and socio-economic factors affecting conservation outcomes during land privatisation in Kenyan Maasailand. PhD. diss., University of Bristol.

Groom, R., and S. Harris. 2008. Conservation on community lands: The importance of equitable revenue sharing. *Environmental Conservation* 35:242–251.

Grootenhuis, J. G. 2000. Wildlife, livestock, and animal disease reservoirs. In *Wildlife Conservation by Sustainable Use*, ed. H. H. T. Prins, J. G. Grootenhuis, and T. T. Dolan, 81–113. Dordrecht, Neth.: Kluwer Academic Publishers.

Grove, R. 1987. Early themes in African conservation: The Cape in the nineteenth century. In *Conservation in Africa: People, Policies, and Practices*, ed. D. M. Anderson and R. Grove, 21–39. Cambridge: Cambridge University Press.

GRUMP. 2004. Global Rural-Urban Mapping Project (GRUMP), Alpha Version: Urban Extents. Socioeconomic Data and Applications Center (SEDAC), Columbia University. http://sedac.ciesin.columbia.edu/gpw. Center for International

Earth Science Information Network (CIESIN), Columbia University; International Food Policy Research Institute (IFPRI); World Bank; and Centro Internacional de Agricultura Tropical (CIAT).

Gulliver, P. H. 1955. *The Family Herds: A Study of Two Pastoral Tribes in East Africa, the Jie and Turkana.* London: Routledge and Kegan Paul.

———. 1975. Nomadic movements: Causes and implications. In *Pastoralism in Tropical Africa,* ed. T. Monod, 369–386. Oxford: Oxford University Press.

Haaland, R. 1995. Sedentism, cultivation, and plant domestication in the Holocene Middle Nile region. *Journal of Field Archaeology* 22:157–174.

Hakansson, N. T. 2004. The human ecology of world systems in East Africa: The impact of the ivory trade. *Human Ecology* 32:561–591.

Hanotte, O., D. G. Bradley, J. W. Ochieng et al. 2002. African pastoralism: Genetic imprints of origins and migrations. *Science* 296:336–339.

Hanski, I., and Y. Cambefort. 1991. *Dung Beetle Ecology.* Princeton: Princeton University Press.

Hao, W. M., and M. H. Liu. 1994. Spatial and temporal distribution of tropical biomass burning. *Global Biogeochemical Cycles* 8:485–503.

Happold, D. C. D. 1995. The interaction between humans and mammals in Africa in relation to conservation: A review. *Biodiversity and Conservation* 4:395–414.

Hardin, G. 1968. Tragedy of the commons. *Science* 162:1243–1248.

Haro, G. O., G. J. Doyo, and J. G. McPeak. 2005. Linkages between community, environmental, and conflict management: Experiences from northern Kenya. *World Development* 33:285–299.

Harris, G., S. Thirgood, G. C. Hopcraft et al. 2009. Global decline in aggregated migrations of large terrestrial mammals. *Endangered Species Research* 7:55–76.

Harrison, S., B. D. Inouye, and H. D. Safford. 2003. Ecological heterogeneity in the effects of grazing and fire on grassland diversity. *Conservation Biology* 17:837–845.

Harrison, Y. A., and C. M. Shackleton. 1999. Resilience of South African communal grazing lands after the removal of high grazing pressure. *Land Degradation and Development* 10:225–239.

Hartnett, D. C., K. R. Hickman, and L. E. F. Walter. 1996. Effects of bison grazing, fire, and topography on floristic diversity in tallgrass prairie. *Journal of Range Management* 49:413–420.

Hartwig, G. W. 1981. Smallpox in Sudan. *International Journal of African Historical Studies* 14:5–33.

Hastenrath, S., and L. Greischar. 1997. Glacier recession on Kilimanjaro, East Africa, 1912–89. *Journal Of Glaciology* 43:455–459.

Hay, S.I., J. Cox, D.J. Rogers et al. 2002. Climate change and the resurgence of malaria in the East African highlands. *Nature* 415:905–909.

Hazzah, L., M.B. Mulder, and L. Frank. 2009. Lions and warriors: Social factors underlying declining African lion populations and the effect of incentive-based management in Kenya. *Biological Conservation* 142:2428–2437.

HDR. 2006. *Human Development Report*. New York: United Nations Development Program.

Heady, H.E., and R.D. Child. 1994. *Rangeland Ecology and Management*. Boulder, Colo.: Westview Press.

Hein, L., and N. De Ridder. 2006. Desertification in the Sahel: A reinterpretation. *Global Change Biology* 12:751–758.

Heine, B., and W.J.G. Mohlig. 1980. *Language and Dialect Atlas of Kenya: Geographical and Historical Introduction*. Berlin: Dietrich Reimer Verlag.

Hemp, A. 2005. Climate change–driven forest fires marginalize the impact of ice cap wasting on Kilimanjaro. *Global Change Biology* 11:1013–1023.

———. 2009. Climate change and its impact on the forests of Kilimanjaro. *African Journal of Ecology* 47:3–10.

Hendrickson, D., J. Armon, and R. Mearns. 1998. The changing nature of conflict and famine vulnerability: The case of livestock raiding in Turkana District, Kenya. *Disasters* 22:185–199.

Hendrix, C.S., and I. Salehyan. 2010. After the rain: Rainfall variability, hydrometeorological disasters, and social conflict in Africa. Paper presented at Climate Change and Security Conference, Trondheim, Norway, June 21–24, 2010. Available at http://ssrn.com/abstract=1641312.

Herendeen, P.S., and B.F. Jacobs. 2000. Fossil legumes from the Middle Eocene (46.0 Ma) Mahenge flora of Singida, Tanzania. *American Journal of Botany* 87:1358–1366.

Herlocker, D.J., and H.J. Dirschl. 1972. Vegetation of the Ngorongoro Conservation Area, Tanzania. *Canadian Wildlife Service Report Series* 19:5–39.

Herskovits, M. 1926. The cattle complex in East Africa. *American Anthropologist* 28:230–272, 361–388, 494–528, 633–664.

Heumann, B.W., J.W. Seaquist, L. Eklundh, and P. Johnsson. 2007. AVHRR-derived phenological change in the Sahel and Soudan, Africa, 1982–2005. *Remote Sensing of Environment* 108:385–392.

Hiernaux, P. 1996. The crisis of Sahelian pastoralism: ecological or economic? Paper 39a, Pastoral Development Network, Overseas Development Institute, London.

———. 1998. Effects of grazing on plant species composition and spatial distribution in rangelands of the Sahel. *Plant Ecology* 138:191–202.

Hiernaux, P., C. L. Bielders, C. Valentin et al. 1999. Effects of livestock grazing on physical and chemical properties of sandy soils in Sahelian rangelands. *Journal of Arid Environments* 41:231–245.

Higgins, S. I., W. J. Bond, and W. S. W. Trollope. 2000. Fire, resprouting, and variability: A recipe for grass-tree coexistence in savanna. *Journal of Ecology* 88:213–229.

Hilborn, R., P. Arcese, M. Borner et al. 2006. Effective enforcement in a conservation area. *Science* 314:1266.

Hildebrand, A. R., G. T. Penfield, D. A. Kring et al. 1991. Chicxulub crater—A possible Cretaceous Tertiary boundary impact crater on the Yucatan Peninsula, Mexico. *Geology* 19:867–871.

Hillman, J. C., and A. K. K. Hillman. 1977. Mortality of wildlife in Nairobi National Park during the drought of 1973–74. *East African Wildlife Journal* 15:1–18.

Hinde, S. L., and H. Hinde. 1910. *The Last of the Masai*. London: William Heinemann.

Hoare, R. 1999. Determinants of human-elephant conflict in a land-use mosaic. *Journal of Applied Ecology* 36:689–700.

———. 2000. African elephants and humans in conflict: the outlook for co-existence. *Oryx* 34:34–38.

Hoare, R., and J. T. du Toit. 1999. Coexistence between people and elephants in African savannas. *Conservation Biology* 13:633–639.

Hobbs, N. T., D. L. Baker, G. D. Bear, and D. C. Bowden. 1996a. Ungulate grazing in sagebrush grassland: Effects of resource competition on secondary production. *Ecological Applications* 6:218–227.

———. 1996b. Ungulate grazing in sagebrush grassland: Mechanisms of resource competition. *Ecological Applications* 6:200–217.

Hobbs, N. T., D. S. Schimel, D. E. Owensby, and D. S. Ojima. 1991. Fire and grazing in the tallgrass prairie: Contingent effects on nitrogen budgets. *Ecology* 72:1374–1382.

Hobbs, N. T., and R. A. Spowart. 1984. Effects of prescribed fire on nutrition of mountain sheep and mule deer during winter and spring. *Journal of Wildlife Management* 48:551–560.

Hobbs, N. T., and D. M. Swift. 1988. Grazing in herds: When are nutritional benefits realized? *American Naturalist* 131:760–764.

Hobbs, R. J., and L. F. Huenneke. 1992. Disturbance, diversity, and invasion: Implications for conservation. *Conservation Biology* 6:324–337.

Hoben, A. 1995. Paradigms and politics: The cultural construction of environmental policy in Ethiopia. *World Development* 23:1007–1021.

Hodgson, J.G., and A.W. Illius. 1996. *The Ecology and Management of Grazing Systems.* Wallingford, U.K.: CAB International.

Hofer, H., K.L.I. Campbell, L.M. East, and S.A. Huish. 1996. The impact of game meat hunting on target and non-target species in the Serengeti. In *The Exploitation of Mammal Populations,* ed. V.J. Taylor and N. Dunstone, 117–146. London: Chapman and Hall.

Hofmann, R.R. 1989. Evolutionary steps of ecophysiological adaptation and diversification of ruminants: A comparative view of their digestive system. *Oecologia* 78:443–457.

Hofmann, R.R., and D.R.M. Stewart. 1972. Grazer or browser: A classification based on stomach structure and feeding habits of East African ruminants. *Mammalia* 36:226–240.

Holdo, R.M., J.M. Fryxell, A.R.E. Sinclair et al. 2011. Predicted impact of barriers to migration on the Serengeti wildebeest population. *PloS ONE* 6:1–7.

Holdo, R.M., R.D. Holt, and J.M. Fryxell. 2009a. Opposing rainfall and plant nutritional gradients best explain the wildebeest migration in the Serengeti. *American Naturalist* 173:431–445.

Holdo, R.M., A.R.E. Sinclair, A.P. Dobson et al. 2009b. A disease-mediated trophic cascade in the Serengeti and its implications for ecosystem C. *PloS Biology* 7:1–12.

Holecheck, J.L., R.D. Pieper, and C.H. Herbel. 2004. *Range Management: Principles and Practices.* Upper Saddle River, N.J.: Pearson Education.

Holekamp, K.E., J.O. Ogutu, H.T. Dublin et al. 1993. Fission of a spotted hyena clan: Consequences of prolonged female absenteeism and causes of female emigration. *Ethology* 93:285–299.

Holmern, T., J. Nyahongo, and E. Roskaft. 2007. Livestock loss caused by predators outside the Serengeti National Park, Tanzania. *Biological Conservation* 135:518–526.

Holmern, T., E. Roskaft, J. Mbaruka et al. 2002. Uneconomical game cropping in a community-based conservation project outside the Serengeti National Park, Tanzania. *Oryx* 36:364–372.

Homewood, K. 2008. *Ecology of Pastoralist Societies.* Oxford: James Currey.

Homewood, K., and D. Brockington. 1999. Biodiversity, conservation, and development in Mkomazi Game Reserve, Tanzania. *Global Ecology and Biogeography* 8:301–313.

Homewood, K., E. Coast, and M. Thompson. 2004. In-migrants and exclusion in east African rangelands: Access, tenure, and conflict. *Africa* 74:567–610.

Homewood, K., and J. Lewis. 1987. Impact of drought on pastoral livestock in Baringo, Kenya, 1983–85. *Journal of Applied Ecology* 24:615–631.

Homewood, K., and W. A. Rodgers. 1984. Pastoralism and conservation. *Human Ecology* 12:431–442.

———. 1987. Pastoralism, conservation, and the overgrazing controversy. In *Conservation in Africa: People, Policies, and Practices*, ed. D. M. Anderson and R. Grove, 111–128. Cambridge: Cambridge University Press.

———. 1991. *Maasailand Ecology: Pastoralist Development and Wildlife Conservation in Ngorongoro, Tanzania*. Cambridge: Cambridge University Press.

Homewood, K., W. A. Rodgers, and K. Arhem. 1987. Ecology of pastoralism in Ngorongoro Conservation Area, Tanzania. *Journal of Agricultural Science* 108:47–72.

Homewood, K., P. C. Trench, and P. Kristjanson. 2009. *Staying Maasai? Livelihoods, Conservation, and Development in East African Rangelands*. London: Springer.

Honey, M. 2008. *Ecotourism and Sustainable Development*. Washington, D.C.: Island Press.

Hopcraft, J. G. C., A. R. E. Sinclair, and C. Packer. 2005. Planning for success: Serengeti lions seek prey accessibility rather than abundance. *Journal of Animal Ecology* 74:559–566.

Horowitz, M. M., and P. D. Little. 1987. African pastoralism and poverty: Some implications for drought and famine. In *Drought and Famine in Africa: Denying Drought a Future*, ed. M. Glantz, 59–82. Cambridge: Cambridge University Press.

Hughes, L. 2002. Moving the Maasai: A colonial misadventure. Ph.D. diss., Oxford University.

———. 2005. Malice in Maasailand: The historical roots of current political struggles. *African Affairs* 104:207–224.

Hulme, D., and M. Murphree, eds. 2001. *African Wildlife and Livelihoods: The Promise and Performance of Community Conservation*. Nairobi: East African Educational Publishers.

Hulme, M., R. Doherty, T. Ngara et al. 2001. African climate change: 1900–2100. *Climate Research* 17:145–168.

Huntley, B. J. 1982. Southern African savannas. In *Ecology of Tropical Savannas*, ed. B. J. Huntley and B. H. Walker, 101–119. Berlin: Springer-Verlag.

Huqqaa, G. N.d. *Gada: The Oromo Traditional, Economic, and Socio-political System, The 37th Gumii Gaayo Assembly*. Addis Ababa: Norwegian Church Aid.

Huston, M. A. 1979. A general hypothesis of species diversity. *American Naturalist* 113:81–101.

———. 1994. *Biological Diversity*. Cambridge: Cambridge University Press.

IEA. 1998. African Mammals Databank—A Databank for the Conservation and Management of the African Mammals, Vols. 1 and 2. Report to the Directorate-

General for Development (DGVIII/A/1) of the European Commission. Project No. B7–6200/94–15/VIII/ENV., Instituto Ecologia Applicata, Rome, Italy.

Igoe, J. 2003. Scaling up civil society: Donor money, NGOs, and the pastoralist land rights movement in Tanzania. *Development and Change* 34:863–885.

Igoe, J., and B. Croucher. 2007. Conservation, commerce, and communities: The story of community-based wildlife management in Tanzania's northern tourist circuit. *Conservation and Society* 5:534–561.

Ikanda, D. K., and C. Packer. 2008. Ritual vs. retaliatory killing of African lions in the Ngorongoro Conservation Area, Tanzania. *Endangered Species Research* 6:67–74.

Iliffe, J. 1995. *Africans: History of a Continent.* Cambridge: Cambridge University Press.

Illius, A. W., and T. G. O'Connor. 1999. On the relevance of nonequilibrium concepts to arid and semiarid grazing systems. *Ecological Applications* 9:798–813.

Imbahale, S., J. M. Githaiga, R. M. Chira, and M. Y. Said. 2008. Resource utilization by large migratory herbivores of the Athi-Kapiti ecosystem. *African Journal of Ecology* 46:43–51.

Isbell, L. A., and T. P. Young. 1993. Human presence reduces predation in a free-ranging vervet monkey population in Kenya. *Animal Behaviour* 45:1233–1235.

IUCN-WCMC. 2005. *Serengeti National Park Tanzania.* Gland, Switz.: International Union for the Conservation of Nature—World Conservation Monitoring Centre.

Jacobs, A. H. 1975. Maasai pastoralism in an historical perspective. In *Pastoralism in Tropical Africa,* ed. T. Monod, 406–425. Oxford: International African Institute.

Jacobs, B. F. 2004. Palaeobotanical studies from tropical Africa: Relevance to the evolution of forest, woodland, and savannah biomes. *Philosophical Transactions of the Royal Society of London, Series B—Biological Sciences* 359:1573–1583.

Jacobs, B. F., J. D. Kingston, and L. L. Jacobs. 1999. The origin of grass-dominated ecosystems. *Annals of the Missouri Botanical Garden* 86:590–643.

James, C. D., J. Landsberg, and S. R. Morton. 1999. Provision of watering points in the Australian arid zone: A review of effects on biota. *Journal of Arid Environments* 41:87–121.

Janis, C. H. 1976. The evolutionary strategy of the Equidae and the origins of rumen and caecal digestion. *Evolution* 30:757–776.

Janssen, T., and K. Bremer. 2004. The age of major monocot groups inferred from 800+ rbcL sequences. *Botanical Journal of the Linnean Society* 146:385–398.

Jaramillo, V. J., and J. K. Detling. 1988. Grazing history, defoliation, and competition: Effects on shortgrass production and nitrogen accumulation. *Ecology* 69:1599–1608.

Jarman, P.J., and A.R.E. Sinclair. 1979. Feeding strategy and the pattern of resource partitioning in ungulates. In *Serengeti: Dynamics of an Ecosystem,* ed. A.R.E. Sinclair and M. Norton-Griffiths, 185–220. Chicago: University of Chicago Press.

Jarvis, A., H.I. Reuter, A. Nelson, and E. Guevara. 2008. Hole-filled SRTM for the globe Version 4, available from the CGIAR-CSI SRTM 90m Database. http://srtm.csi.cgiar.org.

Johanson, D., and M. Edey. 1981. *Lucy: The Beginnings of Humankind.* New York: Simon and Schuster.

Johnson, D.H. 1991. Political ecology in the upper Nile: The twentieth century expansion of the pastoral "common economy." In *Herders, Warriors, and Traders: Pastoralism in Africa,* ed. J.G. Galaty and P. Bonte, 89–117. Boulder, Colo.: Westview Press.

Jones, B., and M. Murphree. 2001. The evolution of policy on community conservation in Namibia and Zimbabwe. In *African Wildlife and Livelihoods: The Promise and Performance of Community Conservation,* ed. D. Hulme and M. Murphree, 38–58. Oxford: James Currey.

Jones, P.G., and P.K. Thornton. 2003. The potential impact of climate change on maize production in Africa and Latin America in 2055. *Global Environmental Change* 13:51–59.

Joubert, D.F., and P.G. Ryan. 1999. Differences in mammal and bird assemblages between commercial and communal rangelands in the succulent Karoo. *Journal of Arid Environments* 43:287–299.

Kaelo, D. 2007. Human-elephant conflict in pastoral areas north of Maasai Mara National Reserve, Kenya. M.Sc. thesis, Moi University, Eldoret.

Kamara, A.B., B. Swallow, and M. Kirk. 2004. Policies, interventions, and institutional change in pastoral resource management in Borana, southern Ethiopia. *Development Policy Review* 22:381–403.

Kangwana, K. 1993. Elephants and Maasai: Conflict and Conservation in Amboseli, Kenya. Ph.D. diss., Cambridge University.

———. 2011. The behavioral responses of elephants to Maasai in Amboseli. In *The Amboseli Elephants,* ed. C.J. Moss, H. Croze, and P.C. Lee, 307–317. Chicago: University of Chicago Press.

Kangwana, K., and C. Browne-Nuñez. 2011. The human context of the Amboseli elephants. In *The Amboseli Elephants,* ed. C.J. Moss, H. Croze, and P.C. Lee, 29–36. Chicago: University of Chicago Press.

Kangwana, K., and R. Ole Mako. 2001. Conservation, livelihoods, and the intrinsic value of wildlife, Tarangire National Park, Tanzania. In *African Wildlife and Livelihoods,* ed. D. Hulme and M. Murphree, 148–159. Oxford: James Currey.

Kanyamibwa, S. 1998. Impact of war on conservation: Rwandan environment and wildlife in agony. *Biodiversity and Conservation* 7:1399–1406.

Kappelman, J., D. T. Rasmussen, W. J. Sanders et al. 2003. Oligocene mammals from Ethiopia and faunal exchange between Afro-Arabia and Eurasia. *Nature* 426:549–552.

Kaser, G., D. R. Hardy, T. Mölg et al. 2004. Modern glacier retreat on Kilimanjaro as evidence of climate change: Observations and facts. *International Journal of Climatology* 24:329–339.

Kaser, G., T. Mölg, N. J. Cullen et al. 2010. Is the decline of ice on Kilimanjaro unprecedented in the Holocene? *Holocene* 20:1079–1091.

Katampoi, K. O., G. O. Genga, M. Mwangi et al. 1990. *Kajiado District Atlas*. Kajiado, Kenya: Arid and Semi-arid Lands Programme.

Keeley, J. E. 2002. Native American impacts on fire regimes of the California coastal ranges. *Journal of Biogeography* 29:303–320.

Keesing, F. 1998. Ecology and behavior of the pouched mouse, *Saccostomus mearnsi*, in central Kenya. *Journal of Mammalogy* 79:919–931.

———. 2000. Cryptic consumers and the ecology of an African savanna. *BioScience* 50:205–215.

Kelbessa, W. 2005. The rehabilitation of indigenous environmental ethics in Africa. *Diogenes* 207:17–34.

Keller, G. 2005. Impacts, volcanism, and mass extinction: Random coincidence or cause and effect? *Australian Journal of Earth Sciences* 52:725–757.

Keller, G., T. Adatte, W. Stinnesbeck et al. 2004. Chicxulub impact predates the K-T boundary mass extinction. *Proceedings of the National Academy of Sciences* 101:3753–3758.

Kellert, S. R., J. N. Mehta, S. A. Ebbin, and L. L. Lichtenfeld. 2000. Community natural resource management: Promise, rhetoric, and reality. *Society and Natural Resources* 13:705–715.

Kelly, R. D., and B. H. Walker. 1977. The effects of different forms of land use on the ecology of the semi-arid region in south-eastern Rhodesia. *Journal of Ecology* 62:553–576.

Kijazi, A. 1997. Principal management issues in the Ngorongoro Conservation Area. In *Multiple Land-Use: The Experience of the Ngorongoro Conservation Area, Tanzania*, ed. D. M. Thompson, 33–44. Gland, Switz.: IUCN.

Kijazi, A., S. Mkumbo, and D. M. Thompson. 1997. Human and livestock population trends. In *Multiple Land-Use: The Experience of the Ngorongoro Conservation Area, Tanzania*, ed. D. M. Thompson, 169–180. Gland, Switz.: IUCN.

Kimani, K., and J. Pickard. 1998. Recent trends and implications of group ranch subdivision and fragmentation in Kajiado District, Kenya. *Geographical Journal* 164:202–213.

Kingdon, J. 1989. *Island Africa*. Princeton: Princeton University Press.

———. 2003. *Lowly Origin*. Princeton: Princeton University Press.

Kingston, J. D., and T. Harrison. 2007. Isotopic dietary reconstructions of Pliocene herbivores at Laetoli: Implications for early hominin paleoecology. *Palaeogeography, Palaeoclimatology, Palaeoecology* 243:272–306.

Kipuri, N., and C. Sørensen. 2008. Poverty, pastoralism, and policy in Ngorongoro. Lessons learned from the ERETO I Ngorongoro Pastoralist Project with implications for pastoral development and the policy debate, ERETO/IIED. ERETO/NPP (Government of Tanzania).

Kipury, N. 1983. *Oral Literature of the Maasai*. Nairobi: East African Educational Publishers.

Kipury, N. O. 1991. Age, gender, and class in the scramble for Maasailand. *Nature and Resources* 27:10–17.

Kirk, M. 1999. The context for livestock and crop-livestock development in Africa: The evolving role of the state in influence property rights over grazing resources in sub-Saharan Africa. In *Property Rights, Risk, and Livestock Development in Africa,* ed. N. McCarthy, B. Swallow, M. Kirk, and P. Hazell, 23–54. Washington, D.C.: International Food Policy Research Institute.

Kissui, B. M. 2008. Livestock predation by lions, leopards, spotted hyenas, and their vulnerability to retaliatory killing in the Maasai steppe, Tanzania. *Animal Conservation* 11:422–432.

Kissui, B. M., A. Mosser, and C. Packer. 2010. Persistence and local extinction of lion prides in the Ngorongoro Crater, Tanzania. *Population Ecology* 52:103–111.

Kiyiapi, J., P. Ntiati, B. Mwongela et al. 2005. A community business: Elerai Ranch and Conservation Area, Kenya. AWF Working Paper Series, African Wildlife Foundation, Nairobi.

Kjekshus, H. 1996. *Ecology Control and Economic Development in East African History.* James London: James Currey; Dar es Salaam: Mkukui na Nyota; Kampala: Fountain Publishers; Athens: Ohio University Press.

Klein, R. G. 1999. *The Human Career: Human, Biological, and Cultural Origins.* Chicago: University of Chicago Press.

Klima, G. 1970. *The Barabaig: East African Cattle Herders.* Prospect Heights, Ill.: Waveland Press.

Knight, R. L., and P. B. Landres. 1998. *Stewardship across Boundaries.* Washington, D.C.: Island Press.

Knoop, W. T., and B. H. Walker. 1985. Interactions of woody and herbaceous vegetation in a southern African savanna. *Journal of Ecology* 73:235–253.

Kolowski, J.M., and K.E. Holekamp. 2006. Spatial, temporal, and physical characteristics of livestock depredations by large carnivores along a Kenyan reserve border. *Biological Conservation* 128:529–541.

————. 2009. Ecological and anthropogenic influences on space use by spotted hyaenas. *Journal of Zoology* 277:23–36.

Konaté, S., X. Le Roux, B. Verdier, and M. LePage. 2003. Effect of underground fungus-growing termites on carbon dioxide emission at the point and landscape-scales in an African savanna. *Functional Ecology* 17:305–314.

Koponen, J. 1988. War, famine, and pestilence in late precolonial Tanzania: A case for a heightened mortality. *International Journal of African Historical Studies* 21:637–676.

Kortenkamp, K.V., and C.F. Moore. 2001. Ecocentrism and anthropocentrism: Moral reasoning about ecological commons dilemmas. *Journal of Environmental Psychology* 21:261–272.

Kristjanson, P.M., M. Radeny, D. Nkedianye et al. 2002. Valuing alternative land-use options in the Kitengela wildlife dispersal area of Kenya. Report prepared for the International Livestock Research Institute, Nairobi.

Kristjanson, P., R.S. Reid, N. Dickson et al. 2009. Linking international agricultural research knowledge with action for sustainable development. *Proceedings of the National Academy of Sciences* 106:5047–5052.

Kshatriya, M. 2005. Effects of fragmentation by fencing on pastoral herd movements in the Kitengela. Report prepared for the International Livestock Research Institute, Nairobi.

KWS. 2010. Amboseli–West Kilimanjaro and Magadi-Natron cross border landscape, wet season 2010, aerial total count. Available at www.kws.org.

Laden, G., and R. Wrangham. 2005. The rise of the hominids as an adaptive shift in fallback foods: Plant underground storage organs (USOs) and australopith origins. *Journal of Human Evolution* 49:482–498.

Lado, L. 1996. Problems of wildlife management and land use in Kenya. *Land Use Policy* 9:169–184.

Lambrechts, C., B. Woodley, A. Hemp et al. 2002. *Aerial Survey of the Threats to Mt. Kilimanjaro Forests.* Dar es Salaam: UNDP.

Lamphear, J. 1976. *The Traditional History of the Jie of Uganda.* Oxford: Clarendon Press.

Lamprey, H.F. 1963. Ecological separation of the large mammal species in the Tarangire Game Reserve, Tanganyika. *East African Wildlife Journal* 1:63–92.

———. 1983. Pastoralism yesterday and today: The overgrazing controversy. In *Tropical Savannas,* vol. 13 of *Ecosystems of the World,* ed. F. Bourlière, 643–666. Amsterdam: Elsevier.

Lamprey, R. H., E. Buhanga, and J. Omoding. 2003. *A Study of Wildlife Distributions, Wildlife Management Systems, and Options for Wildlife-based Livelihoods in Uganda.* Kampala: International Food Policy Research Institute and USAID.

Lamprey, R. H., and F. Mitchelmore. 1996. *Survey of Wildlife in Protected Areas of Uganda.* Kampala: Ugandan Wildlife Authority.

Lamprey, R. H., and R. S. Reid. 2004. Expansion of human settlement in Kenya's Maasai Mara: What future for pastoralism and wildlife? *Journal of Biogeography* 31:997–1032.

Lamprey, R., and R. Waller. 1990. The Loita-Mara region in historical times: Patterns of subsistence, settlement, and ecological change. In *Early Pastoralists of South-western Kenya,* ed. P. Robertshaw, 16–35. Nairobi: British Institute in Eastern Africa.

Landsberg, J., C. D. James, S. R. Morton et al. 2003. Abundance and composition of plant species along grazing gradients in Australian rangelands. *Journal of Applied Ecology* 40:1008–1024.

Lane, C. 1996a. Ngorongoro voices: Indigenous Maasai residents of the Ngorongoro Conservation Area in Tanzania give their views on the proposed general management plan. Report prepared for the FAO, Rome.

———. 1996b. *Pasture Lost: Barabaig Economy, Resource Tenure, and the Alienation of their Land in Tanzania.* Nairobi: Initiatives Publishers.

Laris, P. 2002. Burning the seasonal mosaic: Preventative burning strategies in the wooded savanna of southern Mali. *Human Ecology* 30:155–186.

Laurenroth, W. K., D. G. Milchunas, J. L. Dodd et al. 1999. Effect of grazing on ecosystems of the Great Plains. In *Ecological Implications of Livestock Herbivory in the West,* ed. M. Vavra, W. A. Laycock, and R. D. Pieper, 69–100. Denver: Society of Range Management.

Laws, R. M. 1970. Elephants as agents of habitat and landscape change in East Africa. *Oikos* 21:1–15.

Laws, R. M., I. S. C. Parker, and R. C. B. Johnstone. 1975. *Elephants and Their Habitats: The Ecology of Elephants in North Bunyoro, Uganda.* Oxford: Clarendon Press.

Leach, M. K., and T. J. Givnish. 1996. Ecological determinants of species loss in remnant prairies. *Science* 273:1555–1558.

Leader-Williams, N. 2000. The effects of a century of policy and legal change upon wildlife conservation and utilisation in Tanzania. In *Conservation of Wildlife by*

Sustainable Use, ed. H. H. T. Prins, J. G. Grootenhuis, and T. T. Dolan, 219–245. Boston: Kluwer Academic Publishers.

Leader-Williams, N., J. A. Kayera, and G. L. Overton. 1996. *Tourist Hunting in Tanzania*. Gland, Switz.: IUCN

Leakey, M. D., and J. M. Harris. 1987. *Laetoli: A Pliocene Site in Northern Tanzania*. Oxford: Clarendon Press.

Leakey, M. D., and R. L. Hay. 1979. Pliocene footprints in the Laetolil beds at Laetoli, northern Tanzania. *Nature* 278:317–323.

Leakey, M. G., C. S. Feibel, I. McDougall, and A. Walker. 1995. New 4-million-year-old hominid species from Kanapoi and Allia Bay, Kenya. *Nature* 376:565–571.

Leakey, M. G., F. Spoor, L. N. Leakey, and F. H. Brown. 2001. New hominin discoveries from the Nachukui Formation, west of Lake Turkana. *American Journal of Physical Anthropology, Supplement* 32:96.

Lee, P. C., and M. D. Hauser. 1998. Long-term consequences of changes in territory quality on feeding and reproductive strategies of vervet monkeys. *Journal of Animal Ecology* 67:347–358.

Lee, R. B., and I. DeVore. 1976. *Kalahari Hunter-Gatherers*. Cambridge, Mass.: Harvard University Press.

Legesse, D., C. Vallet-Coulomb, and F. Gasse. 2003. Hydrological response of a catchment to climate and land use changes in tropical Africa: Case study south-central Ethiopia. *Journal of Hydrology* 275:67–85.

Leneman, J. M., and R. S. Reid. 2001. Pastoralism beyond the past. *Development* 44:85–89.

Leonard, W. R., and M. L. Robertson. 1997. Comparative primate energetics and hominid evolution. *American Journal of Physical Anthropology* 102:265–281.

Leuthold, W., and B. M. Leuthold. 1976. Density and biomass of ungulates in Tsavo East National Park, Kenya. *East African Wildlife Journal* 14:49–58.

Lev-Yadun, S., A. Gopher, and S. Abbo. 2000. Archaeology: The cradle of agriculture. *Science* 288:1602–1603.

Levy, M., S. Babu, K. Hamilton et al. 2005. Ecosystem change and human well-being. Chapter 5 of *Millennium Ecosystem Assessment: Conditions and Trends*, vol. 1. Washington, D.C.: Island Press.

Lewis, I. M. 1975. The dynamics of nomadism: Prospects for sedentarization and social change. In *Pastoralism in Tropical Africa*, ed. T. Monod, 426–442. Oxford: Oxford University Press.

Li, Y. P., W. Ye, M. Wang, and X. D. Yan. 2009. Climate change and drought: A risk assessment of crop-yield impacts. *Climate Research* 39:31–46.

Lindsay, W. K. 1987. Integrating parks and pastoralists: Some lessons from Amboseli. In *Conservation in Africa: People, Policies, and Practice*, ed. D. Anderson and R. Grove, 149–167. Cambridge: Cambridge University Press.

Lissu, T. 2000. Policy and legal issues on wildlife management in Tanzania's pastoral lands: The case study of the Ngorongoro Conservation Area. *Law, Social Justice, and Global Development*, online at www2.warwick.ac.uk/fac/soc/law/elj/lgd/2000_1/lissu.

Little, P. D. 1992. *The Elusive Granary: Herder, Farmer, and State in Northern Kenya.* Cambridge: Cambridge University Press.

———. 2003. *Somalia: Economy without a State.* London: International African Institute, in association with Oxford: James Currey; Bloomington: Indiana University Press; Hargeisa: Btec Books.

Little, P. D., and D. W. Brokensha. 1987. Local institutions, tenure, and resource management in East Africa. In *Conservation in Africa: People, Policies, and Practice*, ed. D. Anderson and R. Grove, 193–209. Cambridge: Cambridge University Press.

Little, P. D., K. Smith, B. A. Cellarius et al. 2001. Avoiding disaster: Diversification and risk management among East African herders. *Development and Change* 32:401–433.

Loibooki, M., H. Hofer, K. L. I. Campbell, and M. L. East. 2002. Bushmeat hunting by communities adjacent to the Serengeti National Park, Tanzania: The importance of livestock ownership and alternative sources of protein and income. *Environmental Conservation* 29:391–398.

Loreau, M., S. Naeem, P. Inchausti et al. 2001. Biodiversity and ecosystem functioning: Current knowledge and challenges. *Science* 294:806–808.

Lovatt Smith, D. 1997. *Amboseli: Nothing Short of a Miracle.* Nairobi: Kenway Publications.

Lovejoy, C. O., G. Suwa, L. Spurlock et al. 2009. The pelvis and femur of Ardipithecus ramidus: The emergence of upright walking. *Science* 326:71e1–71e6.

Lovejoy, P. E. 2000. *Transformations in Slavery: A History of Slavery in Africa.* Cambridge: Cambridge University Press.

Lubchenco, J. 1978. Plant species diversity in a marine intertidal community: Importance of herbivore food preference and algal competitive abilities. *American Naturalist* 112:23–39.

Lugard, F. D. 1893. *The Rise of Our East African Empire.* Edinburgh: Blackwood.

Lynn, S. J. 2000. Conservation policy and local ecology: Effects on Maasai land use patterns and human welfare in northern Tanzania. M.Sc. thesis, Colorado State University, Fort Collins.

Mabbutt, J. A. 1984. A new global assessment of the status and trends of desertification. *Environmental Conservation* 11:100–113.

Mace, G., H. Masundire, J. Baillie et al. 2005. Biodiversity. In *Millennium Ecosystem Assessment: Ecosystems and Human Well-Being: Synthesis Report*, 77–122. Millennium Ecosystem Assessment. Washington, D.C.: Island Press.

Mack, R. N., and J. N. Thompson. 1982. Evolution in steppe with few large, hooved mammals. *American Naturalist* 119:157–173.

MacKenzie, J. M. 1987. Chivalry, social Darwinism, and ritualised killing: The hunting ethos in Central Africa up to 1914. In *Conservation in Africa: People, Policies, and Practice*, ed. D. M. Anderson and R. Grove, 41–61. Cambridge: Cambridge University Press.

Mackey, R. L., and D. J. Currie. 2001. The diversity-disturbance relationship: Is it generally strong and peaked? *Ecology* 82:3479–3492.

Maclean, J. E., J. R. Goheen, D. F. Doak et al. 2011. Cryptic herbivores mediate the strength and form of ungulate impacts on a long-lived savanna tree. *Ecology* 92:1626–1636.

Maclennan, S. D., R. J. Groom, D. W. Macdonald, and L. G. Frank. 2009. Evaluation of a compensation scheme to bring about pastoralist tolerance of lions. *Biological Conservation* 142:2419–2427.

Maddock, L. 1979. The "migration" and grazing succession. In *Serengeti: Dynamics of an Ecosystem*, ed. A. R. E. Sinclair and M. Norton-Griffiths, 104–129. Chicago: University of Chicago Press.

Maddox, G. H. 2006. *Sub-Saharan Africa: An Environmental History*. Santa Barbara, Ca.: ABC-CLIO.

Maddox, T. M. 2003. The ecology of cheetahs and other large carnivores in a pastoralist-dominated buffer zone. Ph.D. diss., University College London.

Maglio, V. J. 1978. Patterns of faunal evolution. In *Evolution of African Mammals*, ed. V. J. Maglio and H. B. S. Cooke, 603–619. Cambridge, Mass.: Harvard University Press.

Maglio, V. J., and H. B. S. Cooke, eds. 1978. *Evolution of African Mammals*. Cambridge, Mass.: Harvard University Press.

Malimbwi, R. E., A. Persson, S. Iddi et al. 1992. Effects of spacing on yield and some wood properties of *Pinus patula* at Rongai, northern Tanzania. *Forest Ecology and Management* 53:297–306.

Mango, L. M., A. M. Melesse, M. E. McClain et al. 2011. Land use and climate change impacts on the hydrology of the upper Mara River Basin, Kenya: Results of a modeling study to support better resource management. *Hydrology and Earth System Sciences* 15:2245–2258.

Mapinduzi, A.L., G. Oba, R.B. Weladji, and J.E. Colman. 2003. Use of indigenous ecological knowledge of the Maasai pastoralists for assessing rangeland biodiversity in Tanzania. *African Journal of Ecology* 41:329–336.

Mara-Conservancy. 2011. Monthly report: The Mara Triangle. On-line at http://maratriangle.org.

Marean, C.W. 1997. Hunter-gatherer foraging strategies in tropical grasslands: Model building and testing in the east African Middle and Later Stone Age. *Journal of Anthropological Archaeology* 16:189–225.

Marino, J., C. Sillero-Zubiri, and D.W. Macdonald. 2006. Trends, dynamics, and resilience of an Ethiopian wolf population. *Animal Conservation* 9:49–58.

Marlowe, F.W. 2010. *The Hadza: Hunter-Gatherers of Tanzania.* Berkeley: University of California Press.

Marshall, F. 1990a. Cattle herds and caprine flocks. In *Early Pastoralists of Southwestern Kenya*, ed. P. Robertshaw, 205–260. Nairobi: British Institute of Eastern Africa.

———. 1990b. Origins of specialised pastoralism in East Africa. *American Anthropologist* 92:873–894.

———. 2007. African pastoral perspectives on domestication of the donkey: A first synthesis. In *Rethinking Agriculture: Archaeological and Ethnoarchaeological Perspectives*, ed. T. Denham, J. Iriarte, and L. Vrydaghs, 538–597. Walnut Creek, Ca.: Left Coast Press.

Marshall, F., and E. Hildebrand. 2002. Cattle before crops: The beginnings of food production in Africa. *Journal of World Prehistory* 16:99–143.

Martin, P.S., and R.G. Klein. 1984. *Quaternary Extinctions.* Tucson: University of Arizona Press.

Mati, B.M., S. Mutie, H. Gadain et al. 2008. Impacts of land-use/cover changes on the hydrology of the transboundary Mara River, Kenya/Tanzania. *Lakes and Reservoirs: Research and Management* 13:169–177.

Mburu, J., R. Birner, and M. Zeller. 2003. Relative importance and determinants of landowners' transaction costs in collaborative wildlife management in Kenya: An empirical analysis. *Ecological Economics* 45:59–73.

McBrearty, S., and A.S. Brooks. 2000. The revolution that wasn't: A new interpretation of the origin of modern human behavior. *Journal of Human Evolution* 39:453–563.

McCabe, J.T. 1984. Livestock management among the Turkana: A social and ecological analysis of herding in an east African population. Ph.D. diss., State University of New York, Binghamton.

———. 1987. Drought and recovery: Livestock dynamics among the Ngisonyoka Turkana of Kenya. *Human Ecology* 15:371–389.

————. 1990. Turkana pastoralism: A case against the tragedy of the commons. *Human Ecology* 18:81–103.

————. 1994. Mobility and land use among African pastoralists: Old conceptual problems and new interpretations. In *African Pastoralist Systems: An Integrated Approach*, ed. E. Fratkin, K.A. Galvin, and E.A. Roth, 69–90. Boulder, Colo.: Lynne Rienner Publishers.

————. 1997. Risk and uncertainty among the Maasai of the Ngorongoro Conservation Area in Tanzania: A case study in economic change. *Nomadic Peoples* 1:54–65.

————. 2002. Giving conservation a human face? Lessons from forty years of combining conservation and development in the Ngorongoro Conservation Area, Tanzania. In *Conservation and Mobile Indigenous Peoples: Displacement, Forced Settlement, and Sustainable Development*, ed. D. Chatty and M. Colchester, 61–67. New York: Berghahn Books.

————. 2003. Sustainability and livelihood diversification among the Maasai of northern Tanzania. *Human Organisation* 62:100–111.

————. 2004. *Cattle Bring Us to Our Enemies*. Ann Arbor: University of Michigan Press.

McCabe, J.T., R. Dyson-Hudson, and J. Wienpahl. 1999. Nomadic movements. In *Turkana Herders of the Dry Savanna*, ed. M.A. Little and P.W. Leslie, 108–121. Oxford: Oxford University Press.

McCabe, J.T., P.W. Leslie, and L. DeLuca. 2010. Adopting cultivation to remain pastoralists: The diversification of Maasai livelihoods in northern Tanzania. *Human Ecology* 38:321–334.

McCabe, J.T., N. Molle, and A. Tumaini. 1997a. Food security and cultivation. In *Multiple Land-Use: The Experience of the Ngorongoro Conservation Area, Tanzania*, ed. D.M. Thompson, 397–416. Gland, Switz.: IUCN.

McCabe, J.T., S. Perkin, and C. Schofield. 1992. Can conservation and development be coupled among pastoral people? An examination of the Maasai of the Ngorongoro Conservation Area, Tanzania. *Human Organization* 51:353–366.

McCabe, J.T., E.C. Schofield, and G.N. Pedersen. 1997b. Food security and nutritional status. In *Multiple Land-Use: The Experience of the Ngorongoro Conservation Area, Tanzania*, ed. D.M. Thompson, 285–301. Gland, Switz.: IUCN.

McCann, J.C. 1999. *Green Land, Brown Land, Black Land: An Environmental History of Africa, 1800–1990*. Oxford: James Currey; Portsmouth, N.H.: Heinemann.

McCarthy, N., B. Swallow, M. Kirk, and P. Hazell, eds. 1999. *Property Rights, Risk, and Livestock Development in Africa*. Washington, D.C.: International Food Policy Research Institute; Nairobi: International Livestock Research Institute.

McCauley, D. J., F. Keesing, T. Young, and K. Dittmar. 2008. Effects of the removal of large herbivores on fleas of small mammals. *Journal of Vector Ecology* 33:263–268.

McCauley, D. J., F. Keesing, T. P. Young et al. 2006. Indirect effects of large herbivorous mammals on snakes in an African savanna. *Ecology* 87:2657–2663.

McDougall, I., F. H. Brown, and J. G. Fleagle. 2005. Stratigraphic placement and age of modern humans from Kibish, Ethiopia. *Nature* 433:733–736.

McDowell, R. 1985. *Nutrition of Grazing Ruminants in Warm Climates*. New York: Academic Press.

McNaughton, S. J. 1976. Serengeti migratory wildebeest: Facilitation of energy flow by grazing. *Science* 199:92–94.

———. 1984. Grazing lawns: Animals in herds, plant form, and coevolution. *American Naturalist* 124:863–886.

———. 1985. Ecology of a grazing ecosystem: The Serengeti. *Ecological Monographs* 55:259–294.

———. 1986. Grazing lawns: On domesticated and wild grazers. *American Naturalist* 128:937–939.

———. 1988. Mineral nutrition and spatial aggregation of African ungulates. *Nature* 334:343–345.

———. 1990. Mineral nutrition and seasonal movements of African migratory ungulates. *Nature* 345:613–615.

———. 1992. The propagation of disturbance in savannas through food webs. *Journal of Vegetation Science* 3:301–314.

———. 1994. Biodiversity and function in grazing ecosystems. In *Biodiversity and Ecosystem Function*, ed. E. D. Schulze and H. A. Mooney, 361–383. Berlin: Springer-Verlag.

McNaughton, S. J., and F. F. Banyikwa. 1995. Plant communities and herbivory. In *Serengeti: Dynamics, Management, and Conservation of an Ecosystem*, ed. A. R. E. Sinclair and P. Arcese, 49–70. Chicago: University of Chicago Press.

McNaughton, S. J., F. F. Banyikwa, and M. M. McNaughton. 1997. Promotion of the cycling of diet-enhancing nutrients by African grazers. *Science* 278:1798–1800.

McNaughton, S. J., and N. J. Georgiadis. 1986. Ecology of African grazing and browsing mammals. *Annual Review of Ecology and Systematics* 17:39–65.

McNaughton, S. J., J. L. Tarrants, M. M. McNaughton, and R. H. Davis. 1985. Silica as a defence against herbivory and a growth promoter in African grasses. *Ecology* 66:528–535.

McNaughton, S. J., L. L. Wallace, and M. B. Coughenour. 1983. Plant adaptation in an ecosystem context: Effects of defoliation, nitrogen, and water on growth of an African C_4 sedge. *Ecology* 64:307–318.

McPeak, J., and P. D. Little. 2005. Cursed if you do, cursed if you don't: The contradictory processes of sedentarization in northern Kenya. In *As Pastoralists Settle,* ed. E. Fratkin and E. A. Roth, 87–104. Dordrecht, Neth.: Kluwer Academic Publishers.

Mduma, S., and G. C. Hopcraft. 2008. The main herbivorous mammals and crocodiles in the Greater Serengeti Ecosystem. In *Serengeti III: Human Impacts on Ecosystem Dynamics,* ed. A. R. E. Sinclair, C. Packer, S. A. R. Mduma, and J. M. Fryxell, 497–505. Chicago: University of Chicago Press.

Mduma, S. A. R., A. R. E. Sinclair, and R. Hilborn. 1999. Food regulates the Serengeti wildebeest: A 40-year record. *Journal of Animal Ecology* 68:1101–1122.

MEA. 2005a. *Ecosystems and Human Well-Being: Current State and Trends.* Vol. 1. Millennium Ecosystem Assessment. Washington, D.C.: Island Press.

———. 2005b. *Ecosystems and Human Well-Being: Desertification Synthesis.* Millennium Ecosystem Assessment. Washington, D.C.: World Resources Institute.

Medina, E. 1987. Nutrients: Requirements, conservation, and cycles in the herbaceous layer. In *Determinants of Savannas,* ed. B. H. Walker, 39–65. Oxford: IRL Press.

Meffe, G. K., L. A. Nielsen, R. L. Knight, and D. A. Schenborn. 2002. *Ecosystem Management: Adaptive, Community-Based Conservation.* Washington, D.C.: Island Press.

Meijerink, A. M. J., and W. van Wijngaarden. 1997. Contribution to the groundwater hydrology of the Amboseli ecosystem, Kenya. In *Groundwater/Surface Water Ecotones: Biological and Hydrological Interactions and Management Options,* ed. J. Gibert, J. Mathieu, and F. Fournier, 111–118. Cambridge: Cambridge University Press.

Meinertzhagen, R. 1957. *Kenya Diary 1902–1906.* London: Oliver and Boyd.

Mentis, M. T., and N. M. Tainton. 1984. The effect of fire on forage production and quality. In *Ecological Effects of Fire on South African Ecosystems,* ed. P. de van Booysen and N. M. Tainton, 245–254. Berlin: Springer-Verlag.

Meyer, W. B., and P. S. Schiffman. 1999. Fire season and mulch reduction in a California grassland: A comparison of restoration strategies. *Madroño* 46:25–37.

Mikkelsen, T. S., L. W. Hillier, E. E. Eichler et al. 2005. Initial sequence of the chimpanzee genome and comparison with the human genome. *Nature* 437:69–87.

Milchunas, D. G., and W. K. Laurenroth. 1993. Quantitative effects of grazing on vegetation and soils over a global range of environments. *Ecological Monographs* 63:327–366.

Milchunas, D. G., W. K. Lauenroth, and I. C. Burke. 1998. Livestock grazing: Animal and plant biodiversity of shortgrass steppe and the relationship to ecosystem function. *Oikos* 83:65–74.

Milchunas, D. G., O. E. Sala, and W. K. Lauenroth. 1988. A generalized model of the effects of grazing by large herbivores on grassland community structure. *American Naturalist* 132:87–106.

Mills, A. J. 2006. The role of salinity and sodicity in the dieback of *Acacia xanthophloea* in Ngorongoro Caldera, Tanzania. *African Journal of Ecology* 44:61–71.

Milner-Gulland, E. J., and J. R. Beddington. 1993. The relative effects of hunting and habitat destruction on elephant population dynamics over time. *Pachyderm* 17:75–90.

Milton, S. J. 1994. Growth, flowering, and recruitment of shrubs in grazed and in protected rangeland in the arid Karoo, South Africa. *Vegetatio* 111:17–27.

———. 2004. Grasses as invasive alien plants in South Africa. *South African Journal of Science* 100:69–75.

Milton, S. J., W. R. J. Dean, M. A. Duplessis, and W. R. Siegfried. 1994. A conceptual model of arid rangeland degradation: The escalating cost of declining productivity. *BioScience* 44:70–76.

Mistry, J. 2000. *World Savannas*. London: Prentice-Hall.

Mizutani, F., E. N. Muthiani, P. Kristjanson, and H. Recke. 2005. Impact and value of wildlife in pastoral livestock production system in Kenya: Possibilities for healthy ecosystem conservation and livestock development for the poor. In *Conservation and Development Interventions at the Wildlife/Livestock Interface: Implications for Wildlife, Livestock, and Human Health,* ed. S. A. Osofsky. Occasional Paper of the IUCN Species Survival Commission, No. 30.

Mnaya, B., Y. Kiwango, E. Gereta, and E. Wolanski. Forthcoming. Serengeti and Lake Victoria ecohydrology. In *Serengeti IV*, ed. A. R. E. Sinclair and K. Metzger. Chicago: University of Chicago Press.

Modelski, G., and G. Perry. 2002. Democratization in long perspective. *Technological Forecasting and Social Change* 69:359–376.

Moehlman, P. D. 2005. Endangered wild equids. *Scientific American* 292:86–93.

Moehlman, P. D., V. A. Runyoro, and H. Hofer. 1997. Wildlife population trends in the Ngorongoro Crater. In *Multiple Land-Use: The Experience of Ngorongoro Conservation Area, Tanzania,* ed. D. M. Thompson, 59–80. Gland, Switz.: IUCN.

Mohamed, J. 1999. Epidemics and public health in early colonial Somaliland. *Social Science and Medicine* 48:507–521.

Mol, F. 1996. *Maasai Dictionary: Language and Culture.* Lemek, Kenya: Maasai Centre.

Moleele, N. M., and J. S. Perkins. 1998. Encroaching woody plant species and boreholes: Is cattle density the main driving factor in the Olifants Drift communal grazing lands, south-eastern Botswana? *Journal of Arid Environments* 40:245–253.

Mölg, T., and D. R. Hardy. 2004. Ablation and associated energy balance of a horizontal glacier surface on Kilimanjaro. *Journal of Geophysical Research* 109, D16104: doi:10.1029/2003JD004338.

Mölg, T., D. R. Hardy, N. J. Cullen, and G. Kaser. 2006. Tropical glaciers in the context of climate change and society: Focus on Kilimanjaro (East Africa). In *The Darkening Peaks: Glacial Retreat in Scientific and Social Context*, ed. B. Orlove, E. Wiegandt, and B. Luckman, 168–182. Berkeley: University of California Press.

Mölg, T., G. Kaser, and N. J. Cullen. 2010. Glacier loss on Kilimanjaro is an exceptional case. *Proceedings of the National Academy of Sciences* 107:E68.

Mooney, H. A., and R. J. Hobbs. 2000. *Invasive Species in a Changing World*. Washington, D.C.: Island Press.

Morris, D. L., D. Western, and D. Maitumo. 2008. Pastoralists' livestock and settlements influence game bird diversity and abundance in a savanna ecosystem of southern Kenya. *African Journal of Ecology* 47:48–55.

Mortimore, M., and B. Turner. 2005. Does the Sahelian smallholder's management of woodland, farm trees, rangeland support the hypothesis of human-induced desertification? *Journal of Arid Environments* 63:567–595.

Moss, C. 1988. *Elephant Memories*. New York: Fawcett Cowlitts.

———. 2001. The demography of an African elephant (*Loxodonta africana*) population in Amboseli, Kenya. *Journal of Zoology* 255:145–156.

———. 2010. Amboseli Trust for Elephants newsletter, August. http://elephanttrust.org.

Moss, C. J., H. Croze, and P. C. Lee, eds. 2011. *The Amboseli Elephants: A Long-term Perspective on a Long-lived Mammal*. Chicago: University of Chicago Press.

Muchapondwa, E., F. Carlsson, and G. Kohlin. 2008. Wildlife management in Zimbabwe: Evidence from a contingent valuation study. *South African Journal of Economics* 76:685–704.

Muchiru, A. N. 1992. The role of abandoned Maasai settlements in restructuring savanna herbivore communities, Amboseli, Kenya. M.Sc. thesis, University of Nairobi.

Muchiru, A. N., D. J. Western, and R. S. Reid. 2008. The role of abandoned pastoral settlements in the dynamics of African large herbivore communities. *Journal of Arid Environments* 72:940–952.

———. 2009. The impact of abandoned pastoral settlements on plant and nutrient succession in an African savanna ecosystem. *Journal of Arid Environments* 73:322–331.

Murdock, G. P. 1959. *Africa—Peoples and Their Culture, History*. New York: McGraw Hill.

Murphree, M. W. 2009. The strategic pillars of communal natural resource management: benefit, empowerment and conservation. *Biodiversity and Conservation* 18:2551–2562.

Murray, M. G. 1995. Specific nutrient requirements and migration of wildebeest. In *Serengeti II: Dynamics, Management, and Conservation of an Ecosystem*, ed. A. R. E. Sinclair and P. Arcese, 231–256. Chicago: University of Chicago Press.

Mushi, E. Z., F. R. Rurangirwa, and L. Karstad. 1981. Shedding of malignant catarrhal fever virus by wildebeest calves. *Veterinary Microbiology* 6:281–286.

Mwaikusa, J. T. 1993. Community rights and land use policies in Tanzania: The case of pastoral communities. *Journal of African Law* 37:144–163.

Mwangi, E. 2005. The transformation of property rights in Kenya's Maasailand: Triggers and motivations. CAPRi Working Paper 35, International Food Policy Research Institute, Washington, D.C.

———. 2007. The puzzle of group ranch subdivision in Kenya's Maasailand. *Development and Change* 38:889–910.

Mworia, J. K., W. N. Mnene, D. K. Musembi, and R. S. Reid. 1997. Resilience of soils and vegetation subjected to different grazing intensities in a semi-arid rangeland of Kenya. *African Journal of Range and Forage Science* 14:25–30.

Naughton-Treves, L. 1998. Predicting patterns of crop damage by wildlife around Kibale National Park, Uganda. *Conservation Biology* 12:156–168.

Naughton-Treves, L., and A. Treves. 2005. Socio-ecological factors shaping local support for wildlife: Crop-raiding by elephants and other wildlife in Africa. In *People and Wildlife: Conflict or Coexistence?*, ed. R. Woodroffe, S. Thirgood, and A. Rabinowitz, 252–277. Cambridge: Cambridge University Press.

NCAA. 1996. Summary, Ngorongoro Conservation Area, General Management Plan. Ngorongoro Conservation Area Authority, Ngorongoro, Tanzania.

———. 1999. 1998 aerial *boma* count, 1999 people and livestock census, and human population trend between 1954 and 1999 in the NCA. Ngorongoro Conservation Area Authority, Ngorongoro Crater, Tanzania.

———. 2006. Ngorongoro Conservation Area General Management Plan. Ngorongoro Conservation Area Authority, Arusha, Tanzania.

Neff, J. C., R. L. Reynolds, J. Belnap, and P. Lamothe. 2005. Multi-decadal impacts of grazing on soil physical and biogeochemical properties in southeast Utah. *Ecological Applications* 15:87–95.

Nelson, F. 2010. *Community Rights, Conservation, and Contested Land*. London: Earthscan.

Nelson, F., and A. Agrawal. 2008. Patronage or participation? Community-based natural resource management reform in sub-Saharan Africa. *Development and Change* 39:557–585.

Nelson, F., B. Gardner, J. Igoe, and A. Williams. 2009. Community-based conservation and Maasai livelihoods in Tanzania. In *Staying Maasai? Livelihoods, Conservation, and Development in East African Rangelands*, ed. K. Homewood, P. C. Trench, and P. Kristjanson, 299–333. London: Springer.

Neumann, R. 1995. Ways of seeing Africa: Colonial recasting of African society and landscape in Serengeti National Park. *Ecumene* 2:149–169.

———. 1997. Primitive ideas: Protected area buffer zones and the politics of land in Africa. *Development and Change* 28:559–582.

———. 2002. *Imposing Wilderness: Struggles over Livelihood and Nature Preservation in Africa*. Berkeley: University of California Press.

Newman, J. L. 1995. *The Peopling of Africa*. New Haven, Ct.: Yale University Press.

Newmark, W. D. 1993. The role and design of wildlife corridors with examples from Tanzania. *Ambio* 22:500–504.

———. 1996. Insularization of Tanzanian parks and the local extinction of large mammals. *Conservation Biology* 10:1549–1556.

Newmark, W. D., N. L. Leonard, H. I. Sariko, and D.-G. M. Gamassa. 1993. Conservation attitudes of local people living adjacent to five protected areas in Tanzania. *Biological Conservation* 63:177–183.

Newmark, W. D., D. N. Manyanza, D. G. M. Gamassa, and H. I. Sariko. 1994. The conflict between wildlife and local people living adjacent to protected areas in Tanzania: Human density as a predictor. *Conservation Biology* 8:249–255.

Niamir-Fuller, M. 1999a. Conflict management and mobility among pastoralists in Karamoja, Uganda. In *Managing Mobility in African Rangelands*, ed. M. Niamir-Fuller, 149–183. London: FAO and Beijer Institute of Ecological Economics.

———. 1999b. Introduction. In *Managing Mobility in African Rangelands*, ed. M. Niamir-Fuller, 1–9. London: FAO and Beijer Institute of Ecological Economics.

———. 1999c. Managing mobility in African rangelands. In *Property Rights, Risk, and Livestock Development in Africa*, ed. N. McCarthy, B. Swallow, M. Kirk, and P. Hazell, 102–131. Washington, D.C.: International Food Policy Research Institute; Nairobi: International Livestock Research Institute.

Nicholson, S. E., C. J. Tucker, and M. B. Ba. 1998. Desertification, drought, and surface vegetation: An example from the West African Sahel. *Bulletin of the American Meteorological Society* 79:815–830.

Nicholson, S. E., and X. G. Yin. 2001. Rainfall conditions in equatorial East Africa during the nineteenth century as inferred from the record of Lake Victoria. *Climatic Change* 48:387–398.

Njoka, T. J. 1979. Ecological and socio-cultural trends of Kaputiei group ranches in Kenya. Ph.D. diss., University of California, Berkeley.

Nkako, F. M., C. Lambrechts, M. Gachanja, and B. Woodley. 2005. Maasai Mau Forest Status Report 2005. Report prepared for the Ewaso Njiro South Development Authority, Narok, Kenya.

Nkedianye, D. 2004. Testing the attitudinal impact of a conservation tool outside a protected area: The case for the Kitengela Wildlife Conservation Lease Programme for Nairobi National Park. M.A. thesis, University of Nairobi.

Nkedianye, D., J. de Leeuw, J. O. Ogutu et al. 2011. Mobility and livestock mortality in communally used pastoral areas: The impact of the 2005–2006 drought on livestock mortality in Maasailand. *Pastoralism, Research, Policy, and Practice* 1:1–17.

Nkedianye, D., M. Radeny, P. Kristjanson, and M. Herrero. 2009. Assessing returns to land and changing livelihood strategies in Kitengela. In *Staying Maasai? Livelihoods, Conservation, and Development in East African Rangelands*, ed. K. Homewood, P. Trench, and P. Kristjanson, 115–150. London: Springer.

Nkwame, M. 2009. Wild dogs to be taken to Serengeti. *Tanzania Daily News*, www.dailynews.co.tz/home/?n=5861&cat=home, accessed 24 Dec 2009.

Noe, C. 2003. The dynamics of land-use changes in the Kitendeni wildlife corridor and their impacts on biodiversity. LUCID Working Paper No. 31, University of Dar es Salaam and International Livestock Research Institute, Nairobi.

Norton-Griffiths, M. 1979. The influence of grazing, browsing, and fire on the vegetation dynamics of the Serengeti. In *Serengeti: Dynamics of an Ecosystem*, ed. A. R. E. Sinclair and M. Norton-Griffiths, 310–352. Chicago: University of Chicago Press.

———. 1998. The economics of wildlife conservation policy in Kenya. In *Conservation of Biological Resources*, ed. E. J. Milner-Gulland and R. Mace, 279–293. Chichester, U.K.: Blackwell.

Norton-Griffiths, M., and B. Butt. 2006. Land use economics, Loitokitok Division, Kajiado District, Kenya. Paper prepared for the International Livestock Research Institute, Nairobi, Kenya.

Norton-Griffiths, M., D. Herlocker, and L. Pennycuick. 1975. The patterns of rainfall in the Serengeti ecosystem, Tanzania. *East African Wildlife Journal* 13:347–374.

Norton-Griffiths, M., M. Y. Said, S. Serneels et al. 2008. Land use economics in the Mara area of the Serengeti ecosystem. In *Serengeti III: Human Impacts on Ecosystem Dynamics*, ed. A. R. E. Sinclair, C. Packer, S. A. R. Mduma, and J. M. Fryxell, 379–416. Chicago: University of Chicago Press.

Norton-Griffiths, M., and C. Southey. 1995. The opportunity costs of biodiversity conservation in Kenya. *Ecological Economics* 12:125–139.

Ntiati, P. 2002. Group ranches subdivision study in Loitokitok Division of Kajiado District. LUCID Working Paper No. 7, International Livestock Research Institute, Nairobi.

Nyahongo, J. W., T. Holmern, B. P. Kaltenborn, and E. Roskaft. 2009. Spatial and temporal variation in meat and fish consumption among people in the western Serengeti, Tanzania: The importance of migratory herbivores. *Oryx* 43:258–266.

O'Connor, T. G. 1991. Local extinction in perennial grassland: A life history approach. *American Naturalist* 137:753–773.

———. 1995. Transformation of a savanna grassland by drought and grazing. *African Journal of Range and Forage Science* 12:53–60.

O'Flaherty, R. M. 2003. The tragedy of property: Ecology and land tenure in southeastern Zimbabwe. *Human Organization* 62:178–190.

O'Rourke, J. T., P. J. Terry, and G. W. Frame. 1976. Experimental results of *Eleusine jaegeri* Pilg. control in east African highlands. *East Africa Agriculture and Foresty Journal* 3:253–265.

Oba, G. 1994. Responses of *Indigofera spinosa* to simulated herbivory in a semidesert of North-west Kenya. *Acta Oecologica* 15:105–117.

———. 2011. Colonial resource capture: triggers of ethnic conflicts in the Northern Frontier District of Kenya, 1903–1930s. *Journal of Eastern African Studies* 5:505–534.

Oba, G., and L. M. Kaitira. 2006. Herder knowledge of landscape assessments in arid rangelands in northern Tanzania. *Journal of Arid Environments* 66:168–186.

Oba, G., and D. G. Kotile. 2001. Assessments of landscape level degradation in southern Ethiopia: Pastoralists versus ecologists. *Land Degradation and Development* 12:461–475.

Oba, G., E. Post, P. O. Syvertsen, and N. C. Stenseth. 2000a. Bush cover and range condition assessments in relation to landscape and grazing in southern Ethiopia. *Landscape Ecology* 15:535–546.

Oba, G., N. C. Stenseth, and W. J. Lusigi. 2000b. New perspectives on sustainable grazing management in arid zones of sub-Saharan Africa. *BioScience* 50:35–51.

Oba, G., O. R. Vetaas, and N. C. Stenseth. 2001. Relationships between biomass and plant species richness in arid-zone grazing lands. *Journal of Applied Ecology* 38:836–845.

Odadi, W. O., M. K. Karachi, S. A. Abdulrazak, and T. P. Young. 2011. African wild ungulates compete with or facilitate cattle depending on season. *Science* 333:1753–1755.

Odhiambo, M. O. 2002. An appraisal of the institutional framework for sustaining the development interventions of ERETO. Paper prepared for the Ereto Project, Arusha.

Ogada, D. L., M. E. Gadd, R. S. Ostfeld et al. 2008. Impacts of large herbivorous mammals on bird diversity and abundance in an African savanna. *Oecologia* 156:387–397.

Ogada, M. O., R. Woodroffe, N. O. Oguge, and L. G. Frank. 2003. Limiting depredation by African carnivores: The role of livestock husbandry. *Conservation Biology* 17:1521–1530.

Ogutu, J. O., N. Bhola, and R. Reid. 2005. The effects of pastoralism and protection on the density and distribution of carnivores and their prey in the Mara ecosystem of Kenya. *Journal of Zoology* 265:281–293.

Ogutu, J. O., and H. T. Dublin. 1998. The response of lions and spotted hyaenas to sound playbacks as a technique for estimating population size. *African Journal of Ecology* 36:83–95.

Ogutu, J. O., N. Owen-Smith, H.-P. Piepho, and M. Y. Said. 2011. Continuing wildlife population declines and range contraction in the Mara region of Kenya during 1977–2009. *Journal of Zoology* 285:99–109.

Ogutu, J. O., N. Owen-Smith, H.-P. Piepho et al. Submitted. Changing wildlife populations in Nairobi National Park and adjoining Athi-Kaputiei Plains: Collapse of the migratory wildebeest. *Conservation Biology*.

Ogutu, J. O., H.-P. Piepho, H. T. Dublin et al. 2007. El Niño–Southern Oscillation, rainfall, temperature, and normalized difference vegetation index fluctuations in the Mara-Serengeti ecosystem. *African Journal of Ecology* 46:132–143.

———. 2009. Dynamics of Mara-Serengeti ungulates in relation to land use changes. *Journal of Zoology* 278:1–14.

Ogutu, J. . O., H. P. Piepho, R. S. Reid et al. 2010. Large herbivore responses to water and settlements in savannas. *Ecological Monographs* 80:241–266.

Ogutu, Z. A. 2002. The impact of ecotourism on livelihood and natural resource management in Eselenkei, Amboseli ecosystem, Kenya. *Land Degradation and Development* 13:251–256.

Okello, M. M. 2005a. An assessment of the large mammal component of the proposed wildlife sanctuary site in Maasai Kuku Group Ranch near Amboseli, Kenya. *South African Journal of Wildlife Research* 35:63–76.

———. 2005b. Land use changes and human-wildlife conflicts in the Amboseli area, Kenya. *Human Dimensions of Wildlife* 10:19–28.

Okwi, P. O., G. Ndeng'e, P. Kristjanson et al. 2007. Spatial determinants of poverty in rural Kenya. *Proceedings of the National Academy of Sciences* 104:16769–16774.

Ole-Miaron, J. O. 2003. The Maasai ethnodiagnostic skill of livestock diseases: A lead to traditional bioprospecting. *Journal of Ethnopharmacology* 84:79–83.

Olff, H., and M. E. Ritchie. 1998. Effects of herbivores on grassland plant diversity. *Trends in Ecology and Evolution* 13:261–215.

Olson, J. M., G. Alagarswamy, J. A. Andresen et al. 2008. Integrating diverse methods to understand climate-land interactions in East Africa. *Geoforum* 39:898–911.

Olsson, L., L. Eklundh, and J. Ardo. 2005. A recent greening of the Sahel: Trends, patterns, and potential causes. *Journal of Arid Environments* 63:556–566.

Ostrom, E. 1990. *Governing the Commons: The Evolution of Institutions for Collective Action.* Cambridge: Cambridge University Press.

———. 2003. How types of goods and property rights jointly affect collective action. *Journal of Theoretical Politics* 15:239–270.

———. 2009. A general framework for analyzing sustainability of social-ecological systems. *Science* 325:419–422.

Ottichilo, W. K., J. de Leeuw, and H. H. T. Prins. 2001. Population trends of resident wildebeest [*Connochaetes taurinus hecki* (Neumann)] and factors influencing them in the Masai Mara ecosystem, Kenya. *Biological Conservation* 97:271–282.

Ottichilo, W. K., J. de Leeuw, A. K. Skidmore et al. 2000a. Population trends of large non-migratory wild herbivores and livestock in the Masai Mara ecosystem, Kenya, between 1977 and 1997. *African Journal of Ecology* 38:202–216.

Ottichilo, W. K., J. Grunblatt, M. Y. Said, and P. W. Wargute. 2000b. Wildlife and livestock population trends in the Kenya rangeland. In *Wildlife Conservation by Sustainable Use,* ed. H. H. T. Prins, J. G. Grootenhuis, and T. T. Dolan, 203–218. Dordrecht, Neth.: Kluwer Academic Publishers.

Owen, D. F., and R. G. Wiegert. 1981. Mutualism between grasses and grazers: An evolutionary hypothesis. *Oikos* 36:376–378.

Owen-Smith, N. 1987. Pleistocene extinctions: The pivotal role of megaherbivores. *Paleobiology* 13:351–362.

———. 1988. *Megaherbivores: The Influence of Large Body Size on Ecology.* Cambridge: Cambridge University Press.

Owen-Smith, N., and P. Novellie. 1982. What should a clever ungulate eat? *American Naturalist* 119:151–178.

Packer, C., S. Altizer, M. Appel et al. 1999. Viruses of the Serengeti: Patterns of infection and mortality in African lions. *Journal of Animal Ecology* 68:1161–1178.

Pangle, W. M., and K. E. Holekamp. 2010. Lethal and nonlethal anthropogenic effects on spotted hyenas in the Masai Mara National Reserve. *Journal of Mammalogy* 91:154–164.

Pankhurst, R. 1965. The history and traditional treatment of smallpox in Ethiopia. *Medical History* 9:343–355.

Pankhurst, R., and D. H. Johnson. 1988. The great drought and famine of 1888–92 in northeast Africa. In *The Ecology of Survival: Case Studies from Northeast African History,* ed. D. H. Johnson and D. M. Anderson, 47–70. London: Lester Crook Academic Publishing; Boulder, Colo.: Westview Press.

Parker, I. S. C., and A. D. Graham. 1989a. Elephant decline: Downward trends in African elephant distribution and numbers—Parts I and II. *International Journal of Environmental Studies* 34:13–26.

———. 1989b. Men, elephants, and competition. *Proceedings of the Zoological Society of London* 61:241–252.

Parkipuny, M. S. 1991a. Pastoralism, conservation, and development in the Greater Serengeti Region. Paper No. 26, Drylands Network Programme, International Institute for Environment and Development, London.

———. 1991b. The Maasai of east Africa: A people under the stranglehold of preservation. In *A Stanford Centennial Symposium*. Palo Alto: Stanford Alumni Association.

———. 1997. Pastoralism, conservation and development in the Greater Serengeti Region. In *Multiple Land-Use: The Experience of the Ngorongoro Conservation Area, Tanzania*, ed. D. M. Thompson, 143–168. Gland, Switz.: IUCN.

Parks, S. A., and A. H. Harcourt. 2000. Reserve size, local human density, and mammalian extinctions in U.S. protected areas. *Conservation Biology* 16:800–808.

Parsons, D. A. B., C. M. Shackleton, and R. J. Scholes. 1997. Changes in herbaceous layer condition under contrasting land use systems in the semi-arid lowveld, South Africa. *Journal of Arid Environments* 37:319–329.

Parsons, D. J., and T. Stohlgren. 1989. Effects of varying fire regimes on annual grasslands in the southern Sierra Nevada of California. *Madroño* 36:154–168.

Pasha, I. K. O. 1986. Evolution of individuation of group ranches in Maasailand. In *Range Development and Research in Kenya*, 303–317. Morrilton, Ark.: Winrock International Institute for Agricultural Development; Njoro, Kenya: Egerton College.

Perevolotsky, A., and N. G. Seligman. 1998. Role of grazing in Mediterranean rangeland ecosystems. *BioScience* 48:1007–1017.

Perkin, S. L. 1997. The Ngorongoro Conservation Area: Values, history and land-use conflicts. In *Multiple Land-Use: The Experience of Ngorongoro Conservation Area, Tanzania*, ed. D. M. Thompson, 19–32. Gland, Switz.: IUCN.

Perkins, J. S. 1996. Botswana: Fencing out the equity issue—Cattleposts and cattle ranching in the Kalahari Desert. *Journal of Arid Environments* 33:503–517.

Perkins, J. S., and D. S. G. Thomas. 1993. Spreading deserts or spatially confined environmental impacts? Land degradation and cattle ranching in the Kalahari Desert of Botswana. *Land Degradation and Rehabilitation* 4:179–194.

Peters, C. R., R. J. Blumenschine, R. L. Hay et al. 2008. Paleoecology of the Serengeti-Mara Ecosystem. In *Serengeti III: Human Impacts on Ecosystem Dynamics*, ed. A. R. E. Sinclair, C. Packer, S. A. R. Mduma, and J. M. Fryxell, 47–94. Chicago: University of Chicago Press.

Peterson, A. L. 2001. *Being Human: Ethics, Environment, and Our Place in the World.* Berkeley: University of California Press.

Peterson, D. 2003. *Eating Apes.* Berkeley: University of California Press.

Pickford, M., and B. Senut. 2001. The geological and faunal context of Late Miocene hominid remains from Lukeino, Kenya. *Comptes rendus de l'Académie des Sciences, série II, fascicule A—Sciences de la terre et des planètes* 332:145–152.

Pickford, M., B. Senut, D. Gommery, and J. Treil. 2002. Bipedalism in *Orrorin tugenensis* revealed by its femora. *Comptes rendus Palevol* 1:191–203.

Pimm, S. L., G. J. Russell, J. L. Gittleman, and T. M. Brooks. 1995. The future of biodiversity. *Science* 209:347–350.

Plowright, W. 1982. The effects of rinderpest and rinderpest control on wildlife in Africa. *Symposium of the Zoological Society of London* 50:1–28.

Plowright, W., R. D. Ferris, and G. R. Scott. 1960. Blue wildebeest and the aetiological agent of bovine malignant catarrhal fever. *Nature* 188:1167–1169.

Plummer, T. 2004. Flaked stones and old bones: Biological and cultural evolution at the dawn of technology. *Yearbook of Physical Anthropology* 47:118–164.

Polasky, S., J. Schmitt, C. Costello, and L. Tajibaeva. 2008. Larger-scale influences on the Serengeti Ecosystem: National and international policy, economics, and human demography. In *Serengeti IV: Human Impacts on Ecosystem Dynamics*, ed. A. R. E. Sinclair, C. Packer, S. A. R. Mduma, and J. Fryxell, 347–377. Chicago: University of Chicago Press.

Poole, J. H., and C. J. Moss. 1981. Musth in the African elephant, *Loxodonta africana*. *Nature* 202:830–831.

Potkanski, T. 1997. Pastoral economy, property rights, and traditional mutual assistance mechanisms among the Ngorongoro and Salei Maasai of Tanzania. Paper prepared for the International Institute for Environment and Development (IIED) Drylands Programme, London.

Potts, R. 1996. *Humanity's Descent: The Consequences of Ecological Instability.* New York: William Morrow.

Potts, R., and A. K. Behrensmeyer. 1993. Late Cenozoic terrestrial ecosystems. In *Terrestrial Ecosystems through Time*, ed. A. K. Behrensmeyer, J. D. Damuth, W. A. DiMichele et al., 419–541. Chicago: University of Chicago Press.

Prasad, V., C. A. E. Stromberg, H. Alimohammadian, and A. Sahni. 2005. Dinosaur coprolites and the early evolution of grasses and grazers. *Science* 310:1177–1180.

Pratt, D. J., and M. D. Gwynne. 1977. *Rangeland Management and Ecology in East Africa.* Huntington, N.Y.: R. E. Krieger.

Prendini, L., L. Theron, K. van der Merwe, and N. Owen-Smith. 1996. Abundance and guild structure of grasshoppers (Orthoptera: Acridiodea) in communally grazed and protected savanna. *South African Journal of Zoology* 31:120–129.

Prince, S. D., E. B. De Colstoun, and L. L. Kravitz. 1998. Evidence from rain-use efficiencies does not indicate extensive Sahelian desertification. *Global Change Biology* 4:359–374.

Pringle, R. M., D. F. Doak, A. K. Brody et al. 2010. Spatial pattern enhances ecosystem functioning in an African savanna. *PloS Biology* 8:1–12.

Prins, H. H. T. 1992. The pastoral road to extinction: Competition between wildlife and traditional pastoralism in East Africa. *Environmental Conservation* 19:117–123.

———. 2000. Competition between wildlife and livestock in Africa. In *Wildlife Conservation by Sustainable Use*, ed. H. H. T. Prins, J. G. Grootenhuis, and T. T. Dolan, 51–80. Dordrecht, Neth.: Kluwer Academic Publishers.

Prins, H. H. T., and H. P. Vanderjeugd. 1993. Herbivore population crashes and woodland structure in East Africa. *Journal of Ecology* 81:305–314.

Prothero, D. R., and R. M. Schoch. 2002. *Horns, Tusks, and Flippers: The Evolution of Hoofed Mammals*. Baltimore: Johns Hopkins University Press.

Proulx, M., and A. Mazumder. 1998. Reversal of grazing impact on plant species richness in nutrient-poor vs. nutrient-rich ecosystems. *Ecology* 79:2581–2592.

Ragir, S., M. Rosenberg, and P. Tierno. 2000. Gut morphology and the avoidance of carrion among chimpanzees, baboons, and early hominids. *Journal of Anthropological Research* 56:477–512.

Rainy, M., and J. Rainy. 1989. High noon on the Maasai Mara. *New Scientist* 124:48–53.

Ramsay, J. C. 1965. Kilimanjaro—sources of water supplies. *Tanganyika Notes and Records* 64:92–94.

Reader, J. 1997. *Africa: Biography of a Continent*. London: Hamish Hamilton.

Reed, C. A. 1970. Extinction of mammalian megafauna in the Old World late Quaternary. *BioScience* 20:284–288.

Reed, D. N., T. M. Anderson, J. Dempewolf et al. 2009. The spatial distribution of vegetation types in the Serengeti ecosystem: The influence of rainfall and topographic relief on vegetation patch characteristics. *Journal of Biogeography* 36:770–782.

Rees, P. M., C. R. Noto, J. M. Parrish, and J. T. Parrish. 2004. Late Jurassic climates, vegetation, and dinosaur distributions. *Journal of Geology* 112:643–653.

Reid, H. W., D. Buxton, E. Berrie et al. 1984. Malignant catarrhal fever. *Veterinary Record* 114:582–584.

Reid, R. S. 1992. Livestock-mediated tree regeneration: The impact of pastoralists on woodlands in dry tropical Africa. PhD. diss., Colorado State University, Fort Collins.

Reid, R. S., and J. E. Ellis. 1995. Livestock-mediated tree regeneration: Impacts of pastoralists on dry tropical woodlands. *Ecological Applications* 5:978–992.

Reid, R. S., K. A. Galvin, E. Knapp et al. Forthcoming. Sustainability of the Serengeti-Mara ecosystem for wildlife and people. In *Serengeti IV: Biodiversity,* ed. A. R. E. Sinclair and K. Metzger. Chicago: University of Chicago Press.

Reid, R. S., K. A. Galvin, and R. L. Kruska. 2008a. Global significance of extensive grazing lands and pastoral societies: An introduction. In *Fragmentation in Semi-arid and Arid Landscapes: Consequences for Human and Natural Systems,* ed. K. A. Galvin, R. S. Reid, R. H. Behnke, and N. T. Hobbs, 1–24. Dordrecht, Neth.: Springer.

Reid, R. S., H. Gichohi, M. Y. Said et al. 2008b. Fragmentation of a peri-urban savanna, Athi-Kaputiei Plains, Kenya. In *Fragmentation in Semi-arid and Arid Landscapes: Consequences for Human and Natural Systems,* ed. K. A. Galvin, R. S. Reid, R. H. Behnke, and N. T. Hobbs, 195–224. Dordrecht, Neth.: Springer.

Reid, R. S., D. Nkedianye, M. Y. Said et al. 2009. Evolution of models to support community and policy action with science: Balancing pastoral livelihoods and wildlife conservation in savannas of East Africa. *Proceedings of the National Academy of Sciences,* Early Edition, doi:10.1073/pnas.0900313106.

Reid, R. S., M. Rainy, J. Ogutu et al. 2003. Wildlife, people, and livestock in the Mara ecosystem, Kenya: the Mara count 2002. International Livestock Research Institute, Nairobi, Kenya.

Reid, R. S., M. E. Rainy, C. J. Wilson et al. 2001. Wildlife cluster around pastoral settlements in Africa. Working Paper No. 1, PLE Science Series, International Livestock Research Institute, Nairobi.

Reid, R. S., P. K. Thornton, and R. L. Kruska. 2004a. Loss and fragmentation of habitat for pastoral people and wildlife in East Africa: Concepts and issues. *South African Journal of Grass and Forage Science* 21:171–181.

Reid, R. S., P. K. Thornton, G. J. McCrabb et al. 2004b. Is it possible to mitigate greenhouse gas emissions in pastoral ecosystems of the tropics? *Environment, Development, and Sustainability* 6:91–109.

Reid, R. S., C. J. Wilson, R. L. Kruska, and W. Mulatu. 1997. Impacts of tsetse control and land-use on vegetative structure and tree species composition in southwestern Ethiopia. *Journal of Applied Ecology* 34:731–747.

Renton, A. 2009. Tourism is a curse to us, say Maasai. *Tanzania Daily News* (www .dailynews.co.tz/business/?n=5562), posted 23 November 2009.

Retallack, G. J., J. G. Wynn, B. R. Benefit, and M. L. McCrossin. 2002. Paleosols and paleoenvironments of the middle Miocene, Maboko Formation, Kenya. *Journal of Human Evolution* 42:659–703.

Ribot, J. 2003. Democratic decentralisation of natural resources: Institutional choice and discretionary power transfers in sub-Saharan Africa. *Public Administration and Development* 23:53–65.

Ribot, J. C., A. Agrawal, and A. M. Larson. 2006. Recentralizing while decentralizing: How national governments reappropriate forest resources. *World Development* 34:1864–1886.

Richardson, J. L., and A. E. Richardson. 1972. History of an African rift lake and its climatic implications. *Ecological Monographs* 42:449–534.

Ridley, M. 2003. *Evolution*. Oxford: Blackwell.

Rightmire, G. P., D. Lordkipanidze, and A. Vekua. 2006. Anatomical descriptions, comparative studies, and evolutionary significance of the hominin skulls from Dmanisi, Republic of Georgia. *Journal of Human Evolution* 50:115–141.

Riginos, C. 2009. Grass competition suppresses savanna tree growth across multiple demographic stages. *Ecology* 90:335–340.

Ritchie, M. E. 2008. Global environmental changes and their impact on the Serengeti. In *Serengeti III: Human Impacts on Ecosystem Dynamics,* ed. A. R. E. Sinclair, C. Packer, S. A. R. Mduma, and J. M. Fryxell, 183–208. Chicago: University of Chicago Press.

Ritchie, M. E., D. Tilman, and J. M. H. Knops. 1998. Herbivore effects on plants and nitrogen dynamics in an oak savanna. *Ecology* 79:165–177.

Roba, H. G., and G. Oba. 2009. Community participatory landscape classification and biodiversity assessment and monitoring of grazing lands in northern Kenya. *Journal of Environmental Management* 90:673–682.

Robertshaw, P. 1991. *Early Pastoralists of South Western Kenya*. Nairobi: British Institute of East Africa.

Robinson, J. M. 1991. Phanerozoic atmospheric reconstructions: A terrestrial perspective. *Palaeogeography, Palaeoclimatology, Palaeoecology* 97:51-62.

Roe, D., and J. P. Elliot. 2004. Poverty reduction and biodiversity conservation: Rebuilding the bridges. *Oryx* 38:137–139.

Roelke-Parker, M. E., L. Munson, C. Packer et al. 1996. A canine distemper virus epidemic in Serengeti lions (*Panthera leo*). *Nature* 379:441–445.

Rolston, H. 2003. Feeding people versus saving nature? In *Environmental Ethics: An Anthology,* ed. A. Light and H. Rolston, 451–462. Oxford: Basil Blackwell.

Roque de Pinho, J. 2009. "Staying together": People-wildlife relationships in a pastoral society in transition, Amboseli ecosystem, southern Kenya. PhD. diss., Colorado State University, Fort Collins.

Roques, K. G., T. G. O'Connor, and A. R. Watkinson. 2001. Dynamics of shrub encroachment in an African savanna: Relative influences of fire, herbivory, rainfall, and density dependence. *Journal of Applied Ecology* 38:268–280.

Rose, L., and F. Marshall. 1996. Meat eating, hominid sociality, and home bases revisited. *Current Anthropology* 37:307–338.

Rossel, S., F. Marshall, J. Peters et al. 2008. Domestication of the donkey: Timing, processes, and indicators. *Proceedings of the National Academy of Sciences* 105:3715–3720.

Roth, E.A., and E. Fratkin. 2005. Introduction: The social, health, and economic consequences of pastoral sedentarization in Marsabit District, Northern Kenya. In *As Pastoralists Settle: Social, Health, and Economic Consequences of Pastoral Sedentarization in Marsabit District, Kenya*, ed. E. Fratkin and E.A. Roth, 1–28. New York: Kluwer Academic Publishers.

Runyoro, V.A. 2007. Analysis of alternative livelihoods strategies for the pastoralists of Ngorongoro Conservation Area, Tanzania. PhD. diss., Sokoine University of Agriculture, Morogoro, Tanzania.

Runyoro, V.A., H. Hofer, E.B. Chausi, and P.D. Moehlman. 1995. Long-term trends in the herbivore populations of the Ngorongoro Crater, Tanzania. In *Serengeti II: Dynamics, Management and Conservation of an Ecosystem*, ed. A.R.E. Sinclair and P. Arcese, 146–168. Chicago: University of Chicago Press.

Russell, D., P. Beland, and J.S. McIntosh. 1980. Paleoecology of the dinosaurs of Tendaguru (Tanzania). *Mémoires de la Société géologique de France* 139:169–175.

Russell-Smith, J., D. Lucas, M. Gapindi et al. 1997. Aboriginal resource utilisation and fire management practice in western Arnhem land, monsoonal northern Australia: Notes for prehistory, lessons for the future. *Human Ecology* 25:151–195.

Rutten, M.M.E.M. 1992. *Selling Wealth to Buy Poverty: The Process of Individualisation of Land Ownership among the Maasai Pastoralists of Kajiado District, Kenya, 1890–1990*. Saarbrücken: Breitenbach.

Ryan, K., K. Munene, S.M. Kahinju, and P.M. Kunoni. 2000. Ethnographic perspectives on cattle management in semi-arid environments: A case study from Maasailand. In *The Origin and Development of African Livestock: Archaeology, Genetics, Linguistics, and Ethnography*, ed. R. Blench and K.C. Macdonald, 462–477. New York: Routledge.

Safriel, U., Z. Adeel, D. Niemeijer et al. 2005. Dryland Systems. Chapter 22 in *Millennium Ecosystem Assessment: Conditions and Trends*, vol. 1. Washington, D.C.: Island Press.

Said, M.Y. 2003. Multiscale perspectives of species richness in East Africa. Ph.D. diss., ITC, Enschede, Neth.

Said, M.Y., and R.S. Reid. 2007. Biodiversity. In *Nature's Benefits in Kenya: An Atlas of Ecosystems and Human Well-Being*, 62–79. Washington, D.C.: World Resources Institute, Department of Resource Surveys and Remote Sensing, Central Bureau of Statistics, International Livestock Research Institute.

Salzman, P.C. 1980. *When Nomads Settle: Processes of Sedentarization as Adaptation and Response*. New York: Praeger.

Sandford, G. R. 1919. *An Administrative and Political History of the Masai Reserve.* London: Waterlow and Sons.

Sandford, S. 1983. *Management of Pastoral Development in the Third World.* Chichester, U.K.: John Wiley and Sons.

Sankaran, M., N. P. Hanan, R. J. Scholes et al. 2005. Determinants of woody cover in African savannas. *Nature* 438:846–849.

Sarjeant, W. A. S., and P. J. Currie. 2001. The "Great Extinction" that never happened: The demise of the dinosaurs considered. *Canadian Journal of Earth Sciences* 38:239–247.

Savage, R. J. G. 1978. Carnivora. In *Evolution of African Mammals,* ed. V. J. Maglio and H. B. S. Cooke, 249–267. Cambridge, Mass.: Harvard University Press.

Sayialel, S., and C. J. Moss. 2011. Consolation for livestock loss: A case study in mitigation between elephants and people. In *The Amboseli Elephants,* ed. C. J. Moss, H. Croze, and P. C. Lee, 293–297. Chicago: University of Chicago Press.

Schick, K. D., and N. Toth. 1993. *Making Silent Stones Speak: Human Evolution and the Dawn of Technology.* New York: Simon and Schuster.

Schmidt, P. R. 1997. Archaeological views on a history of landscape change in East Africa. *Journal of African History* 38:393–421.

Scholes, R. J. 1990. The influence of soil fertility on the ecology of southern African dry savannas. *Journal of Biogeography* 17:415–419.

Scholes, R. J., and S. R. Archer. 1997. Tree-grass interactions in savannas. *Annual Review of Ecology and Systematics* 28:517–544.

Scholes, R. J., and B. H. Walker. 1993. *An African Savanna: Synthesis of the Nylsvley Study.* Cambridge: Cambridge University Press.

Schroeder, R. A. 2008. Environmental justice and the market: The politics of sharing wildlife revenues in Tanzania. *Society and Natural Resources,* 21:583–596.

Schüle, W. 1990. Landscapes and climate in prehistory: Interactions of wildlife, man and fire. In *Fire in the Tropical Biota,* ed. J. G. Goldammer, 272–318. Berlin: Springer-Verlag.

Schulte, P., L. Alegret, I. Arenillas et al. 2010. The Chicxulub asteroid impact and mass extinction at the Cretaceous-Paleogene boundary. *Science* 327:1214–1218

Scoones, I. 1991. Wetlands in drylands: Key resources for agricultural and pastoral production in Africa. *Ambio* 20:366–371.

———. 1992. Land degradation and livestock production in Zimbabwe's communal areas. *Land Degradation and Rehabilitation* 3:99–113.

Scott, A. C. 2000. The pre-Quaternary history of fire. *Palaeogeography, Palaeoclimatology, Palaeoecology* 164:297–345.

Scott, A. C., and T. P. Jones. 1994. The nature and influence of fire in Carboniferous ecosystems. *Palaeogeography, Palaeoclimatology, Palaeoecology* 106:91–112.

Seagle, S. W., S. J. McNaughton, and R. W. Ruess. 1993. Simulated effects of grazing on soil nitrogen and mineralization in contrasting Serengeti grasslands. *Ecology* 73:1105–1123.

Segalen, L., J. A. Lee-Thorp, and T. Cerling. 2007. Timing of C4 grass expansion across sub-Saharan Africa. *Journal of Human Evolution* 53:549–559.

Sensenig, R. L., M. W. Demment, and E. A. Laca. 2010. Allometric scaling predicts preferences for burned patches in a guild of East African grazers. *Ecology* 91:2898–2907.

Senut, B., M. Pickford, D. Gommery et al. 2001. First hominid from the Miocene (Lukeino Formation, Kenya). *Comptes rendus de l'Académie des Sciences, série II, fascicule A—Sciences de la terre et des planètes* 332:137–144.

Serneels, S., and E. F. Lambin. 2001. Impact of land-use changes on the wildebeest migration in the northern part of the Serengeti-Mara ecosystem. *Journal of Biogeography* 28:391–407.

Setsaas, T. H., T. Holmern, G. Mwakalebe et al. 2007. How does human exploitation affect impala populations in protected and partially protected areas? A case study from the Serengeti Ecosystem, Tanzania. *Biological Conservation* 136:563–570.

Seymour, C. L., and W. R. J. Dean. 1999. Effects of heavy grazing on invertebrate assemblages in the Succulent Karoo, South Africa. *Journal of Arid Environments* 43:267–286.

Shackleton, C. M. 2000. Comparison of plant diversity in protected and communal lands in the Bushbuckridge lowveld savanna, South Africa. *Biological Conservation* 94:273–285.

Shetler, J. B. 2007. *Imagining Serengeti: A History of Landscape Memory in Tanzania from Earliest Times to the Present.* Athens: Ohio University Press.

Shivji, I. G., and W. B. Kapinga. 1998. *Maasai Rights in Ngorongoro, Tanzania.* London: IIED; Dar es Salaam: HAKIARDHI.

Sillero-Zubiri, C., and D. Gotteli. 1995. Diet and feeding behavior of Ethiopian wolves (*Canis simensis*). *Journal of Mammalogy* 76:531–541.

Simon, N. 1962. *Between the Sunlight and the Thunder.* London: Collins.

Sinclair, A. R. E. 1974. The natural regulation of buffalo populations in East Africa. IV. The food supply as a regulating factor, and competition. *East African Wildlife Journal* 12:291–311.

———. 1977. *The African Buffalo.* Chicago: University of Chicago Press.

———. 1985. Does interspecific competition or predation shape the African ungulate community? *Journal of Animal Ecology* 54:899–918.

———. 1995. Serengeti past and present. In *Serengeti II: Dynamics, Management, and Conservation of an Ecosystem,* ed. A. R. E. Sinclair and P. Arcese, 3–30. Chicago: University of Chicago Press.

Sinclair, A. R. E., and P. Arcese, eds. 1995. *Serengeti II: Dynamics, Management, and Conservation of an Ecosystem*. Chicago: University of Chicago Press.

Sinclair, A. R. E., and J. M. Fryxell. 1985. The Sahel of Africa: Ecology of a disaster. *Canadian Journal of Zoology—Revue canadienne de zoologie* 63:987–994.

Sinclair, A. R. E., J. G. Hopcraft, H. Olff et al. 2008a. Historical and future changes to the Serengeti ecosystem. In *Serengeti III: Human Impacts on Ecosystem Dynamics*, ed. A. R. E. Sinclair, C. Packer, S. A. R. Mduma, and J. M. Fryxell, 7–46. Chicago: University of Chicago Press.

Sinclair, A. R. E., S. A. R. Mduma, and P. Arcese. 2000. What determines phenology and synchrony of ungulate breeding in Serengeti? *Ecology* 81:2100–2111.

———. 2002. Protected areas as biodiversity benchmarks for human impact: Agriculture and the Serengeti avifauna. *Proceedings of the Royal Society of London, Series B—Biological Sciences* 269:2401–2405.

Sinclair, A. R. E., S. Mduma, and J. S. Brashares. 2003. Patterns of predation in a diverse predator-prey system. *Nature* 425:288–290.

Sinclair, A. R. E., S. A. R. Mduma, J. G. C. Hopcraft et al. 2007. Long-term ecosystem dynamics in the Serengeti: Lessons for conservation. *Conservation Biology* 21:580–590.

Sinclair, A. R. E., and M. Norton-Griffiths. 1979. *Serengeti: Dynamics of an Ecosystem*. Chicago: University of Chicago Press.

Sinclair, A. R. E., C. Packer, S. A. R. Mduma, and J. M. Fryxell, eds. 2008b. *Serengeti III: Human Impacts on Ecosystem Dynamics*. Chicago: University of Chicago Press.

Singer, F. J., and K. A. Schoenecker. 2003. Do ungulates accelerate or decelerate nutrient cycling? *Forest Ecology and Management* 181:189–204.

Sitati, N. W., M. J. Walpole, and N. Leader-Williams. 2005. Factors affecting susceptibility of farms to crop raiding by African elephants: Using a predictive model to mitigate conflict. *Journal of Applied Ecology* 42:1175–1182.

Sitati, N. W., M. J. Walpole, R. J. Smith, and N. Leader-Williams. 2003. Predicting spatial aspects of human-elephant conflict. *Journal of Applied Ecology* 40:667–677.

Skarpe, C. 1991. Impact of grazing in savanna ecosystems. *Ambio* 20:351–356.

Smith, A. B. 1992. *Pastoralism in Africa: Origins and Development Ecology*. London: Hurst and Co.; Athens: Ohio University Press; Johannesburg: Witwatersrand University Press.

———. 2005. *African Herders: Emergence of Pastoral Traditions*. Lanham, Md.: Altamira Press.

Smith, R. J., D. J. Muir, M. J. Walpole et al. 2003. Governance and the loss of biodiversity. *Nature* 426:67–70.

Sobania, N. W. 1991. Feasts, famines, and friends: Nineteenth-century exchange and ethnicity in the Eastern Lake Turkana region. In *Herders, Traders, and Warriors: Pastoralism in Africa,* ed. J. G. Galaty and P. Bonte, 118–142. Boulder, Colo.: Westview Press.

Söderström, B., and R. S. Reid. 2010. Abandoned pastoral settlements provide concentrations of resources for savanna birds. *Acta Oecologica* 36:184–190.

Sørensen, C., M. Ole Moita, and L. Ole Kosyando. 2005. Community views on best practices and lessons learned from Ereto I. Ereto Project Report, DANIDA, Endulen, Tanzania.

Spinage, C. A. 1973. A review of ivory exploitation and elephant population trends in Africa. *East African Wildlife Journal* 11:281–289.

———. 1992. The decline of the Kalahari wildebeest. *Oryx* 26:147–150.

Stebbins, G. L. 1981. Coevolution of grasses and herbivores. *Annals of the Missouri Botanical Garden* 68:75–86.

Steinhart, E. I. 2000. Elephant hunting in 19th century Kenya: Kamba society and ecology in transformation. *International Journal of African Historical Studies* 33:335–349.

Stephens, P. A., C. A. d'Sa, C. Sillero-Zubiri, and N. Leader-Williams. 2001. Impact of livestock and settlement on the large mammalian wildlife of Bale Mountains National Park, southern Ethiopia. *Biological Conservation* 100:307–322.

Stewart, D. R. M., and D. R. P. Zaphiro. 1963. Biomass and density of wild herbivores in different East African habitats. *Mammalia* 27:483–496.

Stoner, C., T. Caro, S. Mduma et al. 2006. Changes in large herbivore populations across large areas of Tanzania. *African Journal of Ecology* 45:202–215.

Strait, D. S., and B. A. Wood. 1999a. Biogeographic implications of early hominid phylogeny. *American Journal of Physical Anthropology,* supplement 28:259.

———. 1999b. Early hominid biogeography. *Proceedings of the National Academy of Sciences* 96:9196–9200.

Stringham, T. K., W. C. Frueger, and P. L. Shaver. 2003. State and transition modelling: An ecological process approach. *Journal of Range Management* 56:106–113.

Stromberg, C. A. E. 2011. Evolution of grasses and grassland ecosystems. *Annual Review of Earth and Planetary Sciences* 39:517–544.

Struhsaker, T. 1973. Recensus of vervet monkeys in Masai-Amboseli Game Reserve, Kenya. *Ecology* 54:930–932.

Sues, H. D., and R. R. Reisz. 1998. Origins and early evolution of herbivory in tetrapods. *Trends in Ecology and Evolution* 13:141–145.

Sugimoto, A., D. E. Bignell, and J. A. MacDonald. 2000. Global impact of termites on the carbon cycle and atmospheric trace gases. In *Termites: Evolution, Sociality,*

Symbiosis, Ecology, ed. T. Abe, D.E. Bignell, and M. Higashi, 409–435. Dordrecht, Neth.: Kluwer Academic Publishers

Sullivan, S. 1999. The impacts of people and livestock on topographically diverse open wood- and shrublands in arid north-west Namibia. *Global Ecology and Biogeography Letters* 8:257–277.

Sullivan, S., and R. Rohde. 2002. On non-equilibrium in arid and semi-arid grazing systems. *Journal of Biogeography* 29:1595–1618.

Surovell, T., N. Waguespack, and P.J. Brantingham. 2005. Global archaeological evidence for proboscidean overkill. *Proceedings of the National Academy of Sciences* 102:6231–6236.

Sutherst, R.W. 1987. Ecoparasites and herbivore nutrition. In *The Nutrition of Herbivores,* ed. J.B. Hacker and J.H. Ternouth, 191–209. Sydney: Academic Press Australia.

Sutton, J.E.G. 1993. Becoming Maasailand. In *Being Maasai,* ed. T. Spear and R. Waller, 38–60. London: James Currey.

Svejcar, T.J. 1989. Animal performance and diet quality as influenced by burning on tallgrass prairie. *Journal of Range Management* 37:392–397.

Swallow, B., and A.B. Kamara. 1999. The dynamics of land use and property rights in semi-arid East Africa. In *Property Rights, Risk, and Livestock Development in Africa,* ed. N. McCarthy, B. Swallow, M. Kirk, and P. Hazell, 243–275. Washington, D.C.: International Food Policy Research Institute; Nairobi: International Livestock Research Institute.

Tablino, P. 1999. *The Gabra.* Nairobi: Paulines Publications Africa.

Talbot, L.M. 1972. Ecological consequences of rangeland development in Masailand, east Africa. In *The Careless Technology: Ecology and International Development,* ed. M.T. Farvar and J.P. Milton, 694–711. Garden City, N.Y.: Natural History Press.

Talbot, L.M., and M.H. Talbot. 1962. Food preferences of some East African wild ungulates. *East African Agricultural and Foresty Journal* 27:131–138.

Tansley, A.G., and R.S. Adamson. 1925. Studies of the vegetation of the English chalk. III. The chalk grasslands of the Hampshire Sussex border. *Journal of Ecology* 13:177–223.

Taylor, R. 2009. Community based natural resource management in Zimbabwe: The experience of CAMPFIRE. *Biodiversity and Conservation* 18:2563–2583.

Thieme, H. 1997. Lower Palaeolithic hunting spears from Germany. *Nature* 385:807–810.

Thirgood, S., C. Mlingwa, E. Gereta et al. 2008. Who pays for conservation? Current and future financing scenarios for the Serengeti ecosystem. In *Serengeti III: Human Impacts on Ecosystem Dynamics,* ed. A.R.E. Sinclair, C. Packer, S.A.R. Mduma, and J.M. Fryxell, 443–470. Chicago: University of Chicago Press.

Thirgood, S., A. Mosser, S. Tham et al. 2004. Can parks protect migratory ungulates? The case of the Serengeti wildebeest. *Animal Conservation* 7:113–120.

Thomas, C. D., A. Cameron, R. E. Green et al. 2004. Extinction risk from climate change. *Nature* 427:145–148.

Thomas, D. S. G., D. Sporton, and J. S. Perkins. 2000. The environmental impact of livestock ranches in the Kalahari, Botswana: Natural resource use, ecological change, and human response in a dynamic dryland system. *Land Degradation and Development* 11:327–341.

Thompson, D. M., ed. 1997. *Multiple Land-Use: The Experience of the Ngorongoro Conservation Area, Tanzania*. Gland, Switz.: IUCN.

Thompson, D. M., and K. Homewood. 2002. Entrepreneurs, elites, and exclusion in Maasailand: Trends in wildlife conservation and pastoralist development. *Human Ecology* 30:107–138.

Thompson, D. M., S. Serneels, D. Kaelo, and P. C. Trench. 2009a. Maasai Mara—Land privatization and wildlife decline: Can conservation pay its way? In *Staying Maasai: Livelihoods, Conservation, and Development in East African Rangelands*, ed. K. Homewood, P. Kristjanson, and P. Trench, 77–114. London: Springer.

Thompson, L. G., H. H. Brecher, E. Mosley-Thompson et al. 2009b. Glacier loss on Kilimanjaro continues unabated. *Proceedings of the National Academy of Sciences* 106:19770–19775.

Thompson, L. G., E. Mosley-Thompson, M. E. Davis et al. 2002. Kilimanjaro ice core records: Evidence of Holocene climate change in tropical Africa. *Science* 298:589–593.

Thompson, S. C. G., and M. A. Barton. 1994. Ecocentric and anthropocentric attitudes toward the environment. *Journal of Environmental Psychology* 14:149–157.

Thomson, J. 1885. *Through Maasailand*. London: Frank Cass.

Thornton, P. K., S. B. BurnSilver, R. B. Boone, and K. A. Galvin. 2006. Modelling the impacts of group ranch subdivision on agro-pastoral households in Kajiado, Kenya. *Agricultural Systems* 87:331–356.

Thornton, P. K., P. G. Jones, G. Alagarswamy, and J. A. Andresen. 2009. Spatial variation of crop yield response to climate change in East Africa. *Global Environmental Change* 19:54–65.

Thornton, P. K., P. G. Jones, P. J. Ericksen, and A. J. Challinor. 2011. Agriculture and food systems in sub-Saharan Africa in a 4°C+ world. *Philosophical Transactions of the Royal Society, Series A* 369:117–136.

Thurow, T. L., and A. J. Hussein. 1989. Observations on vegetation responses to improved grazing systems in Somalia. *Journal of Range Management* 42:16–19.

Tiffen, M., M. Mortimore, and F. Gichuki. 1994. *More People, Less Erosion*. Nairobi: ACTS Press.

Tiki, W., G. Oba, and T. Tvedt. 2011. Human stewardship or ruining cultural landscapes of the ancient Tula wells, southern Ethiopia. *Geographical Journal* 177:62–78.

Tignor, R. L. 1976. *The Colonial Transformation of Kenya: The Kamba, Kikuyu, and Maasai from 1900 to 1939*. Princeton: Princeton University Press.

Tilman, D., J. Fargione, B. Wolff et al. 2001. Forecasting agriculturally driven global environmental change. *Science* 292:281–284.

Tobler, M. W., R. Cochard, and P. J. Edwards. 2003. The impact of cattle ranching on large-scale vegetation patterns in a coastal savanna in Tanzania. *Journal of Applied Ecology* 40:430–444.

Todd, S. W. 2006. Gradients in vegetation cover, structure, and species richness of Nama-Karoo shrublands in relation to distance from livestock watering points. *Journal of Applied Ecology* 43:293–304.

Todd, S. W., and M. T. Hoffman. 1999. A fence-line contrast reveals effects of heavy grazing on plant diversity and community composition in Namaqualand, South Africa. *Plant Ecology* 142:169–178.

Tolsma, D. J., W. H. O. Ernst, and R. A. Verwey. 1987. Nutrients in soil and vegetation around two artifical waterpoints in eastern Botswana. *Journal of Applied Ecology* 24:991–1000.

Trapnell, C. G. 1959. Ecological results of woodland burning experiments in northern Rhodesia. *Journal of Ecology* 47:126–168.

Trapnell, C. G., M. T. Friend, G. T. Chamberlain, and H. F. Birch. 1976. The effects of fire and termites on a Zambian woodland soil. *Journal of Ecology* 64:577–588.

Trauth, M. H., J. C. Larrasoana, and M. Mudelsee. 2009. Trends, rhythms, and events in Plio-Pleistocene African climate. *Quaternary Science Reviews* 28:399–411.

Trauth, M. H., M. A. Maslin, A. Deino, and M. R. Strecker. 2005. Late Cenozoic moisture history of East Africa. *Science* 309:2051–2053.

Trauth, M. H., M. A. Maslin, A. L. Deino et al. 2007. High- and low-latitude forcing of Plio-Pleistocene East African climate and human evolution. *Journal of Human Evolution* 53:475–486.

Tripati, A. K., C. D. Roberts, and R. A. Eagle. 2009. Coupling of CO_2 and ice sheet stability over major climate transitions of the last 20 million years. *Science* 326:1394–1397.

Trollope, W. S. W. 1982. Ecological effects of fire in South African savannas. In *Ecology of Tropical Savannas*, ed. B. J. Huntley and B. H. Walker, 292–306. Berlin: Springer-Verlag.

Trollope, W. S. W., and L. Trollope. 1995. Report on the fire ecology of the Ngorongoro Conservation Area in Tanzania with particular reference to the Ngorongoro

Crater. Report prepared for the Department of Livestock and Pasture Science, University Fort Hare, Alice, South Africa.

Trollope, W. S. W., L. A. Trollope, and P. Morkel. 2002. Report on prescribed burning as a means of controlling the incidence of ticks, the current condition of the Lerai Forest and the infestation of forbs in the Ngorongoro Crater, Tanzania. Report prepared for the Department of Livestock and Pasture Science, University of Fort Hare, Alice, Sout Africa.

Tucker, C. J., H. E. Dregne, and W. W. Newcomb. 1991. Expansion and contraction of the Sahara Desert from 1980 to 1990. *Science* 253:299–301.

Turner, A., and M. Anton. 2004. *Evolving Eden.* New York: Columbia University Press.

Turner, M. D. 1999a. Spatial and temporal scaling of grazing impact on the species composition and productivity of Sahelian annual grasslands. *Journal of Arid Environments* 41:277–297.

———. 1999b. The role of social networks, indefinite boundaries, and political bargaining in maintaining the ecological and economic resilience of the transhumance systems of Sudano-Sahelian West Africa. In *Managing Mobility in African Rangelands,* ed. M. Niamir-Fuller, 97–123. London: Intermediate Technology Publications.

Turton, D. 1987. The Mursi and national park development in the Lower Omo Valley. In *Conservation in Africa: People, Policies, and Practices,* ed. D. M. Anderson and R. Grove, 169–186. Cambridge: Cambridge University Press.

———. 1988. Looking for a cool place: The Mursi, 1890s to 1980s. In *The Ecology of Survival: Case Studies from Northeast African History,* ed. D. H. Johnson and D. M. Anderson, 261–282. London: Lester Crook Academic Publishing; Boulder, Colo.: Westview Press.

———. 1991. Movement, warfare, and ethnicity in the lower Omo Valley. In *Herders, Traders, and Warriors: Pastoralism in Africa,* ed. J. G. Galaty and P. Bonte, 145–169. Boulder, Colo.: Westview Press.

UNESCO. 2007. World Heritage Committee report, 31st Session, Christchurch, New Zealand, 23 June–2 July 2007. UNESCO, Paris.

Uno, K. T., T. E. Cerling, J. M. Harris et al. 2011. Late Miocene to Pliocene carbon isotope record of differential diet change among East African herbivores. *Proceedings of the National Academy of Sciences* 108:6509–6514.

van der Hammen, T. 1983. The paleoecology and palaeogeography of savannas. In *Tropical Savannas,* vol. 13 of *Ecosystems of the World,* ed. F. Bourlière, 19–35. Amsterdam: Elsevier.

van de Vijver, C., C. A. Foley, and H. Olff. 1999a. Changes in the woody component of an East African savanna during 25 years. *Journal of Tropical Ecology* 15:545–564.

van de Vijver, C., P. Poot, and H. H. T. Prins. 1999b. Causes of increased nutrient concentrations in post-fire regrowth in an East African savanna. *Plant and Soil* 214:173–185.

van Langevelde, F., C. van de Vijver, L. Kumar et al. 2003. Effects of fire and herbivory on the stability of savanna ecosystems. *Ecology* 84:337–350.

Van Soest, P. J. 1994. *Nutritional Ecology of the Ruminant.* Ithaca, N.Y.: Cornell University Press.

van Wijngaarden, W. 1985. *Elephants, Trees, Grass, Grazers: Relationships between Climate, Soils, Vegetation, and Large Herbivores in a Semi-arid Savanna Ecosystem (Tsavo, Kenya).* Enschede, Neth.: ITC Publications.

Veblen, K. E., and T. P. Young. 2010. Contrasting effects of cattle and wildlife on the vegetation development of a savanna landscape mosaic. *Journal of Ecology* 98:993–1001.

Veitia, R. A., and S. Bottani. 2009. Whole genome duplications and a "function" for junk DNA? Facts and hypotheses. *PloS ONE* 4:7.

Verlinden, A. 1997. Human settlements and wildlife distribution in the southern Kalahari of Botswana. *Biological Conservation* 82:129–136.

Verschuren, D., K. R. Laird, and B. F. Cumming. 2000. Rainfall and drought in equatorial east Africa during the past 1,100 years. *Nature* 403:410–414.

Vesey-Fitzgerald, D. R. 1960. Grazing succession among East African game animals. *Journal of Mammalogy* 41:161–172.

Vetter, S. 2005. Rangelands at equilibrium and non-equilibrium: recent developments in the debate. *Journal of Arid Environments* 62:321–341.

Virani, M. Z., C. Kendall, P. Njoroge, and S. Thomsett. 2011. Major declines in the abundance of vultures and other scavenging raptors in and around the Masai Mara ecosystem, Kenya. *Biological Conservation* 144:746–752.

Vitousek, P. M., P. R. Ehrlich, A. H. Ehrlich, and P. A. Matson. 1986. Human appropriation of the products of photosynthesis. *BioScience* 36:368–373.

Vitousek, P. M., H. A. Mooney, J. Lubchenco, and J. M. Melillo. 1997. Human domination of Earth's ecosystems. *Science* 277:494–499.

Vrba, E. S., G. H. Denton, T. C. Partridge, and L. H. Burckle. 1995. *Paleoclimate and Evolution with Emphasis on Human Origins.* New Haven, Ct.: Yale University Press.

Waithaka, J. M. 1994. The ecological role of elephants in restructuring plant and animal communities in different eco-climatic zones in Kenya and their impacts on land-use patterns. Ph.D. diss., Kenyatta University, Nairobi.

Walker, J. D., and J. W. Geissman. 2009. *Geologic Time Scale: Geologic Society of America.* doi:10.1130/2009.CTS004R2C.

Waller, R. 1988. Emutai: Crisis and response in Maasailand 1883–1902. In *The Ecology of Survival: Case Studies from Northeast African History,* ed. D. H. Johnson and

D. M. Anderson, 73–112. London: Lester Crook Academic Publishing; Boulder, Colo.: Westview Press.

———. 2004. "Clean" and "dirty": Cattle disease and control policy in colonial Kenya, 1900–40. *Journal of African History* 45:45–80.

Waller, R., and N. Sobania. 1994. Pastoralism in historical perspective. In *African Pastoralist Systems: An Integrated Approach*, ed. E. Fratkin, K. A. Galvin, and E. A. Roth, 45–68. Boulder, Colo.: Lynne Rienner Publishers.

Wanitzek, U., and H. Sippel. 1998. Land rights in conservation areas in Tanzania. *GeoJournal* 46:113–128.

Ward, D., B. T. Ngairorue, J. Kathena et al. 1998. Land degradation is not a necessary outcome of communal pastoralism in arid Namibia. *Journal of Arid Environments* 40:357–371.

Waters, G., and J. Odero. 1986. *Geography of Kenya and the East African Region*. London: Macmillan.

Watson, J. P. 1967. A termite mound in an iron-age burial ground in Rhodesia. *Journal of Ecology* 55:663–669.

WCED. 1987. *Our Common Future: Report of the World Commission on Environment and Development*. Oxford: Oxford University Press.

WDPA. 2009. WDPA Consortium 2009 World Database on Protected Areas, www.unep-wcmc.org/wdpa/. UNEP-World Conservation Monitoring Centre (UNEP-WCMC), Cambridge.

Webb, J. L. A. 2005. Malaria and the peopling of early tropical Africa. *Journal of World History* 16:269–291.

Wendorf, F., and R. Schild. 1984. The emergence of food production in the Egyptian Sahara. In *From Hunters to Farmers: The Causes and Consequences of Food Production in Africa*, ed. J. D. Clark and S. A. Brandt, 93–101. Berkeley: University of California Press.

West, P., J. Igoe, and D. Brockington. 2006. Parks and peoples: The social impact of protected areas. *Annual Review of Anthropology* 35:251–277.

Westerberg, L. O., K. Holmgren, L. Borjeson et al. 2010. The development of the ancient irrigation system at Engaruka, northern Tanzania: Physical and societal factors. *Geographical Journal* 176:304–318.

Western, D. 1973. The structure, dynamics, and changes of the Amboseli ecosystem. Ph.D. diss., University of Nairobi.

———. 1975. Water availability and its influence on the structure and dynamics of a savannah large mammal community. *East African Wildlife Journal* 13:265–286.

———. 1982a. Amboseli National Park: Enlisting local landowners to conserve migratory wildlife. *Ambio* 11:302–308.

———. 1982b. The environment and ecology of pastoralists in arid savannas. *Development and Change* 13:183–211.

———. 1989. Conservation without parks: Wildlife in the rural landscape. In *Conservation for the Twenty-first Century*, ed. D. Western and M. C. Pearl, 158–165. Oxford: Oxford University Press.

———. 1994. Ecosystem conservation and rural development: The case of Amboseli. In *Natural Connections: Perspectives in Community-Based Conservation*, ed. D. Western and J. J. R. Grimsdell, 15–52. Washington, D.C.: Island Press.

———. 2005. The ecology and changes of the Amboseli Ecosystem: Recommendations for planning and conservation. Report prepared for the African Conservation Centre, Nairobi.

———. 2006. A half a century of habitat change in Amboseli National Park, Kenya. *African Journal of Ecology* 45 302–310.

———. 2009. The future of Maasailand: Its people and wildlife. In *Staying Maasai: Livelihoods, Conservation, and Development in East African Rangelands*, ed. K. Homewood, P. Kristjanson, and P. Trench, v–viii. New York: Springer.

Western, D., and T. Dunne. 1979. Environmental aspects of settlement site decisions among pastoral Maasai. *Human Ecology* 7:75–98.

Western, D., and V. Finch. 1986. Cattle and pastoralism: Survival and production in arid lands. *Human Ecology* 14:77–94.

Western, D., and H. Gichohi. 1993. Segregation effects and the impoverishment of savanna parks: The case for ecosystem viability analysis. *African Journal of Ecology* 31:269–281.

Western, D., R. Groom, and J. Worden. 2009. The impact of subdivision and sedentarization of pastoral lands on wildlife in an African savanna ecosystem. *Biological Conservation* 142:2538–2546.

Western, D., and W. K. Lindsay. 1984. Seasonal herd dynamics of a savanna elephant population. *African Journal of Ecology* 22:229–244.

Western, D., and D. Maitumo. 2004. Woodland loss and restoration in a savanna park: A 20-year experiment. *African Journal of Ecology* 42:111–121.

Western, D., and D. L. Manzolillo Nightingale. 2004. *Environmental Change and the Vulnerability of Pastoralists to Drought: A Case Study of the Maasai in Amboseli, Kenya*. Nairobi: UNEP.

Western, D., S. Russell, and I. Cuthill. 2009b. The status of wildlife in protected areas compared to non-protected areas of Kenya. *PloS ONE* 4:1–6.

Western, D., and D. M. Sindiyo. 1972. The status of the Amboseli rhino population. *East African Wildlife Journal* 10:43–57.

Western, D., and C. van Praet. 1973. Cyclical changes in the climate and habitat of an East African ecosystem. *Nature* 241:104–106.

Western, D., and M. Wright. 1994. The background to community-based conservation. In *Natural Connections: Perspectives in Community-Based Conservation*, ed. D. Western and J.J.R. Grimsdell, 1–14. Washington, D.C.: Island Press.

Westoby, M. 1979–80. Elements of a theory of vegetation dynamics in arid rangelands. *Israel Journal of Botany* 28:169–194.

Weyerhauser, F.J. 1985. Survey of elephant damage to baobabs in Tanzania's Lake Manyara National Park. *African Journal of Ecology* 23:235–243.

White, T.D., G. WoldeGabriel, B. Asfaw et al. 2006. Asa Issie, Aramis, and the origins of *Australopithecus. Nature* 440:883–889.

Whyte, I.J., and S.C.J. Joubert. 1988. Blue wildebeest population trends in the Kruger National Park and the effects of fencing. *South African Journal of Wildlife Research* 18:78–87.

Wiegand, T., and S.J. Milton. 1996. Vegetation change in semiarid communities Simulating probabilities and time scales. *Vegetatio* 125:169–183.

Wilkinson, D.M. 1999. The disturbing history of intermediate disturbance. *Oikos* 84:145–147.

Williams, A.P., and C. Funk. 2011. A westward extension of the warm pool leads to a westward extension of the Walker circulation, drying eastern Africa. *Climate Dynamics* 37:2417–2435.

Williams, S.D. 1998. Grévy's zebra: Ecology in a heterogeneous environment. Ph.D. diss., University College, London.

Wilson, C.J., R.S. Reid, N.L. Stanton, and B.D. Perry. 1997. Ecological consequences of controlling the tsetse fly in southwestern Ethiopia: Effects of land-use on bird species diversity. *Conservation Biology* 11:435–447.

Wilson, R.T. 1980. Wildlife in northern Darfur, Sudan: A review of its distribution and status in the recent past and at present. *Biological Conservation* 17:85–101.

Wing, S.L., and H.-D. Sues. 1993. Mesozoic and early Cenozoic terrestrial ecosystems. In *Terrestrial Ecosystems through Time*, ed. A.K. Behrensmeyer, J.D. Damuth, W.A. DiMichele et al. , 327–416. Chicago: University of Chicago Press.

Wittemyer, G., P. Elsen, W.T. Bean et al. 2008. Accelerated human population growth at protected area edges. *Science* 321:123–126.

Wolanski, E., and E. Gereta. 2001. Water quantity and quality as the factors driving the Serengeti ecosystem, Tanzania. *Hydrobiologia* 458:169–180.

Wolanski, E., E. Gereta, M. Borner, and S. Mduma. 1999. Water, migration, and the Serengeti ecosystem. *American Scientist* 87:526–533.

Wolff, C., G.H. Haug, A. Timmermann et al. 2011. Reduced interannual rainfall variability in east Africa during the last ice age. *Science* 333:743–747.

Wood, B. 2002. Palaeoanthropology: Hominid revelations from Chad. *Nature* 418:133–135.

———. 2011. Did early *Homo* migrate "out of" or "in to" Africa? *Proceedings of the National Academy of Sciences* 108:10375–10376.

Wood, B., and M. Collard. 1999. The genus *Homo*. *Science* 284:65–71.

Wood, B., and N. Lonergan. 2008. The hominin fossil record: Taxa, grades, and clades. *Journal of Anatomy* 212:354–376.

Wood, T. G. 1988. Termites and the soil environment. *Biology and Fertility of Soils* 6:228–236.

Woodroffe, R., and J. R. Ginsberg. 1998. Edge effects and the extinction of populations inside protected areas. *Science* 280:2126–2128.

Woods, M. 2007. Engaging the global countryside: Globalization, hybridity, and the reconstitution of rural place. *Progress in Human Geography* 31:485–507.

Worden, J. S. 2007. Fragmentation and settlement pattern in Maasailand: Implications for pastoral mobility, drought vulnerability, and wildlife conservation in an East African savanna. Ph.D. diss., Colorado State University, Fort Collins.

Worden, J., V. Mose, and D. Western. 2010. Aerial census of wildlife and livestock in eastern Kajiado. Available at www.amboseliconservation.org.

Worden, J. S., R. S. Reid, and H. Gichohi. 2003. Land-use impacts on large wildlife and livestock in the swamps of the Greater Amboseli Ecosystem. LUCID Working Paper No. 27, International Livestock Research Institute, Nairobi.

Wrangham, R. W., J. H. Jones, G. Laden et al. 1999. The raw and the stolen: Cooking and the ecology of human origins. *Current Anthropology* 40:567–594.

Yibarbuk, D., P. J. Whitehead, J. Russell-Smith et al. 2001. Fire ecology and Aboriginal land management in central Arnhem Land, northern Australia: A tradition of ecosystem management. *Journal of Biogeography* 28:325–343.

Young, T. P. 2006. Declining rural populations and the future of biodiversity: Missing the forest for the trees? *Journal of International Wildlife Law and Policy* 9:319–334.

Young, T. P., and K. Lindsay. 1988. Role of even-aged population structure in the disappearance of Acacia xanthophloea woodlands. *African Journal of Ecology* 26:69–72.

Young, T. P., T. A. Palmer, and M. E. Gadd. 2005. Competition and compensation among cattle, zebras, and elephants in a semi-arid savanna in Laikipia, Kenya. *Biological Conservation* 122:351–359.

Young, T. P., N. Partridge, and A. Macrae. 1995. Long-term glades in *Acacia* bushland and their edge effects in Laikipia, Kenya. *Ecological Applications* 5:97–108.

Zavaleta, E. S., and K. B. Hulvey. 2004. Realistic species losses disproportionately reduce grassland resistance to biological invaders. *Science* 306:1175–1177.

Zeidane, M. O. 1999. Le rôle des institutions provisoires dans la gestion des ressources à propriété commune des terres arides en Mauritanie. In *Managing Mobility in African Rangelands*, ed. M. Niamir-Fuller, 47–75. London: FAO and Beijer International Institute of Ecological Economics.

Zhivotovsky, L. A., N. A. Rosenberg, and M. W. Feldman. 2003. Features of evolution and expansion of modern humans, inferred from genomewide microsatellite markers. *American Journal of Human Genetics* 72:1171–1186.

INDEX

Page numbers in italics indicate figures, illustrations, and maps.

119, 120, 124, 125, *126,* 147, 151–52,
155–59. *See also* degradation of environment; irreversible loss
deforestation, 164, 171, 181; in Amboseli
ecosystem, 185, 192–93, 194–95; in
Serengeti-Mara region, 164, 171, 181
degradation of environment: overview and
definition of, 113, 124; pastoralists as
enriching savannas and, 113, 124;
pastoralists as incompatible with wildlife
and, 147, 151–52, *153,* 155–59, 161; wildlife
and, 113, 124, 147, 151–52, *153,* 155–59,
161. *See also* decline of environment;
irreversible loss
desertification (overgrazing). *See* grazing;
overgrazing (desertification)
deserts, in Africa, *20,* 21, *246*
development, 182, 208, 217–18, 219–20,
257–58
diffuse competition, 66
dik diks, 37, 43, 83
DiMichele, W. A., 75
Dinka herders of South Sudan, *24,* 24–25,
58, 111–12
dinosaurs, 74, *76,* 78–80, 82
direct competition, 66
diseases: animals' effects on animals and, 44;
dinosaurs and, 79; ecosystems and, 14, 15;
rainfall change predictions and, 241;
Serengeti-Mara region as affected by, 44,
67, 177; transmission between livestock
and wildlife of, 57, 66, 67, 170, 172, 215,
225–26; as wildlife stressors, 13, 45, 66,
106, 107–9. *See also specific diseases*
disturbance-driven savannas, 19, *20,*
245, 269n2
diversify (enrich), 123, 125, *126,* 131. *See also*
simplify (impoverish) or simplifying
response
diversity of approaches, and middle ground,
258–60, *259*
Djibouti: bimodal rainfall in, 27, *27;* colonial
period and, 110; east Africa and, *5,* 15;
language groups of, *7;* soft-boundary
savannas and, 72; tribal groups of, *9,*
267n7; wildlife populations missing data
for, 280n57

domestic dogs, 54, 65–66, 67, 120, 150, 172.
See also wild dogs
donors and private sector support, and
middle ground, 262–63
Dorobo (Ndorobo) people, 104, 176, *259*
dromedary (single-humped) camel herds, 55
droughts: in Amboseli ecosystem, 188–89;
in Amboseli National Park, 189; in
Kaputiei Plains, 210; pastoralists and, 52,
53, 56, 61, 112, 180, 181–82; rainfall
change predictions and, 242; in
Serengeti-Mara region, 180, 181; wildlife
and, 180, 181–82
drylands: development in, 118; of east Africa,
25; farming in, 116; global view of, 21–22,
125–26; human populations' development
as part of, 14–15; in U.S., 125–26
Dublin, H. T., 35, 109, 178
dung: dung beetles and, 38–40, *40, 76,* 78;
settlements' effects on savannas and,
36, 122–23, 135–38, 140. *See also* soil
fertility
Du Toit, J. T., 67, 130–31, 160
dystrophic (nutrient-poor) savannas, 29, *29,*
30, 46, 242–43. *See also* soil fertility

east Africa: overview and use of term for, *5,*
15; bimodal rainfall in historical context
in, 49–50; cattle types and, 282n30;
colonial period in, *106,* 110–15, 278n31;
conservation in historical context and, 2;
elevation changes in Africa, 26–27, *28,*
270n16; farmers in, 32; future scenarios
and, 201; human population movement in,
100; hunter-gatherers in, 32; ivory trade
and, 103; locations within, 3, *5, 6;*
pastoralists populations in, 30–31,
116–17; slave trade and, 103; tribal groups
in, 4, *9, 10, 11,* 100, 267n7. *See also*
Africa; future for pastoralists and wildlife
in savannas; language groups in east
Africa; savannas in east Africa; *specific
countries, herder groups,* and *wildlife parks
and reserves*
East Coast fever (ECF), 226, 236
ecocentric perspective, 7–8, 114, 152, *153,*
155, 161

forage *(continued)*
  goats and, 64, 150; rhinos and, 21, 43, 224;
    sheep and, 64, 150; shrubs as, 37–38, 42,
    55, 129; wildebeest and, 43, 228. *See also*
    plant-eating wildlife (herbivores)
Ford, J., 107, 109
forests (woodlands). *See* savanna woodlands;
    trees and grass balance; woodlands
    (forests)
Frank, D. A., 129
Frank, L., 59
Friends of Nairobi National Park, 218,
    219
Fryxell, J. M., 127
Fulani (Fulbe) herders of west Africa, 18, 23
Funk, C., 242
future for pastoralists and wildlife in
    savannas: overview of, 238, 251–53,
    264–66; Amboseli ecosystem and,
    199–201; climate change and, 50, 181–82,
    241–43; community-based resource
    management and, 249; costs and benefits
    of wildlife for pastoralists and, 249–50,
    252, 253, 255, 265; current trends as
    context for, 240–46, 247, 248–51;
    diversity of approaches and, 239–40;
    farming and, 245, 248; globalization and,
    243; hard-boundary savannas and, 245;
    human populations growth and, 244–45;
    imagined scenario for, 264–66; Kaputiei
    Plains and, 217–20; land rights for
    pastoralists and, 248–49; NCA and,
    233–37; Ngorongoro Maasai herders and,
    235–37; NGOs' influence and, 243;
    non-pastoral supplemental activities for
    pastoralists and, 116, 118, 247–48;
    paradigms and paradigm shifts and, 112,
    253, 261, 266; peace committees and, 244;
    political corruption and, 171, 226, 244,
    257, 265, 267; power for locals over land
    and profits and, 4, 243, 248–149; rights
    for wildlife and, 243; savannas and, 245,
    246, 247, 295n19; Serengeti-Mara region
    and, 180–83; social conflicts and, 243–44;
    soft-boundary savannas and, 245. *See also*
    middle ground; pastoralists (pastoral
    lifestyle or pastoral herders)

Gabra herders of Kenya, *10*, 23, *24*, 31, 50,
    270n10
Galla (Oromo) herders of Ethiopia, 8, *9*, 133,
    149, 154, 267n7
Galvin, K. A., 23, 49, 51, 226, 236, 242
gazelles: diseases and, 107; evolution
    timeline and, 83; forage for, 228; Grant's
    gazelle and, 38, 42–43, 91, 107, 135, 141,
    198, 247; migrations and, 165; parks and,
    188; Thomson's gazelle and, *39*, 42, 91,
    107, 130, 135, 141, 168
Georgiadis, N. J., 138
Ghana, 67
Gichohi, H., 131, 143, 211, 219
Gillson, L., 109
Ginsberg, J. R., 67
giraffes: as browsers, 37, *38*, 41; deforesta-
    tion's effects on, 195; evolution timeline
    and, 80, 83; farmer's crop destruction by,
    185; forage for, 43; group grazing and, 43;
    parks and, 188; predation and, 45;
    scavenger (proto-)humans and, 91; trees
    and grass balance in savannas and, 46
globalization, *71*, 208, 210, 217, 243
global viewpoints: of drylands, 21–22,
    125–26; of pastoralists, 22–23, 269n6,
    270n20; population statistics, 12–13; of
    rainfall, 25, 26, *27;* of savannas, 18–19,
    21–22; of wildlife, 21
goats: overview of, 47–48, 54–56, 59;
    bleeding to use blood as food and, 50–51;
    as browsers, *33*, 150, 155; diseases and,
    107; evolution timeline and, 83; food
    production through livestock timeline
    and, 96; forage for, 64, 150; shoat herds
    and, 54, 55; wildlife attacks on livestock
    and, 149
Goldman, M., 238
Gordon, R. G., Jr., 267n7
governance structure of pastoralists, and
    middle ground, 260–61
government policies (postcolonial policies):
    benefit sharing for pastoralists and,
    255–56; community-based resource
    management and, 249; conservation
    models and, 174–75, *259;* empowerment
    and power over land and profits for

pastoralists in context of, 180, 254–55; farming and, 248; in future scenarios for pastoralists and wildlife in savannas, 261–62; group ranches and, 214; humans as ecosystem engineers and, 118; Kaputiei Plains as affected by, 118, 213–14; paradigms as culturally constructed and, 261; pastoralists as affected by, 226; pastoralists' settlements and, 118–19; policymakers' paradigm shift and, 261; political corruption and, 244; power for locals over land and profits and, 254–55; privatization of land and, 214; social conflicts and, 244. *See also* colonial period

Grant's gazelle, 38, 42–43, 91, 107, 135, 141, 198, 247

grasses: biomass as edible and, 41–42, 90; evolution timeline and, 8, 84, 275n19; homestead, 122–23; resilience of, 127–30; as resistant to fires, 37; survival and success in savannas for, 37, 42, 82; wildlife types dependent on, 37. *See also* plant-eating wildlife (herbivores); savannas; trees and grass balance

grazing: diversity in plants and, 128–29; elephants and, 18, 34–35, 43, 45–46, 189; elephants' effects on trees and, 33, *33,* 33–34, 46, 194, 195; enriching response data and, 125, *126,* 128–29, 131, 139–40, 145, 151–52, 158; equilibrium dynamics and, 127; evolution timeline and, 82, 83, 91, 125; homestead grasses and, 122–23, 189; humped response and, 124–25, *126,* 128–29, 131–33, 135, 141, 151–52, 158; livestock and wildlife competition for, 150, 208, 212; livestock and wildlife group, *3,* 43, 48, 173; in Nairobi city for livestock, 210; in NCA for livestock, 135; near settlements by wildlife, 140–43; in North America, 46, 125–26, *126,* 128, 129, 132; opening up savannas by, 64; overgrazing theories, 127; plants as tolerant of, 46; plant species as affected by, 128–29, 131; poison as cause of death of wildlife and, 138; savannas as affected by, 125, 128, 131–32, 281n19; simplifying response and, 125; soils as tolerant of, 46,

125; "tragedy of the commons" and, 127, 216; wildlife group, *3,* 43, 48, 173; wildlife's effects on savannas and, 46, 130, 132, 134, 138–39, 140–43. *See also* grasses; overgrazing; plant-eating wildlife (herbivores); plant species

grazing down, 142, 189. *See also* grazing

grazing lawns: fires and, 134, 211; in Kaputiei Plains, 211; livestock and, 122, 134, 173, 211; Nairobi National Park and, 211; NCA and, 228; pastoralists and livestock as enriching savannas and, 43, 122–23, 130, 134, 141, 145, 173; predation and, 130, 134, 141, 142, 211; wildlife and, 43, 122–23, 130, 134, 141, 145, 173. *See also* grazing; grazing down

Great Rift Valley (Rift Valley). *See* Rift Valley (Great Rift Valley)

Greenberg, J. H., 3–4, *7,* 32, 267n7

Grime, J. P., 124

group grazing, *3,* 43, 48, 173. *See also* grazing

group ranches: overview of, *106;* in Amboseli ecosystem, 115, 117, 186, 197–98, 200; in Amboseli National Park, 186, 197–98; government policies and, 214; humans as ecosystem engineers and, 115, 197–98; Kaputiei Maasai herders and, 207, 208; in Kaputiei Plains, 115, 186, 207, 208, 214; in Mara region of Kenya, 166; in Serengeti-Mara region, 174; tourism on former, 238; water points and, 214; wildlife loss due to, 197–98. *See also* land

Grumeti Game Reserve, Tanzania, 166, *167,* 168, 171

Hadza (Kindiga) hunter-gatherers of Tanzania, 32, 100, 224, 231, 234

Hakansson, N. T., 103–4

hard-boundary savannas: future for pastoralists and wildlife and, 245; Nairobi city and, *70,* 202, 208; Nairobi National Park and, *70,* 208, 218; pastoralists as enriching savannas and, 130; savannas distribution predictions and, 245, *246,* 247, 295n19; in Serengeti-Mara region, 169–70; soft- to hard-boundary savannas

hard-boundary savannas *(continued)*
transition and, *69*, 119, *144*, 159–61;
South Sudan and, 68; water points and,
159–60. *See also* mixed-boundary
savannas; soft-boundary savannas
Hardin, G., 127, 216
Haro, G. O., 244
Harris, J. M., 92
hartebeest, 83, 92, 93, 107, 142–43, 188, 213,
216, 272n43
Heine, B., 267n7
herbivores (plant-eating wildlife). *See* forage;
grasses; grazing; plant-eating wildlife
(herbivores); plant species; wildlife
herders. *See* pastoralists (pastoral lifestyle or
pastoral herders); pastoralists and livestock as enriching savannas; pastoralists
and livestock as incompatible with wild-
life; *specific countries and herder groups*
Herero herders of southern Africa, *24*, 25
Herskovits, M., 127
Hiernaux, P., 131–32, 157
highway development, 164, 182, 208, 257–58
Hildebrand, A. R., 79
Hildebrand, E., 101
Himba herders of Namibia, 23, 155, 270n10
hippos, 21, 45, 83, 107, 109, 123, 141
Hoare, R., 67, 105, 160
Hobbs, N, T., 150
Holekamp, K. E., 172–73
Homewood, K., 56, 221, 229
honey gatherers, 35, 184, 192
Hook, R. W., 75
Hughes, L., 111, 195
human-centric perspective, 152, *153*, 155, 161
human evolution. *See* (proto-)human
evolution
human population conflicts with wildlife:
Amboseli ecosystem, 149, 188–90;
anthropocentric perspective and, 4, 7–8;
buffaloes and, 146–47, 149, 172; colonial
period and, 102; ecocentric perspective
and, 7–8, 114; elephants and, 59, 66–68,
102, 104, 149, 172; farming and, 102, 104;
hyenas and, 58, 149; leopards and, 149,
163–64; lions and, 58, 149, 227–28;
Maasai herders and, 59, 146–47; Maasai

Mara National Reserve and, 59; in Mara
region of Kenya, 59, 146–47; Ngorongoro
Maasai herders and, 227–28; pastoralists
as incompatible with wildlife and, 59,
146–47, 149; perspectives on, 8; poison as
cause of death of wildlife and, 149, 173,
196, 201, 261; rhinos, 149; in Serengeti-
Mara region, 172; water points and, 3,
67–68, 150, 159–60, 190, 211. *See also*
wildlife attacks on livestock
human populations: development of, 4,
13–15; diseases in context of wildlife's
balance with, 61, *65*, 65–66; in environments outside savannas, 12–15; future
scenarios for, 201, 244–45; in Kaputiei
Plains, 209, 213, 219; land use as affected
by, 68, *69, 70, 71,* 72; movement of, 96,
98, 99–100; nature as separate from, 2,
7–8, 112; NCA and, 224; population
density effects and, 68, *69, 70, 71,* 72;
populations size statistics and, 67; in
Serengeti-Mara region, 169; size of,
67–68; wildlife as affected by, 61–62,
64–68, *65, 69, 70, 71,* 72; wildlife's
balance with, 2, *3,* 6–9, *7, 31,* 48, 61, *65,*
65–66, 123, 185; zoning for wildlife and,
219, 221, 261. *See also* hunter-gatherers;
pastoralists (pastoral lifestyle or pastoral
herders); *specific diseases* and *herder groups*
human populations and wildlife balance:
overview of, 240; Amboseli ecosystem
and, 149, 188–90; diseases and, 61, *65,*
65–66; farming and, 15, 35, 49, 59, 62–64,
63, 67–68, 102, 160–61; hunting game
and, 160–61; Maasai Mara National
Reserve and, *144,* 170; Mara region of
Kenya and, 143–44, *144,* 159–60; NCA
and, 227–29, 234; pastoralists and, *7,*
8–9, *31,* 48; pastoralists as enriching
savannas and, 143–44, *144;* resources
and, *65,* 66; Serengeti-Mara region and,
144, 170, 173, 182. *See also* human
populations
humans as ecosystem engineers: overview of,
95; bimodal rainfall and, *97,* 99, 100;
colonial period and, *106,* 110–15, 278n31;
conservation reserves and, 115; diseases as

stressors on humans timeline and, 105, *106*, 107, 108; diseases as stressors on wildlife timeline and, *106*, 107–9; evolution timeline in context of, 92–94; fires due to diseases as stressors on wildlife and, 109; food production through livestock timeline and, 96, *97*, *98*, 99–103; government policies and, 118; group ranches and, 115, 197–98; human population movements and, 96, *98*, 99–100; hunter-gatherers and, 96, 99, 101; ironwork technology and, 100, 103; ivory trade timeline and, 103–5, *106*, 278n19; language groups of east Africa and, *7*, 99–100; marginalization of pastoralists and, 110–19; parks and, 104, 109, 114, 115, 119, 120; in Serengeti-Mara region, 176–78; slave trade timeline and, 103, 105, *106*; soft-boundary savannas and, 101–2; soft- to hard-boundary savannas transition and, 119; water needs and, 115, 116, 119; water points and, 177, 214; wildlife as affected by, 102, 119–21, 280n57; woodlands as affected by, 108–9, 229

humped response (intermediate disturbance hypothesis), 124–25, *126*, 128–29, 131–33, 135, 141, 151–52, 158. *See also* enriching response; no response to grazing data

hunter-gatherers: overview of, *10*, 100–101; fourth stage of humans as effective ecosystem engineers and, 93–94; in Amboseli ecosystem, 190; in east Africa, 32; fires and, 35, 62, *63*, 92; humans as ecosystem engineers and, 96, 99, 101; in Kaputiei Plains, 205; in NCA, 221; in Serengeti-Mara region, 164–65; wildlife as affected by, 62, *63*, 102

hunting game: in Amboseli National Park, 189; colonial period and, *106*, 113–15; for food by African, 104, 107, 148–49, 170–71, 189, 209, 212; for food by pastoralists, 148–49, 170–71, 212; human populations and wildlife balance and, 160–61; in Kenya, 148, 170–71, 179, 250, 256; in Rwanda, 149, 231, 250; in Serengeti-Mara region, 166, 170–71; in

South Africa, 113; in Sudan, 120; in Tanzania, 149, 174, 231, 250; for trophies, 2, *106*, 113–15, 179, 250, 254, 256. *See also* hunter-gatherers

hyenas: adaptation to pastoralists and livestock by, 142, 172–73; evolution timeline and, 80, 84, 88; pastoralists as enriching savannas and, 142; poison as cause of death and, 173; population statistics for, 168; scavenger (proto-) humans and, 91; wildlife attacks on livestock and, 36, 58–59, 64, 65, 135, 141, 149, 172

Ikorongo Game Reserve, Tanzania, 166, *167*, 171

impoverish (simplify or simplifying response), 123, 125, *126*, 129, 131–32, 148, 151–52. *See also* enrich (diversify)

impoverishment of human populations (poverty): Kaputiei Maasai herders and, 209, 215, 218–19; in Nairobi, 217; NCA and, 4, 221, 224–27, 235, 237; parks and, 4, 226, 235, 237; pastoralists and, 1, 4, 127, 215; privatization of land and, 215; soil fertility as cause of, 30

Indigofera spinosa, 55, 132

individual ownership (privatization) of land. *See* group ranches; privatization (individual ownership) of land

industrial enterprises, *71*, 188, 208, 210, 217

Intergovernmental Pael on Climate Change, 241

intermediate disturbance hypothesis (humped response), 124–25, *126*, 128–29, 131–33, 135, 141, 151–52, 158. *See also* enriching response; no response to grazing data

Intertropical Convergence Zone (ITCZ), 25

Iraqw people of Tanzania, *7*, 50, 100

ironwork technology, 100, 103

irreversible loss: overview and term of use, 124; of savanna productivity due to pastoralists and livestock, 147–48, 151–59; of wildlife due to pastoralists and livestock, 114, 160. *See also* decline of environment; degradation of environment

148, 170–71, 179, 250, 256; irreversible
loss of savanna productivity, 155; ivory
trade and, 104; Laikipia ranch and, *24,* 59,
110–11, 114, 142, 205–6, 247, *259,* 260;
language groups of, *7,* 100; livestock
populations and, 55, 156; maps and
location of, 3, *5, 6,* 15; measurements of
change in savannas and, 153; nomadic
herders in, 13; pastoralists in, 2, *3,* 111;
plant diversity and, 131; political
corruption and, 244; raiding by
pastoralists in, 57–58, 244; rainfall change
predictions and, 241–42; settlements'
effects in, 139; sheep herds and, 54–55;
tribal groups of, *10,* 100, 267n7; water
points and, 66–67; wildlife in, 4, 30, *31,*
62, 66–67, 109, 249–50; woodlands and,
33–34, 109. *See also* Maasailand; *specific
herder groups* and *wildlife parks and reserves*
Kenya Wildlife Service, 198, 218, 219, 249, *259*
Khoisan language group, *7,* 32, 96, 99,
100–101
Kijazi, A., 227
Kilimanjaro National Park, Tanzania, 184,
185, *187. See also* Mt. Kilimanjaro
Kindiga (Hadza) hunter-gatherers of
Tanzania, *10,* 32, 100, 224, 231, 234
Kipury, N., 215
Kirk, M., 117
Kitengela (Kaputiei or Athi-Kaputiei)
Plains. *See* Kaputiei Maasai herders of
Kenya; Kaputiei (Kitengela or Athi-
Kaputiei) Plains
Kitengela town, 208, 210. *See also* Kaputiei
(Kitengela or Athi-Kaputiei) Plains
Kolowski, J. N., 172
Konaté, S., 40
Koponen, J., 105
Kruger National Park, South Africa, 216, 218
Kruska, R. L.: African types of land and
land use map by, *20,* 269n2; elevation
changes in Africa map by, 26–27, *28,*
270n16; global view of pastoralists and,
269n6; language groups of east Africa
and, *7;* locations in east African region
map by, *5;* locations in northern Kenya to
northern Tanzania map by, *6;* rainy

season patterns in Africa map by, 26, *27;*
savannas, forests, deserts, conservation
areas, cropland, and urban areas in Africa
map by, 269n2; tribal groups of Eritrea,
Ethiopia, Djibouti and Somalia map by, *9,*
267n7; tribal groups of Sudan and South
Sudan map by, *11,* 267n7; tribal groups of
Uganda, Kenya, Tanzania, Rwanda and
Burundi map by, *10,* 267n7
Kshatriya, M., 216
Kuria of Tanzania, 58, 164

Laetoli, Tanzania, 73, 88, 176, 230
Laikipia ranch, Kenya, *24,* 59, 110–11, 114,
142, 205–6, 247, *259,* 260
Lake Baringo, Kenya, 56
Lake Manyara, Tanzania, 35, 108, 141, 194
Lake Turkana region, Kenya, *97,* 104, 136.
See also Turkana herders of Kenya
Lamphear, J., 108
Lamprey, H., 42–43, 120, 127, 148–49, 178
land: overview of use and types of, 19, *20,* 21;
communal ownership of, 23, *24,* 25, 60,
106, 110, 116, 249; development of, 217–18,
219–20; moderate use in Serengeti-Mara
region of, 227; pastoralists' loss of, *106,*
110–11, 115–16, 117, 143, 165, 189–90,
195–96, 205–6; regulations for use of, 68;
relocation as cause for loss of, *106,* 110–11,
165, 195–96, 205–6; as resource, 8; social
relationships, among pastoralists and, 60,
116; urban-elite's ownership of, 117, 203,
214, 254; value inherent in, 8; wildlife
conservation and, 4, 12. *See also* power for
locals over land and profits; privatization
(individual ownership) of land
land rights: as contested in Kenya, 171, 178,
181; government policies and, 117, 213–14;
indigenous groups', 181, 213, 234, 237; in
NCA, 117, 232–33, 237; Ngorongoro
Maasai herders and, 117, 232–33, 237; parks
and, 4, 12; for pastoralists, 15, 60–61, 64,
244, 248–49, 261, 264–65; pastoralists' loss
of, 189–90; power for locals and, 68, 72,
115, 180, 235, 260; privatization of land
effects to ensure, 117, 213–14; villages in
Tanzania, 232. *See also* land; rights

fires and, 35, 109, 133, 134, 178; group ranches and, 115, 117, 174, 186, 197–98; hard-boundary savannas and, 169–70, 174–75; human population conflicts with elephants and, 149; humans as ecosystem engineers and, 178; hunting for food and, 107, 148, 149; irreversible loss of savanna productivity and, 154, 157; ivory trade and, 104; land management strategies and, 175, 182–83; land rights for, 110–11, 195–96, 232; language groups and, 100; livestock and, 50, *51*, 150, 196–97; livestock and wildlife competition for grazing and, 150; marginalization of pastoralists and, 110–11, 114–15; measurements of change in savannas and, 153, 157; NCA and, 179; pastoralists' effects on wildlife and, 3, 150; poaching offensive by, 188; population statistics and, 169; privatization of land and, 174, 182, 197; raiding by pastoralists in, 58; relocation and land loss for, *106*, 110–11, 195–96; retaliatory killing of wildlife by, 149, 190, 199; as seasonally moving herders, 23–24, *24;* settled families with nomadic herds and, *10*, *24*, 24–25; settlements effects and, *136;* settlements' effects on wildlife distribution and abundance and, 140; trees and grass balance and, 178; warrior headdresses with feathers and, 148; wildlife attacks and, 59, 146–47; wildlife grazing effects and, 102, 132. *See also* Kaputiei Maasai herders of Kenya; Maasai Mara National Reserve, Kenya

Maasailand: conservation and, 221; conservation models and, 174; diseases in, 205; droughts and, 242; economics in reserves and, 200; farming and, 119, 245; future in, 201, 242, 245, 250–51, *259;* group ranches in, 117, 207; hard-boundary savannas and, 169; hunter-gatherer shift to pastoralists and, 102; livestock livelihood in, 196–97; pastoralists and, 225; pastoralists and livestock as enriching savannas and, 240, 250–51; power for locals over land and

profits and, 235; privatization of land in, 117; profits, *259;* raiding in, 57–58; settlements' effects on wildlife distribution and abundance and, 140; soft boundaries and, 169; trees around settlements in, 139; water points and, 3; wildlife and, 3, 15, 140, 141. *See also* Amboseli ecosystem; Kaputiei (Kitengela or Athi-Kaputiei) Plains; Ngorongoro Conservation Area (NCA); Serengeti-Mara region in Kenya and Tanzania

Maasai Mara National Reserve, Kenya: overview of, 166; adaptation of wildlife to pastoralists and livestock in, 142, 172; conservation models and, 174, 179–80; economic costs for parks and, 165; human populations and wildlife balance in, *144*, 170; irrigated farming and, 171; pastoralists and livestock in, 169–70; relocation and land loss for pastoralists and, 165, 231; tourism benefits from "honey pot effect" and, 169; wildlife attacks in, 59; wildlife diversity in, 43, 143–44. *See also* Maasai herders of Kenya and Tanzania; Mara region of Kenya

MacKenzie, J. M., 113

McBrearty, S., 93

McCabe, J. T., 52, 54, 57, 225

McCauley, D. J., 44

McNaughton, S. J., 43, 129, 130, 140

McPeak, J., 118, 247–48

Madagascar, 26–27, *28*, 30, 270n16

Maddox, T. M., 173

Maitumo, D., 34, 194

malaria, 100, 111. *See also* diseases

Mali, 132, 155

malignant catarrhal fever (MCF), 66, 150, 170, 172, 215, 225–26, 236

Mara region of Kenya: cattle and, *51*, 282n30; colonial period and, 109; fires due to diseases as stressors on wildlife and, 109; group ranches in, 117, 166, 174; human population conflicts with wildlife and, 59, 146–47; human populations and wildlife balance in, 143–44, *144*, 159–60; humans as ecosystem engineers and, 178; nutrient

Mara region of Kenya *(continued)*
hotspots and, 140; parks' exclusion
policies and, 143–44, *144;* pastoralists as
enriching savannas and, 146–47; poison
as cause of death of wildlife and, 149, 173;
private conservancies in, 238; settlements'
effects on savannas and, 135–38, *136;*
soft- to hard-boundary savannas
transition and, 159; soil fertility and, 33;
trees and grass balance and, 35; wildlife
in, *39,* 59, 109, 119, 143–44, *144,* 146–47,
149, 159–60, 173; woodlands spread due
to diseases and, 109. *See also* Maasai
herders of Kenya and Tanzania; Maasai
Mara National Reserve, Kenya
Mara River of Kenya and Tanzania, 159, 160,
164, 166, 168, 171, 176, 180–81. *See also*
Serengeti-Mara region in Kenya and
Tanzania
marginalization, of pastoralists, 110–15, 118
Marshall, F., 49–50, 96, 101, 135, 176
Maswa Game Reserve, Tanzania, 166, 169
Mau Forest, Kenya, 164, 166, 167–68,
171, 181
Mauritania, 22, 23, 270n10
MCF (malignant catarrhal fever), 66, 150,
170, 172, 215, 225–26, 236
Mduma, S., 45, 171
methods and resources, 16–18
middle ground: overview of, 12, 15, 251–52;
adaptive stewardship and management for
pastoralists and, 257–58; benefit sharing
for pastoralists, 255–56; collaboration
and, 258; diversity of approaches and,
258–60, *259;* donors and private sector
support and, 262–63; empowerment for
pastoralists, 253–55; equity for pastoralists
and, 256–57; examples of, 258–60, *259;*
future scenarios, 251–53, 264–66;
governance structure of pastoralists and,
260–61; government policies and
policymakers in future scenarios and,
261–62; imagined scenario for, 264–66;
NGOs and, 262; pluralistic and
crosscultural conservation ethic for,
251–52; scientists' support and, 263–64.
See also future for pastoralists and wildlife

in savannas; pastoralists (pastoral lifestyle
or pastoral herders); savannas; wildlife
migrations of wildlife: overview of, 165–66,
168, 195; in Amboseli ecosystem, 168;
birds and, 195; elephants and, 196; grazing
intensity during, 130; hard-boundary
savannas and, 169–70; in Kaputiei Plains,
168, 207, 211–13; Loita migration and,
168–69, 170; in Nairobi National Park,
207, 211–12; NCA and, 224; patterns for,
44–45, 272n43; in Serengeti-Mara region,
130, 165–66, 168–69, 169–70; in South
Sudan, 120, 244; in Sudan, 120; as wildlife
tool, 44–45, 272n43. *See also* migrations
of wildlife; movement; movement
patterns, for pastoralists; wildlife; wildlife
evolution
migratory patterns, for wildlife, 44–45,
272n43
Mills, A. J., 229
Milton, S. J., 157–58
miombo forests, 19–20, 28, 41, *81,* 82
mixed-boundary savannas: costs and benefits
of wildlife for pastoralists and, 68; in
Kaputiei Plains, *71,* 210–11; livestock and,
68, *69, 71,* 72; pastoralists and, 68, *69, 71,*
72; savannas distribution predictions and,
245, *246,* 247; tourism and, 68, 72
mixed farmers (sedentary herder-farmers).
See agro-pastoralists (settled herders with
nomadic herds); group ranches;
pastoralists (pastoral lifestyle or pastoral
herders); sedentary herder-farmers
(mixed farmers)
Mkomazi Game Reserve, Tanzania, 115, 257
Moehlman, P. D., 65, 227–28
Mohlig, W. J. G., 267n7
Mongolian herders of Mongolia, 22–23, 197
monomodal rainfall, in Africa, 26, *27. See
also* rainfall and rainy season patterns
Moss, C. J., 188, 190, 198–99
Mt. Kilimanjaro: overview and location of,
184, *187,* 188; ecosystems and, 206;
farming at base of, 119; future in, 242;
icefields and glaciers on, 184, 185, 191–92;
pastoralists and, 190; rainfall and, 185;
wildlife on slopes of, 186, 196. *See also*

Amboseli National Park, Kenya;
Kilimanjaro National Park
movement patterns, for pastoralists, 23–24,
24, 52, 96, *98*, 214, 216, 236, 270n9. *See
also* migrations of wildlife
Mpala and Mpala Research Centre, Kenya,
41, 43
Murdock, G. P., 4, *7*, *9*, *10*, *11*, 30, 267n7
Mursi herders of Ethiopia, *7*, 8, *9*, 107
Muslim societies, 50, 103
Mwangi, E., 200

Nairobi, Kenya: overview of, 203, *204*, 205;
grazing for livestock in, 210; hard-
boundary savannas and, *70*, 202, 208;
impoverishment and, 217; industrial
enterprises and, 188; Kaputiei Plains land
development by, 217–18, 219–20;
privatization of land and, 213, 217–18;
urban-elite landowners from, 203
Nairobi National Park, Kenya: overview and
history of, 114, 202; costs and benefits of
wildlife for pastoralists and, 250;
ecosystems and, 206; fires and, 211;
grazing lawns and, 211; hard-boundary
savannas and, *70*, 208, 218; livestock and
wildlife competition for grazing and, 212;
migrations of wildlife in, 207, 211–12;
privatization of land and, 217–18;
soft-boundary savannas and, *71*, 208; soil
fertility and, 211; urban-elite landowners
and, 203; water for wildlife in, 207;
wildlife and, 207, 210–11, 212, 213, 250;
woodlands and, 206. *See also* Nairobi,
Kenya
Namibia, 23, 30, 131–32, 154, 155, 156,
270n10
Narok District and reserve, 110, 174, 179,
186, 195, 214, 257
national parks: humans as ecosystem
engineers and, 104, 109, 114, 115, 119, 120.
See also specific national parks
nature, as separate from human populations,
2, 7–8, 112
Naughton-Treves, L., 256
NCA (Ngorongoro Conservation Area). *See*
Ngorongoro Conservation Area (NCA)

NCAA (Ngorongoro Conservation Area
Authority), 221, 224, 226, 227, 232–35.
See also Ngorongoro Conservation Area
(NCA)
Ndorobo (Dorobo) people, 104, 176, *259*
Nelson, F., 249
Neumann, R., 114
Ngoroko herders of Kenya, 57
Ngorongoro Conservation Area (NCA):
overview of, 166, *167*, 221–22, *222*;
agro-pastoralists and, 4, 224; central
management issue and, 221, 232, 234;
conservation lease programs and, 235;
conservation models and, 179, 221, 227,
237; diseases transmission between
livestock and wildlife and, 225, 236;
economic costs for parks and, 165, 235;
ecosystems and, 222, 224; elevation
changes in, *28*, 224; Empakaai Crater
and, 179, 222, 229–30; farming in, 227,
234; fires and, 135, 227, 228, 230; future
for, 233–37; grazing lawns and, 228;
grazing of livestock and, 135; historical
context for, 230–33; human populations
and, 224; human populations and wildlife
balance and, 227–29, 234; hunter-
gatherers and, 221; impoverishment of
human populations and, 4, 221, 224–27,
235, 237; land rights in, 117, 232–33, 237;
Lerai Forest and, 224, 229; livestock
balance with wildlife and, 236–37;
livestock restrictions in parks and, 4,
221–22, 226, 233–34, 236–37; livestock-
wildlife group grazing and, *173*;
migrations of wildlife and, 224; in modern
times, 222, *223*, 224–30; NCAA and,
224, 226, 227, 232–35; Ngorongoro
Crater and, 119, 222, *223*, 231–32; power
for locals over land and profits and, 226,
235; (proto-)human evolution and, 91,
176; rainfall change predictions and, 241;
relocation of pastoralists from, 231–32;
salt levels in soil and, 228, 229; scavenger
(proto-)humans and, 91; sedentary
herder-farmers and, 230; soil fertility and,
228, 229; tourism and, 221, 224–26, 228,
232, 234–35; tourism companies and, 224,

"pure" pastoralists (nomadic herders), 23, 24, 270nn9–10. *See also* pastoralists (pastoral lifestyle or pastoral herders)

quagga, 113

raiding by pastoralists, 36, 57–58, 108, 111–12, 243–44
rainfall and rainy season patterns: bimodal rainfall and, 49–50, *97*, 99, 100, 209–10; climate change and, 238, 241–42; global view of, 25, 26, *27;* monomodal rainfall in, 26, *27;* predictions for changes in, 241; rainfall-driven savannas and, 19, *20*, 245, 269n2; savanna types and, 33, *33;* in west Africa, 49
Rainy, J., 43, 141
Rainy, M., 43
ranches: commercial, 22, 48, 55; commercial wildlife, *71*, 72; fenced, *24*, 25, 47–48, 54. *See also* group ranches
relocation and land loss, for pastoralists, *106*, 110–11, 165, 195–96, 205–6, 228–29, 231–32
Rendille herders of Kenya, *10*, 23, *24*, 31, 50, 60, 108, 112, 270n10
Renton, A., 254
reserves and wildlife parks (parks). *See* wildlife parks and reserves (parks); *specific wildlife parks and reserves*
Reshiat (Daasanach) of Ethiopia, *9*, 32, 58, 60
rhinos: black, 160; diseases and, 107; evolution timeline and, 80, 83, 93; farmers' effects on, 160; forage for, 21, 43, 224; group grazing and, 43; human population conflicts with, 149; poaching in Ngorongoro and, 233; population statistics and, 196; predation and, 45; reproductive rates of, 160; settlements' effects on, 142–43, 160
Rift Valley (Great Rift Valley): ecological diversity in, 91; escarpment of, 17–18, 208; evolution timeline and, *81*, 82–83, 209; lakes in, *87*, 89; map of, 5; pastoralists' marginalization in context of, 111, 195, 206; (proto-)human evolution and, *97*, 99, 100; wildlife population statistics and, 30. *See also* Kaputiei (Kitengela or Athi-Kaputiei) Plains

rights: grazing, 60–61, 261; human, 4, 12, 15, 64; indigenous people's, 1, 176, 178–79, 181, 213, 231, 234–35, 237; NGOs and, 262; for pastoralists, 15, 60–61, 244, 248, 261, 264–65; power for locals over land and profits and, 68, 72, 115, 180, 218, 235, 260; water, 181; for wildlife, 15, 243, 251, 256. *See also* land rights
rinderpest: control of, 109; diseases as stressors on wildlife and, *106*, 107–9, *177–78;* eradication of, 13, 45, 109, 150; pandemic of, 64, 66, 95, *106*, 230; wildebeest migration as affected by, 109
Ritchie, M. E., 181–82, 241
Rodgers, W. A., 221, 229
Runyoro, V. A., 135, 224, 227–29, 233–34
Rutten, M. M. E. M, 208, 214, 215
Ruwenzori Mountains, 30, *31*, 191
Rwanda: bimodal rainfall in, 27, *27;* colonial period and, 110; doctors per population statistics, 13; farming and, 245; hard-boundary savannas and, 68; hunting for food by pastoralists and, 149; hunting game in, 149, 231, 250; ironwork technology and, 100; language groups of, *7;* map and location of, *5*, 15; political power connection with livestock and, 50; soft-boundary savannas and, 72; tribal groups of, *10*, 267n7; wildlife populations in, 120

safety nets, and survival strategies, 59–61, 242
the Sahara, 12, 19, 26–27, 30, 96, *97*, 127
the Sahel, 12, 23, 30, 127, 156, 158
Said, M., 62, 117, 119
salt levels: in soil, 194–95, 228, 229; in water, 44, 168
Samburu herders of Kenya, *10*, 23–24, *24*, 24–25, 107, 112
Sankaran, M., 29, *29*
savanna evolution: overview of, 73–74; first stage of "deep time" and, 74, 75, *76;*

77–79; second stage in evolution of wildlife and savannas and, 80, *81,* 82–85; third stage in evolution from 5.3–1.8 Ma and, *28,* 91; elevation changes in Africa and, 26–27, *28,* 270n16; grass survival and success and, 37, 42, 82; tree survival and success in, 37. *See also* evolution timeline; savannas

savannas: overview and types of, 19–20, *20;* African types of, 19–20, *20,* 245, 269n2, 269n4; bimodal rainfall in, 27, *27;* biomass as edible and, 41–42, 90; definition and description of, *20,* 32–36, 77; fires' effects on, *33,* 34–36, 35, 132–35, 144, 148, 151, 161; future for, 245, *246,* 247, 295n19; future scenarios and, 201; global view of, 18–19, 21–22; grazing of elephants effects on trees and, 33–34, 45–46; irreversible loss of productivity due to pastoralists and livestock and, 147–48, 151–59; measurements of change caused by pastoralists in, 151–55, 157; migratory patterns for wildlife and, 44–45, 272n43; multiple species association, 43; open up through forest reduction and, 33–35, 64, 124, 133, 178; overgrazing by wildlife, 45; polarization and, 119; productivity of, 42; savanna woodlands, 74, 80, 269n2, 269n4; settlements' effects on, 67, 135–43, *136,* 140–43, 143–44, *144;* shrubs and, 19, 33; soil fertility and, *29,* 29–30, 33–34; termites' effects on, 40–41; trees and grass balance in, 33, *33,* 34; wet savannas and, 41, 42, 102, 116–17, 139, 243, 245, 247. *See also* boundaries in savannas, and soft- to hard-boundary transition and; future for pastoralists and wildlife in savannas; grasses; savannas in east Africa; savanna woodlands

savannas in east Africa: overview of use of term for, 1–12, 267n4; as ancestral home of human species, 32, 86, 239; anthropocentric perspective of conflicts in historical context and, 4, 7–8; bimodal rainfall in, 26–27, *27;* diversity and opportunity in, *20,* 25–32; ecocentric perspective of conflicts in historical context and, 7–8; human populations in environments outside, 12–15; as human populations' ancient home, 1, 267n4; parks and, 2, 4; pastoralists' effects due to restrictions in parks and, 4; pastoralists' population statistics and, 4; pastoral lifestyle common to, 2–3, *3;* soil fertility in, *29,* 29–30, 270n18; wildlife conservation supporters and, 4; wildlife populations in, 2, 4, 30, *31,* 42–43. *See also* future for pastoralists and wildlife in savannas; middle ground; pastoralists (pastoral lifestyle or pastoral herders); savanna evolution; savannas; wildlife; wildlife parks and reserves

savanna woodlands: overview and definition of, 269n2, 269n4; evolution timeline and, 52, 73, 80; fires in, 92, 133; grazing of elephants effects on trees and, 33–34, 45–46; miombo forests as, 19–20, 28, 41, *81,* 82; (proto-)human evolution and, 73, 84–85, 269n4; trees and grass balance in savannas, 33, *33;* trees' survival and success in, 37. *See also* savanna evolution; savannas; savannas in east Africa; woodlands (forests)

Sayialel, K., 188

Sayialel, S., 188, 199

scavenger (proto-)humans, 1, 32, 74, 90, 91, 92–93, 230, 239. *See also* (proto-)human evolution

Scholes, R. J., 41–42

scientists' support, for middle ground, 263–64

Scoones, I., 152

seasonally moving herders (transhumant pastoralists), 18, 23–24, *24,* 54, 270n9. *See also* pastoralists (pastoral lifestyle or pastoral herders)

security/insecurity, as challenge for pastoralists, 52, *53,* 57–59

sedentary herder-farmers (mixed farmers): overview of, *24,* 25; "enriched" savannas and, 123; future for, 247; in Kaputiei Plains, 197, 209–10; NCA and, 230; pastoralists lifestyle changes and, 116; in

response and, 126, *126,* 129; *Indigofera
spinosa* and, 55, 132; no response to
grazing and, 129; predation and, 134;
savannas and, 19, 33; settlements and, 137,
138; simplifying response and, 132;
Suaeda monoica and, 188, 194; termite
mounds and, 41; wildlife's effects on, 130.
See also plant-eating wildlife (herbivores);
plant species

simplify (impoverish) or simplifying
response or simplifying response, 123,
125, *126,* 129, 131–32, 148, 151–52. *See also*
enrich (diversify)

Sinclair, A. R. E. (Tony), 43, 45, 127, 177–78

single-humped (dromedary) camel herds, 55

slave trade, 103, 105, *106*

small mammals. *See* wildlife; *specific species*

smallpox, 105, 107, 108, 205

Smith, A. B., 22, 101

Sobania, N. W., 58, 60, 111, 238

social conflicts, and pastoralists, 36, 57–58,
108, 111–12, 243–44

social relationships, among pastoralists, *53,*
56, 59–61, 116, 197, 214–16, 235. *See also*
pastoralists (pastoral lifestyle or pastoral
herders)

soft-boundary savannas: future for
pastoralists and wildlife and, 245; humans
as ecosystem engineers and, 101–2;
Nairobi National Park and, *71,* 208;
pastoralists as enriching savannas and,
69, 130–32; savannas distribution
predictions and, 245, *246,* 247; soft- to
hard-boundary savannas transition and,
69, 119, *144,* 159–61; South Sudan and,
72. *See also* hard-boundary savannas;
mixed-boundary savannas

soil fertility: overview of, *29,* 29–30, 270n18;
climate change and, 242–43; diversity in
plants and, 129; fires and, 125; grazing
affect and, 46, 125; human populations
needs and, 30; impoverishment of human
populations and, 30; in Mara region of
Kenya, 33; in Nairobi National Park, 211;
Nairobi National Park and, 211; in NCA,
228, 229; nutrient-poor savannas and, 29,
29, 30, 46; nutrient-rich savannas and, *29,*

29–30, 46, 129–30, 209; pastoralists as
incompatible with wildlife and, 151–52,
153, 155–57, 159; salt levels and, 194–95,
228, 229; savannas and, *29,* 29–30,
33–34, 270n18; wildebeest migration as
cause of loss of soil and, 45; wildlife's
effects on, *29,* 30. *See also* dung

Somalia: bimodal rainfall in, 26–27, *27;*
colonial period and, 110; diseases in
human populations, 105; hunting wildlife
and, 149; irreversible loss of savanna
productivity and, 154, 155; language
groups of, *7;* map and location of, *5,* 15;
pastoralists in, 23–24, *24,* 61, 149; plant
species affects by grazing, 131; as
seasonally moving herders, 23–24, *24;*
subsistence herders in, 22–23; tribal
groups of, *9,* 267n7; wildlife populations
missing data for, 280n57

South Africa: bimodal rainfall and, 26;
decline of environment and, 157–58;
evolution timeline and, *76,* 78; hunting
game for trophies in, 113; irreversible loss
of savanna productivity and, 155–56;
livestock population decline in, 156; parks
and, 216, 218; pastoralists and livestock as
enriching savannas, 64, 126, 131, 139;
(proto-)human evolution and, 92;
savannas and, 30, 158; termites and, 41;
trees and grass balance, 34; wildlife
populations and, 216

southern Africa: cattle populations and, 156;
colonial period conservation efforts and,
113; conservation in historical context
and, 2; irreversible loss of savanna
productivity and, 155–56; rinderpest and,
107; sedentary herder-farmers in, *24,* 25;
settlement effects and, 139; slave trade
and, 103. *See also* Africa; *specific countries,
herder groups,* and *locations*

southern Ethiopia, 3, 58, 60, 116–17, 119,
153, 245

South Sudan: bimodal rainfall in, 27, *27;*
language groups of, *7;* overview and
location of, *5;* colonial period and,
110–12; hard-boundary savannas and, 68;
human population movements and, 96,

199, 224, 226, 228, 232, 255–56, 258; conservation models and, 115–16, 174; costs and benefits of wildlife for pastoralists and, 249, 252, 255; ecotourism and, 116, 182, 200, 206; enriching savannas by pastoralists and, 65, *65;* future scenarios and, 252, 255; globalization and, 243; group ranches converted for, 238; "honey pot effect" and, 169; joint operations with pastoralists in, 258, *259;* land management strategies by pastoralists and, 175; Maasai Mara National Reserve and, 169; mixed-boundary savannas effects and, 68, 72; NCA and, 221, 224–26, 228, 232, 234–35; Ngorong-oro Maasai herders and, 225, 235; photographic, 250; political corruption and, 244; power for locals over land and profits and, 68, 72, 115, 180, 199–200, 218; "wildlife Africa" myth and, 165

"tragedy of the commons," 127, 216

transhumant pastoralists (seasonally moving herders), 18, 23–24, *24,* 54, 270n9

Trapnell, C. G., 34

trees and grass balance: in Amboseli ecosystem, 194; diseases' effects on, 108–9; fires effects on, 151, 177–78; in savannas, 33, *33,* 34; in Serengeti-Mara region, 177–78. *See also* grasses; woodlands (forests)

tribal groups. *See* pastoralists (pastoral lifestyle or pastoral herders); pastoralists and livestock as enriching savannas; pastoralists and livestock as incompatible with wildlife; *specific countries and tribal groups*

trypanosomosis disease, 56, 61, 109, 133, 150, 172, 178

Tsavo National Park, Kenya, 33–34, 109, 114

tsetse flies: livestock and, 123, 179; trypanosomosis disease and, 56, 61, 109, 133, 150, 172, 178; woodlands and, 109, 133, 178

Tswana of southern Africa, *24,* 25

Turkana herders of Kenya: bleeding livestock to use blood as food and, 50–51; cattle herds and, 57; climate as control for livestock levels and, 127–28; dromedary camel herds and, 55; on fires, 133; human population movements in east Africa in prehistorical and historical periods and, 100; movement patterns of pastoralists for survival and, *24,* 52, 54; as nomadic herders, 23, *24,* 270n10; raiding by pastoralists against, 57–58; setting the scene, 17; settlement effects and, 135; sheep herds and, 54–55; trees around settlements and, 36, 139. *See also* Lake Turkana region, Kenya

Turton, D., 8, 58

Tvedt, T., 153–54

Uganda: bimodal rainfall in, 27, *27;* colonial period and, 110; colonial period control of resources and, 112; farming and, 245; farming as commercial enterprise and, 119; hard-boundary savannas and, 68; hunting for food by pastoralists and, 148–49; ironwork technology and, 100; language groups of, *7;* map and location of, *5,* 15; parks and, 119, 120, 149; political corruption and, 244; raiding by pastoralists in, 58; rainfall change predictions and, 241–42; settlement effects and, 139; soft-boundary savannas and, 72; tribal groups of, *10,* 267n7; wildlife populations in, 120; woodlands spread due to diseases and, 109

UNESCO, 221, 231

ungulates, 42, 80, 83. *See also* wildlife

United States: bison and, 46, 125–26, 128, 129, 140, 169; commercial ranches in, 22; drylands in, 125–26; fires and burning shrublands in, 134; grazing effects and, 46, 125–26, *126,* 129; ivory trade and, 104; parks in, 46, 129. *See also* North America

urban-elite landowners, 117, 203, 214, 254. *See also* land

Vanderjeugd, H. P., 108

Van Praet, C., 191, 194

Van Wijngaarden, W., 33–34, 193

Vesey-Fitzgerald, D. R., 142

Vrba, E. S., 88

wildlife parks and reserves *(continued)*
exclusion policies, 110–11, 143, 189–90;
historical context for, 2; human populations' effects on, 67; impoverishment of
human populations and, 4, 226, 235, 237;
land rights for pastoralists and, 64;
livestock restrictions in, 4, 221–22, 226,
233–34, 236–37; parks and people model
and, 143; pastoralists as enriching savannas
and, 115, 143–45; pastoralists as incompatible with wildlife and, 115, 149, 156;
pastoralists populations in, 4; polarization
in savannas and, 119; power for locals
over land and profits and, 179, 199–200,
238; soft-boundary savannas and, *69, 71,
72. See also* conservation models; middle
ground; migrations of wildlife; protected
or conservation areas; savannas; wildlife;
wildlife evolution; *specific countries,* and
wildlife parks and reserves
Williams, A. P., 242
Wilson, R. T., 120
Wittemyer, G., 67
WMAs (Wildlife Management Areas), 175
Wood, B., 85
woodlands (forests): in Africa, *20, 21,*
269n2; in Amboseli ecosystem, 186,
192–93; in Amboseli National Park,
188; biomass as edible and, 41–42, 90;
deforestation and, 164, 171, 181; die-offs
of, 108, 193–95, 229; diseases' effects
on, 108–9; evolution timeline and,
76–80, 84, 86; humans as ecosystem
engineers and, 108–9, 229; miombo
forests and, 19–20, 28, 41, *81,* 82;
Nairobi National Park and, 206; open
up of savannas through reduction of,
33–35, 64, 124, 133, 178; raiding
protection from, 36; as resistant to fires,
37; salt levels in soil effects on, 194–95,

229; in Serengeti-Mara region, 109, 165,
171, 177–78; swamplands and, 194–95;
trees around settlements and, 36, 139;
trees' effects on water and, 186, 192–93;
tree survival and success in, 37; tsetse
flies and, 109, 133; wood cutting by tree
fellers and, *33,* 35, 184, 192–93. *See also*
savanna woodlands; trees and grass
balance
Woodroffe, R., 67
Worden, J., 198
World Heritage Site, 221, 231. *See also*
Ngorongoro Conservation Area (NCA)

Yellowstone National Park, U.S., 46, 129
Young, T. P., 43, 44, 108, 142, 194, 245

Zambia, *29, 30, 34,* 133
zebras: diseases and, 107, 109; evolution
timeline and, 80, 83; grazing and, 41, 189,
195, 203; human populations and wildlife
balance and, 236; migrations and, 165,
206, 211–12; Ngorongoro and, 224; parks
and, 188; population statistics for, 168,
198, 213, 227; predation and, 45;
settlements and, 141; water needs for, 43,
189, 203, 224; wildlife diversity and, 188
Zimbabwe: CAMPFIRE program in, 236,
252, 254–55; elephants and, 45–46, 67;
farming's effects on wildlife and, 160;
grazing effects and, 131–32; livestock
populations and, 156; parks and, 45–46;
plants and, 28, 131–32; settlements' effects
on wildlife and, 160
zoning for human populations and wildlife,
219, 221, 261. *See also* conservation
models; future for pastoralists and wildlife
in savannas; human populations; wildlife;
wildlife parks and reserves (parks)
Zulu people, *24, 25*

University of California Press gratefully acknowledges the following generous donors to the Authors Imprint Endowment Fund of the University of California Press Foundation.

Wendy Ashmore
Clarence & Jacqueline Avant
Diana & Ehrhard Bahr
Nancy & Roger Boas
Robert Borofsky
Beverly Bouwsma
Prof. Daniel Boyarin
Gene A. Brucker
William K. Coblentz
Joe & Wanda Corn
Liza Dalby
Sam Davis
William Deverell
Frances Dinkelspiel & Gary Wayne
Ross E. Dunn
Carol & John Field
Phyllis Gebauer
Walter S. Gibson
Jennifer A. González
Prof. Mary-Jo DelVecchio Good & Prof. Byron Good
The John Randolph Haynes & Dora Haynes Foundation / Gilbert Garcetti
Daniel Heartz
Leo & Florence Helzel / Helzel Family Foundation
Prof. & Mrs. D. Kern Holoman
Stephen & Gail Humphreys
Mark Juergensmeyer
Lawrence Kramer
Mary Gibbons Landor

Milton Keynes UK
Ingram Content Group UK Ltd.
UKHW030621120824
446774UK00002B/11/J

9 780520 273559